Horticultural Crop Breeding

NIPA GENX ELECTRONIC RESOURCES & SOLUTIONS P. LTD.
New Delhi-110 034

About the Author

Prof. Bhimasen Naik (b.1961) hails from an agri-horticultural family of Western Odisha (Village: Jamtalia, Tehsil: Sundargarh, District: Sundargarh). He obtained B.Sc. (Ag. & A.H.) and M.Sc. (Ag.) in Genetics and Plant Breeding from Chandra Shekhar Azad University of Agriculture and Technology, Kanpur in 1981 and 1983, respectively, and Ph.D. from Utkal University, Bhubaneswar in1999. He has been awarded ICAR Junior Research Fellowship and CSIR Senior Research Fellowship during his studies. He is a Fellow of the Indian Society of Genetics & Plant Breeding, New Delhi; and life member of the Crop Improvement Society of India, Ludhiana; the Indian Society of Plant Breeders, Coimbatore; the Society for Advancement of Rice Research, Hyderabad; the Indian Society of Oilseeds Research, Hyderabad and the Society for Plant Biochemistry & Biotechnology, New Delhi. He has worked on breeding and genetics of ginger, turmeric, mungbean, linseed and rice. He has more than 36 years of research and teaching experience at Odisha University of Agriculture and Technology. He has 56 research papers published in peer-reviewed national and international journals, 23 abstracts of symposia, 05 books, 01 book chapter (Springer-Nature) and 20 popular articles to his credit. He has guided one M.Sc. (Ag.) and one Ph.D. student. He is associated in the development of five cultivars in ginger (Suruchi), turmeric (Surama, Ranga and Rashmi) and linseed (Arpita); collaborated in development of three cultivars in rice (Ashutosh, Gobinda and Hasanta) and registered one genetic stock of mungbean (BSN 1, INGR 00011).

Horticultural Crop Breeding
Principles and Practices

Bhimasen Naik
Former Professor of Plant Breeding and Genetics
Odisha University of Agriculture and Technology
Chiplima Campus, Sambalpur-768 025, Odisha

a division of
NIPA GENX ELECTRONIC RESOURCES & SOLUTIONS P. LTD.
New Delhi-110 034

nipa®
a division of
NIPA GENX ELECTRONIC
RESOURCES & SOLUTIONS P. LTD.

101,103, Vikas Surya Plaza, CU Block
L.S.C.Market, Pitam Pura, New Delhi-110 034
Ph : +91 11 27341616, 27341717, 27341718
E-mail:newindiapublishingagency@gmail.com
www: www.nipabooks.com

For customer assistance, please contact
Phone: + 91-11-27 34 17 17 Fax: + 91-11- 27 34 16 16
E-Mail: feedbacks@nipabooks.com

ISBN: 978-93-90591-78-7

Composed and Designed by NIPA.

To
Prasanti (Wife)
Rashmi (Daughter), Namrata (Daughter)
Soumen (Son-in-law)
and
Abhipsa (Grand daughter)

Acknowledgement

I thank all the authors of the books/chapters/reviews/articles mentioned in the bibliography that I have referred to while writing this undergraduate textbook.

I thank Dr. Bertrand C.Y. Collard for his generous permission to adapt their review papers "Collard BCY, Jahufer MZZ, Brouwer JB and Pang ECK 2005. An introduction to markers, quantitative trait loci (QTL) mapping and marker-assisted selection for crop improvement: The basic concepts. *Euphytica* 142: 169-196" for the 'Chapter 4 Molecular Genetic Basis I: Markers, QTL Mapping and Marker Assisted Selection', and "Verdeprado H, Kretzschmar T, Begum H, Raghavan C, Joyce P, Lakshmanan P, Cobb JN and Collard BCY 2018. Association mapping in rice: basic concepts and perspectives for molecular breeding, *Plant Production Science* DOI: 10.1080/1343943X.2018.1483205" for the 'Chapter 5 Molecular Genetic Basis II: Association Mapping', and suggestions for improvement of the text.

I also thank Dr. H. Shimelis for his prompt permission to adapt their review paper "Shimelis H and Laing M 2012. Timelines in conventional crop improvement: pre-breeding and breeding procedures. *Australian Journal of Crop Science* 6 (11):1542-1549" for the 'Chapter 15 Timelines in Conventional Crop Improvement: Pre-breeding and Breeding Procedures'.

I acknowledge the assistance received in the form of constructive comments from my colleagues Dr. Atanu Seni, Dr. (Mrs.) Rini Pal, Dr. Dipankar Mandal, Dr. Purandar Mandal, Dr. Biswa Ranjan Nayak, Dr. Kaushik Kumar Panigrahi and Dr. Pamirelli Ranjith. They have gone through some chapters critically and put forth their suggestions for improvement.

B.S. Naik

Preface

I got an opportunity to be associated with the students of the College of Horticulture, Chiplima as Associate Dean for about one year. I realized two points: (i) absence of a suitable textbook for B.Sc. (Hons) Horticulture written as per the syllabus recommended by the 5th Deans' Committee and (ii) most of the textbooks are not oriented towards horticultural crops. So, a textbook entitled "Horticultural Crop Breeding: Principles and Practices" was endeavoured to fulfill these two requirements.

The course "Principles of Plant Breeding" is in the second semester of B.Sc. (Hons) Horticulture programme. The two important prerequisite courses of "Principles of Genetics and Cytogenetics", and "Elementary Statistics and Computer Application" are taught in the first semester. With this academic background of the students the book is written. The whole of the book is spread over 29 chapters grouped into three sections. Section I deals with overview and biological foundations, Section II with breeding principles (also called 'theory') and Section III with breeding practices (also called 'methods', 'techniques', 'practice', 'procedures', 'schemes', 'plans' or 'approaches'). Breeding practices may be conventional (also called 'traditional', 'classical' or 'general') or non-conventional (also called 'unconventional', 'non-traditional', 'modern' or 'special'). Conventional breeding practices utilize variations present in nature or created by intra-specific (inter-varietal) hybridization for development of cultivars. On the other hand, non-conventional breeding practices utilize variations created by artificial mutagenesis, wide (distant) and somatic hybridization, polyploidy, somaclonal variation and genetic engineering (also called 'recombinant DNA technology' or 'transgenic technology'). Besides, mission-oriented breeding practices such as breeding for biotic and abiotic stresses have been discussed. The author does not claim any originality of the book—it draws heavily from many text/reference books, review articles and some other references. Like others, this textbook is a history of the subject. It is an introductory level textbook – a first course in plant breeding. So, the size is limited so that it can be taught in one semester. A student in the class once asked me why she was reading the course "Principles of Plant Breeding". The question was very pertinent. Hence, the textbook should be introductory as well as capable of inculcating in the mind of students the probable way of application in practical crop breeding programme.

I acknowledge the assistance of Mr. Subrat Kumar Chhanda for computer-typing of the manuscript. I congratulate M/S New India Publishing Agency, New Delhi for their support and publishing in a short time. I cherish the encouragement and cooperation received from my wife Prasanti during the preparation of the manuscript.

Though the book is primarily written for B.Sc. (Hons) Horticulture students, the counterparts of B.Sc. (Hons) Agriculture also may be benefitted. It may serve as a reference book for post-graduate students and horticultural plant breeders. Every effort has been made to render the book simple, lucid and correct in presentation. However, error if any is mine; I would highly appreciate if it is brought to my notice for rectification in next edition.

I seek suggestions from learned teachers and students for improvement of the book.

B.S. Naik
Chiplima, Sambalpur, Odisha

Contents

Section II: Breeding Principles

Section III: Breeding Practices

Part A: Conventional Breeding Practices

Part B: Non-Conventional Breeding Practices

Section I
Overview and Biological Foundations

1

Horticultural Crop Breeding An Overview

'To keep agricultural production sustainable, efforts must be made in the field of cultivation methods as well as crop breeding. The two fields complement each other.'

— VIB

1.1 Introduction

The word '**horticulture**' was first used in 1600's, and is derived from two Latin words, '**hortus**'(garden) and '**cultûra**'(cultivation), and it means '**the culture** (cultivation) **of garden plants**'. Relf (1992) defined "horticulture as an applied science and art of growing fruits, vegetables, herbs and spices, nuts, tubers, aromatic and medicinal plants, plantation crops, mushrooms and ornamental plants which include shrubs, flowering plants and turfs stimulating minds, spirits and emotions of individuals, enriching health and wellness of communities, and linking human beings to nature and its landscaped exteriors for wellness and wellbeing". Fruits and vegetables are nourishing and protective foods. It has been recognized that fruits are a rich source of antioxidants which protect us from dreadful diseases. Vegetables and fruits contain vitamin E, K, and β–carotene in addition to minerals like zinc, copper and manganese. To meet the ever-increasing demand of the country for fruits, vegetables, tubers, spices, flowers, ornamentals etc., we must strive hard to deliver better cultivars in as many aspects as possible to the society so that human beings will be in sound health physically as well as mentally. To achieve this goal plant breeding in general, and horticultural plant breeding in particular will come to our help.

1.2 Plant Breeding as a Dynamic Science

1.2.1 Definition

'**Plant Breeding**' consists of two words. '**Plant**' usually means angiosperms and '**breeding**' is derived from '**breed**' meaning lexically '**to produce progeny** or **offspring**'. It is the applied branch of botany and deals with the improvement of crops. It is also known as 'crop breeding', 'crop improvement' or 'plant improvement'. Plant breeding has been defined in different ways by different authors:

1.2.1.1 Definitions during Classical Period (1900-1980)

1. Plant breeding is defined as the **evolution** directed by the will of man (N.I. Vavilov 1935, 1951).
2. Plant breeding is merely a continuation of the natural **evolution** of the crop plant species changing its course in the direction of greater use to mankind (G.L. Stebbins 1957).
3. Plant breeding is the **genetic adjustment** of plants to the service of man (Frankel 1958).
4. Plant breeding is the art and the science of **changing and improving** the heredity of plants (J.M. Poehlman 1959).
5. Plant breeding is the art and science of **changing** plants genetically (R.W. Allard 1960).
6. Plant breeding is continuous improvement of crop plants as the result of **evolution** directed by man for his own ends (W. Williams 1964).
7. Plant breeding is defined as the art and science of **improving** the genetic pattern of plants in relation to their economic use (D.C. Smith 1966).
8. Plant breeding is the **genetic adjustment** of plants to the social, cultural, economic and technological aspects of the environment (Frankel 1968).
9. Plant breeding is a **technology** of developing superior crop plants for various purposes (R. Riley 1978).
10. Plant breeding is a **technology** which has the practical purpose of producing material objects—the variety or cultivar—which offer advantages on pre-existing varieties in terms of absolute yield, stability of yield, agronomic convenience or quality of the market produce (R. Riley 1979).
11. Plant breeding is the current phase of crop **evolution** (N.W. Simmonds 1979).

1.2.1.2 Definitions during Modern Period (1980 onwards)

1. Plant breeding has been an **integral part of the daily life** of humankind or people living in the countryside for the past 10,000 years. After a brief sojourn in splendid isolation, plant breeding, once again is taking its place in our imperfect but wonderful world of nature and people in nature (D. N. Duvick 1996).
2. Plant breeding is an applied, **multidisciplinary science**. It is the application of genetic principles and practices associated with the development of cultivars more suited to the needs of humans than the ability to survive in the wild. It uses knowledge from agronomy, botany, genetics, cytogenetics,

molecular genetics, physiology, pathology, entomology, biochemistry, and statistics (R.H.J. Schlegel 2003).

3. The ultimate outcome of plant breeding is **mainly improved cultivars**. Therefore, plant breeding is primarily an organismal science even though it is eminently suited to translate information at the molecular level (DNA sequences, protein products) into economically important phenotypes. The traditional definition of plant breeder includes only those scientists who develop new cultivars and improved germplasm, however many feel this definition should be expanded to include scientists who contribute to crop improvement through breeding research (C. Ransom, C. Drake, K. Ando and J. Olmstead 2006).
4. Plant breeding is a **deliberate effort** by humans to nudge nature, with respect to the heredity of plants, to an advantage (G. Acquaah 2012).
5. Plant breeding can be broadly defined as **alterations caused in plants** as a result of their use by humans, ranging from unintentional changes resulting from the advent of agriculture to the application of molecular tools for precision breeding (F. Breseghello and A.S.G. Coelho 2013).
6. Plant breeding is defined as application of **genetic principles** to upgrade the productivity of crop plants per unit area, time and input (energy) coupled with tolerance to biotic and abiotic stresses and adaptability to cropping system and socio-economic reach of the farmers of the region/state (M. Ram 2014).

During classical period (1900-1980), different authors called plant breeding as human directed evolution, genetic adjustment, changing and/or improving, or technology for development of superior crop plants. In modern period (1980 onwards), some authors emphasised scientific aspects whereas, others practical aspects. Out of 17 definitions, those of Schlegel 2003 and Ram 2014 are better ones.

If we combine: plant breeding may be defined as the art, multidisciplinary science and technology of developing genetically better cultivars with the application of genetic principles and practices to upgrade the productivity of crop plants per unit area, time and input (energy) coupled with tolerance to biotic and abiotic stresses and adaptability to cropping system and socio-economic reach of the farmers of the region/state.

1.2.2 Art, Science and Technology

Plant Breeding is an art, a science and a technology as described below:

1. Art refers to human imagination, creativity and skill. Plant breeding is an art because selection of superior plants requires human skill, imagination and experiences.
2. Science refers to systematic knowledge of a subject gained through human observations and experiments involving certain principles. Plant breeding is a science because development of suitable varieties or hybrids involves genetic principles, sequential steps and experimentation.
3. Technology refers to development of a useful commercial product/ service involving scientific principles and human skill. In Plant breeding, the aim is to develop superior varieties or hybrids for commercial cultivation. So, it is a technology.

Thus, plant breeding is an art, a science and a technology of developing genetically superior plants in terms of their economic utility for the mankind.

1.2.3 Agricultural Plant Breeding versus Horticultural Plant Breeding

Based on crops, **Plant Breeding** can be divided into **'Agricultural Plant Breeding'** (also called **'Agricultural Crop Breeding'**) and **'Horticultural Plant Breeding'** (also called **'Horticultural Crop Breeding'**). Agricultural crop breeding deals with genetic improvement of agricultural crops (field crops) such as cereals, millets, pulses, oilseeds, fibre crops, sugar crops etc. Horticultural crop breeding deals with genetic improvement of horticultural crops such as vegetables, fruits, ornamental plants etc.

Knowledge on the methods of reproduction of crops is of primary importance from the standpoint of plant breeding. Besides, their annual or perennial nature also decides the methods of breeding to be adopted to get desired improvement. The agricultural crops are mostly annual (naturally self-pollinated/often cross-pollinated/ naturally cross-pollinated) and pose little problem for breeding. The horticultural crops may be asexual / sexual (naturally self-pollinated / often cross-pollinated / naturally cross-pollinated / dioecious cross-pollinated) and annual or perennial. Based on modes of reproduction and pollination, and habit, they can be classified as:

A. Annual Horticultural Crops

1. Naturally (Normally) Self-Pollinated Crops (e.g., tomato, cowpea)
2. Often Cross-Pollinated Crops (e.g., okra, chilli)
3. Naturally (Normally) Cross-Pollinated Crops (e.g., radish, pumpkin)

B. Perennial Horticultural Crops

1. Crops exclusively propagated by vegetative means (e.g., ginger, turmeric)
2. Crops propagated by both vegetative and sexual means (e.g., mango, guava)
3. Crops exclusively propagated by sexual means (e.g., coconut, arecanut)

C. Dioecious Horticultural Crops

1. Annual Crops (e.g., pointed gourd, little gourd, spine gourd, spinach)
2. Perennial Crops (e.g., asparagus, papaya, date palm, palm)

Common breeding methods can be applied to the naturally self-pollinated and often cross-pollinated annual crops. However, the parents chosen for breeding under often cross-pollinated crop need to be selfed for few generations to select desirable, homozygous lines. Again, common breeding methods can be applied to naturally cross-pollinated and dioecious cross-pollinated crops. But, in latter case both male and female parents must be improved and maintained.

1.2.3.1 Limitations of Horticultural Breeding

In the case of perennial horticultural crops due to their perennial nature, breeders face a lot of problems (limitations) such as:

a) Being heterozygous, on cross-fertilization, it is difficult to predict the performance of their hybrids.
b) In crops like garlic, ginger, turmeric, which either do not produce flowers or the flowers are sterile, hybridization is not possible.
c) Fruit crops like citrus produce apomictic seedlings, which complicate the isolation of hybrid seedling from the apomicts.
d) Hybrid seedlings require longer periods (say 6 to 10 years) to bear the economic crops.
e) Existence of self- and cross-incompatibility in fruit crops like apple makes it difficult to hybridize the selected parents.
f) Relatively larger land is required to grow and evaluate hybrids, involving high cost of maintenance.
g) The crossed fruits yield only one hybrid seed per fruit as in mango, coconut and cashew giving very small number of hybrid progenies for evaluation and selection.
h) Complex floral biology, e.g. presence of male and perfect flowers as in mango, cashew and avocado makes emasculation and crossing very laborious.

i) Hybridization becomes a difficult task in cases where crossed fruits yield sterile seeds.

Despite these difficulties, superior genotypes developed through breeding can be easily preserved by vegetative propagation in most of these crops except in seed propagated crops. Most of them are propagated by asexual means but some are exclusively propagated by sexual means.

1.2.4 Branches of Horticultural Breeding

There are many branches of horticultural breeding that are based on technique and experimental material used and plant trait improved. These branches are defined below:

1. **Fruit Breeding:** It deals with genetic improvement of various fruit crops.
2. **Vegetable Breeding:** It deals with genetic improvement of various vegetable crops.
3. **Spice Breeding:** It deals with genetic improvement of various spice and condiment crops.
4. **Tuber Crop Breeding:** It deals with genetic improvement of various tuber crops.
5. **Ornamental Crop Breeding:** It deals with genetic improvement of various flowers and ornamental crops.
6. **Plantation Crop Breeding:** It deals with genetic improvement of various plantation crops.
7. **Medicinal Crop Breeding:** It deals with genetic improvement of various medicinal crops.
8. **Stress Breeding (Resistance Breeding)**: It deals with development of crops resistant to biotic (diseases, insect pests and parasitic weeds) and abiotic (drought, salinity, alkalinity, acidity, frost, cold, heat, metal toxicity, nutrient deficiency etc.) stresses.
9. **Mutation Breeding:** It deals with genetic improvement of crops for various economic characters using physical and chemical mutagens.
10. **Quality Breeding:** It deals with genetic improvement of crops in relation to quality attributes such as market quality, industrial quality and nutritional quality.
11. **Ploidy Breeding:** It deals with genetic improvement of crops through manipulation of chromosome number.
12. **Introgressive Breeding:** It deals with genetic improvement of crops by transferring desirable genes from wild species into cultivated ones.

13. **Molecular Breeding:** It deals with genetic improvement of crops for various economic characters through indirect selection for molecular markers (DNA markers). It utilizes molecular techniques such as RFLP, AFLP, CAPS, RAPD, SSR etc. for selection of superior plants.
14. **Transgenic Breeding:** It deals with genetic improvement of crops for various economic characters through biotechnology (genetic engineering). It utilises transgenes bypassing sexual process.

1.3 Involvement of other Disciplines

To start a judicious breeding programme, a plant breeder should have good knowledge of the following disciplines:

1. **Genetics and Cytogenetics:** Plant breeding is applied branch of genetics. It involves various genetical principles. So, knowledge of genetics and cytogenetics is essential.
2. **Morphology and Taxonomy:** Some knowledge of angiospermic morphology and taxonomy is essential to a plant breeder for utilization of morphological variation and transfer of resistance genes from wild species.
3. **Plant Physiology:** For development of abiotic resistant and physiologically efficient cultivars, knowledge of plant physiology is essential.
4. **Plant Pathology:** For developing disease resistant cultivars, knowledge of plant pathology is essential.
5. **Entomology:** For development of insect resistant or tolerant cultivars, a plant breeder needs some knowledge of entomology.
6. **Crop Production Technology:** A breeder should be able to raise a good crop. So, knowledge of the principles of crop production is essential.
7. **Soil Science:** To develop salinity, alkalinity or acidity resistant cultivars a breeder should have knowledge of soil science.
8. **Biochemistry:** For quantification of vitamins and minerals in fruits and vegetables, some knowledge of biochemistry is required.
9. **Agricultural Engineering:** Development of crop cultivars suitable for mechanical harvesting needs knowledge of agricultural engineering. In many vegetable crops including tomato and potato, cultivars are available which can be harvested with machine.
10. **Statistics and Biometrics:** A plant breeder must evaluate germplasm and elite lines in field experiments. So, he/she should have good knowledge of statistical methods and experimental designs. Again, he/she must carry out biometrical analyses, namely, correlation, path coefficient, discriminant

function, diallel, partial diallel, line × tester, stability, D^2 statistics etc. Hence, he/she should have good knowledge of biometrical techniques.

11. **Computer:** Now computer programmes are available for various biometrical analyses. Again, computers are useful in compiling information and plant modelling studies. So, some knowledge of computer handling is essential for a plant breeder.
12. **Plant Biotechnology:** Plant Biotechnology is a combination of plant tissue culture and genetic engineering. Biotechnology is a useful tool for development of transgenic crop plants and molecular breeding. Thus, knowledge of plant biotechnology is essential for a plant breeder.

In short, a plant breeder must be a versatile person with a broad spectrum of knowledge of the above disciplines. He / She is not necessarily an expert in all these, but should be in a position to apply the knowledge and experience of the said disciplines in developing cultivars.

1.4 Activities of Plant Breeding

The products of plant breeding activities are improved cultivars. They should reach farmers to harness fruit of plant breeding. It involves five main steps, *viz.*, (i) creation of genetic variation, (ii) selection, (iii) evaluation, (iv) seed multiplication and (v) seed distribution.

1.4.1 Creation of Genetic Variation

Genetic variation is a pre-requisite in plant breeding programme. It can be created by domestication, germplasm collection, plant introduction, hybridization (inter varietal, distant, somatic), mutation, polyploidy, somaclonal variation and genetic engineering.

1.4.2 Selection

Selection is identification and isolation of plants having the desirable combinations of characters. It is based on phenotype. The efficiency of this activity determines the success of breeding programme. Various breeding methods have been designed to increase the efficacy of selection. Selection finally yields an improved line/strain/population.

1.4.3 Evaluation

The newly selected lines/strains/populations are tested for yield as well as other traits and compared with the existing best cultivars called **checks**. Evaluation is a stepwise process, ordinarily conducted at several locations for three or more years under the respective All India Co-Ordinated Crop Improvement Project.

Multilocation trials are also conducted within the state. If the new line/strain/ population is found superior to the checks, it is released and notified as a new cultivar, and its seed can now be multiplied and, more importantly, certified by a seed certification agency for quality.

1.4.4 Seed Multiplication

This step concerns with a large-scale production of certified seed of the released and notified cultivar. Seed production is usually done by seed production agencies in a stepwise manner (foundation and certified seed), and the seed is certified by a seed certification agency.

1.4.5 Seed Distribution

Certified seed is ultimately sold to the farmers who use it for commercial cultivation. This activity alone makes it possible to reap the economic benefit from the above four activities of plant breeding.

1.5 History of Plant Breeding

Richards (1994) recognises four distinct phases in the history of plant breeding (crop/plant improvement):

1. The era of farmer managed land races **(Phase-I)**
2. The 'hit or miss' phase of breeding—selection for yield and elimination of defects **(Phase-II)**
3. The **first designer phase** in which breeders focused on modified plant architecture—crop ideotypes for mono-cropping in well controlled farm environment **(Phase-III)**
4. The **second designer phase** in with transgenic crops are "constructed" **(Phase-IV)**

1.5.1 Phase I

Ever since the early man settled from shifting to sedentary agriculture, consciously or unconsciously he tried to improve crops to produce more from the piece of land he possessed. Initially, it was a slow process involving **selection** of suitable crop/crop varieties solely for food purpose. In ancient times, our forefathers in various parts of the world **domesticated** (brought into cultivation) a few hundred species from thousands available. In the process they transformed them consciously or unconsciously into crops. Through a long history of trial and error, a relatively few plant species have become the mainstay of agriculture including horticulture. This process of domestication involved the identification of certain useful wild species combined with a process of selection that brought about changes in appearance, quality and productivity. **The selection of naturally occurring variants is the basis of plant breeding**. Over thousands of years, this technique resulted in the development of modern crops. The discovery of techniques for **vegetative propagation** (asexual reproduction), such as by using natural offshoots, rooted stem cuttings, or various grafting techniques, made it possible to fix genetic variants. It was the technique used for many fruit trees enabling identical plants to be cultivated in orchards. As humans carried improved crops to new locations, opportunities opened to increase genetic variation from natural intercrosses with new wild populations.

1.5.2 Phase II

In the 18th and 19th centuries, a conscious attempt was made to predict the performance of plants that could be expected from one generation to the next. The concept of **pedigree** (ancestry) led to refinement in selection process by keeping records and the assessment of lineage. Furthermore, it became obvious that variation could be managed by controlling the mating process. This new type of selection was named **'pedigree selection'** and it was found to increase the efficiency of the process. **'Progeny testing'**, evaluating the genetic worth by performance of the progeny, increased the efficiency of this process. The **commercial plant breeding** began in the second half of the 19th century by seed producers. It involved hybridization (crossing) between selections to control genetic recombination followed by selection of improved types. This is still the basis of **conventional** (also called **traditional, classical,** or **general**) plant breeding methods. *Interestingly, much of this early type of plant breeding was carried out without a clear understanding of the genetic mechanism involved in inheritance.*

1.5.3 Phase III

The modern plant breeding methods have their bases in the genetic principles. The science of genetics began with rediscovery of Mendel's laws of inheritance in 1900. This was immediately followed by a steady development of scientific breeding based on genetic principles. Based on sound genetic principles, plant breeding marched on by achieving favourable accumulations and combinations of genes through hybridization. A gene discovered in a wild plant could be transferred to a suitable adapted cultivar by the technique of **backcross**. In the early 20th century, it was demonstrated that **hybrid vigour** or **heterosis** (extra vigour) could be exploited in plant breeding. This resulted in the development of **hybrid maize** (corn) which led to enormous increase in productivity. The success with maize prompted exploitation of heterosis not only in cross-pollinated crops, but also in self-pollinated crops such as **tomato** and rice. The 20th century has, therefore, seen a tremendous increase in crop yield. Detailed analyses have indicated that *half of the increase can be attributed to improvement introduced through plant breeding* and half through agronomy (crop production technology), each dependent on the other.

During second half of 20th century, high yielding varieties (**HYV**), especially of wheat and rice, were developed using **semi-dwarf** ***Norin 10*** and ***Dee-geo-woo-gen*** lines, respectively. These HYVs were semi-dwarf, lodging-resistant, thick-stemmed, responsive to irrigation and added chemical fertilizers, generally resistant to pests and diseases, and photoperiod-insensitive. *The developed wheat cultivars saved over one billion people from starvation in India and Pakistan*. The IR-8, the "Miracle Rice" and subsequently developed rice cultivars boosted the rice production in the world. Thus, in true sense the era of **Green Revolution** was heralded.

1.5.4 Phase IV

Recombinant DNA technology, also called **transgenic technology** or **genetic engineering**, is the most powerful and revolutionary of the new genetics developed in the second half of the 20th century. It is possible to isolate stretches of DNA (deoxy ribonucleic acid) from one organism, store in a bacterial host, select unique combinations, then incorporate them into DNA of other species, where it can be expressed. Refinement in the technique makes it possible to create mutations in desired genes, further increasing variability. Other uses of molecular biology known as **genomics** involve the detailed mapping of the DNA and the identification of useful stretches of the molecule. This makes it possible to improve the efficiency of selection (**marker-assisted selection**), because it is directly based on the genes rather than the organism, where the effects may be confounded by environment and genetic interactions. *The use of genetic engineering in agriculture and horticulture to create genetically modified foods is viewed by many as the natural culmination of the* ***green revolution.***

1.6 Major Contribution by Some Indian Plant Breeders

In India remarkable contribution has been made in the field of plant breeding. The significant contribution of some Indian plant breeders is briefly presented below:

1. **T.S. Venkatraman** (1884-1963): He was an eminent sugarcane breeder. He transferred thick stem and high sugar content from tropical noble cane to north Indian canes. This process is known as noblization of sugarcane. He developed several cultivars of sugarcane having high yield and high sugar content. He was Director of Sugarcane Breeding Institute, Coimbatore.

2. **B.P. Pal** (1906 -1989)**:** He was a pioneer wheat and **flower** breeder. He is known for his contribution to the breeding of superior disease resistant N.P. cultivars of wheat. He was the first Director General of ICAR.
3. **M.S. Swaminathan** (b.1925): He is a famous plant breeder responsible for green revolution in India. He developed high yielding cultivars of wheat and rice. He was Director General of ICAR and later Director General of IRRI, Philippines from 1982-1988. He coined the much quoted "There is a horticulture remedy for every nutritional malady". Voted one among the three great Asians-Mahathma Gandhi, Rabindranath Tagore and M.S. Swaminathan- he becomes history and a household name.
4. **Pushkarnath**: A famous **potato** breeder who developed several high yielding cultivars of potato. He was Director of CPRI, Shimla.
5. **N.G.P. Rao** (1927-2016): He was eminent sorghum breeder. He developed several high yielding hybrids of grain sorghum. He was chairman of ASRB and Vice-Chancellor of Marathwada Agricultural University, Parbhani.
6. **K. Ramiah** (1892-1988)**:** He was a renowned rice breeder. He developed several high yielding cultivars of rice. He was Director of CRRI (now NRRI), Cuttack.
7. **Ram Dhan Singh** (1891-1977): He was an inborn plant breeder who developed several superior cultivars of wheat in Punjab. His famous cultivar was C 591.
8. **D.S. Athwal** (1928-2017)**:** He was a famous pearl millet breeder. He worked in PAU, Ludhiana and developed several superior cultivars of pearl millet (bajra).
9. **Basiswar Sen (Boshi Sen)** (1887-1971): He was an eminent maize breeder. He developed several superior cultivars of maize for hill region of Uttaranchal. He established the Vivekananda Parvatiya Krishi Anusandhan Shala at Almora,Uttaranchal and was its Director till his death.
10. **Dharampal Singh**: He, an eminent oilseed breeder, released several cultivars of oilseed (rapeseed and mustard) from Kanpur. He was Vice-Chancellor of G.B. Pant University of Agriculture and Technology, Pantnagar.
11. **C.T. Patel** (1917-1990): He was a famous cotton breeder who developed world's first cotton hybrid in 1970 for commercial cultivation in India. He is known as father of hybrid cotton.
12. **V. Santhanam** (b.1925): He is a famous cotton breeder who has developed several high yielding cultivars of upland cotton and two cultivars of Egyptian cotton. He was the first Project Coordinator for All India Cotton Improvement Project and later worked in FAO as cotton consultant.

13. **D.C. Mohanty** (1939-2016): He was a pioneer **spices** breeder who developed several ginger and turmeric cultivars. He released the first ginger cultivar 'Suprabha' in India. He worked at High Altitude Research Station of OUAT at Pottangi in Odisha.
14. **K.V. Peter** (b.1948): He is associated with development and release of improved and high yielding cultivars/hybrids in tomato, brinjal, chilli, bittergourd, melons, amaranths and cowpea. Sources of resistance to bacterial wilt in tomato, chilli and brinjal, aphids in cowpea and viral leaf curl in chilli were located by him and are now used in breeding programmes. Prof. Peter provided managerial support to the Indian Institute of Spices Research, Calicut to possess the World's largest germplasm of black pepper and cardamom. He is a prolific writer and academic editor. He has authored/ edited 74 books published both in India and abroad. Director, World Noni Research Foundation, Chennai (2007); Vice-Chancellor, Kerala Agricultural University (2001-2006); Director of Research, Kerala Agricultural University (1999-2001); Director, Indian Institute of Spices Research, Calicut (1991-1999); Professor of Horticulture, KAU (1979-2008); Assistant Professor, GBPUAT, Pantnagar (1975-1979).
15. **Hari Har Ram** (b.1948): He released and commercialized 74 cultivars / hybrids of major vegetables and soybean at GB Pant University of Agriculture and Technology, Pantnagar and Krishidhan Vegetable Seeds India Pvt Ltd, Pune. He headed product research and development on major vegetable crops like hot-pepper, tomato, okra, eggplant, gourds, cucumber, onion, peas and beans in Krishidhan Vegetable Seeds India Pvt Ltd, Pune, and taught genetics/plant breeding/vegetable breeding to PG students, conducted research on breeding/genetics/seed production of vegetables and soybean at GBPUAT, Pantnagar in various capacities starting as Asst. Professor and ending as Professor and Head Vegetable Science/Programme Coordinator-Vegetable Research/Coordinator-Seed Science and Technology, General Manager Seed Production at Univ. Farm and Nodal Officer, Pantnagar Centre for Plant Genetic Resources. He authored/edited 15 books out of which (1) Vegetable Breeding-Principles and Practices, (2) Crop Breeding and Biotechnology and (3) Genetic Resources and Seed Enterprises are being used as textbooks at post graduate level in State Agricultural Universities.
16. **K.L. Chadha** (b.1936): His research pertains to nutrition, use of plant growth regulators and variety improvement in important fruit crops like mango, grape, pineapple, citrus and stone fruits. He is credited with establishing the Central Mango Research Station at Lucknow from scratch and he gave a new direction to research programmes under the All-India Coordinated Fruit

Improvement Project. He served as Project Coordinator (Mango, Grape, Guava, Apple, Stone Fruits and Walnut) and Head, Central Mango Research Station, Lucknow. He was Director, IIHR, Bengaluru from 20.05.1981 to 21.01.1986. Thereupon, he was appointed as the first Horticulture Commissioner, Govt. of India, New Delhi, followed by his appointment as the first Deputy Director General (Horticulture) at ICAR headquarters in New Delhi. He added wealth of his experience through a number of publication (30 books and bulletins) including 'Advances in Horticulture' - a monumental publications in 13 volumes covering 9410 pages. Being the doyen of horticulture, Dr. Chadha is aptly referred to as the 'Father of Modern Horticulture'. In the words of Dr. M.S. Swaminathan "Dr K.L. Chadha has provided outstanding leadership to horticultural renaissance of India - he has rendered invaluable service for which we are all very grateful" a phenomenon which has led to the Horticulture (Golden) Revolution in India.

17. **Kirti Singh** (b.1934): His research pertains to vegetable science. He was Dean of Agriculture, Narendra Dev University of Agriculture and Technology, Faizabad (UP), 1977-84 and Vice-Chancellor, 1984-89; Vice-Chancellor, Himachal Pradesh Agricultural University, Palampur, 1989-93; Indira Gandhi Agricultural University, Raipur, 1993-95; Member, 1995-98 and Chairman, ASRB, 1998-99; Advisor, Govt. of Nepal for Establishment of Agricultural University, 1995, FAO consultant in Cambodia, 1999.

18. **G. Kalloo**: He developed several vegetable cultivars. He was Director of Indian Institute of Vegetable Research, Varanasi (UP) during Sept 1999 - Sep 2001, DDG (Horticulture), ICAR, New Delhi and Vice Chancellor, JNKVV, Jabalpur, MP. Out of his number of books, Genetic Improvement of Vegetable Crops, Genetic Improvement of Tomato and Vegetable Breeding (3volumes) are popular among PG students.

1.7 Goals of Horticultural Breeding

The goals or objectives of breeding differ in different crops at different places depending on local preferences. However, the general goals are discussed below:

1. **Higher productivity**: Higher yield per unit area is one important character across fruits, vegetables and flowers. The economic part may differ in different groups–may be fruits, seeds or flowers.
2. **Better quality:** Quality in general perception includes all those parameters that affect its marketing, shelf life/vase life, edibility and nutrition. Fruit shape, size, colour, firmness, flavour, pigments, vitamins, proteins, essential amino acids and minerals are quality attributes in various vegetable and

fruits. Longer keeping quality and vase life are important quality parameters in flowers.

3. **Maturity duration**: Crop rotational efficiency is enhanced by cultivation of short duration cultivars. Early maturing cultivars are available in cauliflower, radish, knolkhol, tomato, peas, potato, onion etc.

4. **Resistance against insect pests and diseases**: Pest attack often affects plant part that is directly consumed by human being. Spraying of chemicals for pest control often leads to pesticide contamination causing health hazards and increases the cost of cultivation. So, breeding for resistant cultivars in vegetables and fruits is required. In some vegetables like tomato, potato and brinjal pest resistant transgenic cultivars have been developed.

5. **Photo-insensitivity**: Development of photo-thermo-insensitive cultivars has considerably increased the area of cultivations in several vegetables. Breeding of high temperature tolerant strains in temperate vegetables has enabled their cultivation in intermediate as well as sub-tropical areas. Conversely, cold tolerance in tropical vegetables is helping in development of strains for temperate areas.

6. **Longer blooming period**: Longer blooming or flowering period is one important character in flowers and ornamentals.

7. **Determinate growth:** Determinate growth along with short duration is desirable. Determinate types are physiologically more efficient. In tomato, peas etc. determinate strains have been found.

8. **Drought resistance:** Vegetables are considered crops of assured irrigation area. Cultivars with resistance to moisture stress have been developed but number is few. Green peas, some leafy vegetables and melons in cucurbits offer scope for cultivation under rainfed farming. Breeding drought resistant cultivars is needed in vegetables to make them important component of rainfed farming.

9. **Reduction in anti-nutritional factors:** Trypsin inhibitors, oxalic acid, phenolic contents, nitrate content, bitter principles of cucurbits are some undesirable molecules in vegetables. Breeding efforts are used to reduce or eliminate them.

10. **Cold tolerance:** It is especially desired in summer vegetables at seedling or early vegetative growth stage. For instance, cold tolerance in chilli, tomato and brinjal help in their nursery raising during mid-winter or else the seedling have to be provided protection. Tolerance to frost is required even in winter vegetables.

1.8 Horticultural Breeding in India

1.8.1 Histroy

Organised agricultural research in India dates back to 1871 when the Government of India created the Department of Agriculture. The first scientist appointed in the department in 1892 was an agricultural chemist. In 1905, the Imperial Agricultural Research Institute was established in Pusa, now in Bihar. This was the first agricultural research institute in the country. The buildings of this institute were damaged by an earthquake in 1934. Therefore, it was shifted to New Delhi in 1936. Its name was changed to the Indian Agricultural Research Institute in 1946.

Agricultural colleges were established at Kanpur, Pune, Sabour, Llyalpur (Faisalabad, now in Pakistan) and Coimbatore between the years 1901 and 1905. The main objective of these colleges was to impart agricultural education and training. About this time, several Agricultural Research Farms were established in each province. But the progress in agricultural research was disappointingly slow. In view of this, the Imperial Council of Agricultural Research was established in 1929, its name was changed to the Indian Council of Agricultural Research (ICAR) in 1946.

The Indian Central Cotton Committee was established in 1921. The committee carried out many notable research in breeding and cultivation of cotton, e.g., development of 70 improved cultivars of cotton. Encouraged by these results, 9 central commodity committees were set up on jute, sugarcane, tobacco, oilseed, **coconut, arecanut, spices, cashewnut** and lac. Later, these committees were merged with the ICAR.

In 1956, a Project for Intensification of Regional Research on Cotton, Oilseed and Millets (PIRRCOM) was initiated to intensify research on these crops. The PIRRCOM was located at 17 different centres spread over throughout the country. It focussed on cotton, castor, groundnut, *Brassica* spp., til, toria, taramira, jowar and bajra.

The All India Coordinated Maize Improvement Project started in 1957 with the objective of exploiting heterosis. The first hybrid maize cultivars developed under the project were released in 1961. The phenomenal success of this project prompted the ICAR to initiate coordinated project for the improvement of other crops as well. The All India Coordinated Vegetable Improvement Project (AICVIP) was established in 1970-71, which was upgraded in 1987 to the Project Directorate of Vegetable Research. The Indian Institute of Horticultural Research (IIHR) was established at Bangalore (Bengaluru) in 1967. The ICAR was reorganised twice in 1966 and 1973 with a view to improve its efficiency. The first agricultural university was established in 1960 at Pantnagar, Nainital (U.P).

Subsequently, such universities were established in other states of the country. Presently, the 47 State Agricultural Universities (SAUs), 3 Central Agricultural Universities (CAUs) and 4 Deemed Universities have the responsibility of education, research and extension in the different areas of agriculture. In addition, 10 Research Institutes, 6 Directorates, 7 National Research Centres (NRCs), 13 All India Coordinated Research Projects (AICRPs) and 6 Network Projects/ Outreach Programmes of ICAR are engaged in improvement of horticultural crops.

1.8.2 Major Achievements

Relatively less attention was given in India to horticulture in the first six five-year plans. In the past, plant breeders have not paid so much attention to vegetables and fruits, let alone flowers and ornamental plants, as they have given to agricultural crops.

1.8.2.1 Vegetable Breeding

More than 70 types of vegetables are grown on a national level in the country. But, maximum emphasis is on vegetables like tomato, brinjal, chilli, cauliflower, cabbage and peas. Not much effort is made to enlarge the area under minor and exotic vegetables, even though they hold the key to export market. Before the launching of research projects for improvement of vegetable crops, introductions were made from various countries. For example, cultivars 'Arkel' and 'Bonneville' of garden pea were introduced from the United Kingdom and the United States of America, respectively. Cultivars 'Sioux' of tomato (from the USA), 'Sugarbaby' of watermelon, 'Early Grano' of onion (from the USA) and 'Pusa Chetki' of radish (improvement from material introduced from Denmark) became popular in India. Production of vegetables and fruits has increased in India after independence.

The real impact of vegetable research on enhanced productivity was seen with the release of **first tomato hybrid 'Karnataka' in 1973** and **first capsicum hybrid 'Bharat'** by the Indo-American Hybrid Seed Company based at Bangalore (Bengaluru). Several hybrids of tomato, cabbage, cauliflower, capsicum, okra (lady's finger), chilli, brinjal (egg plant), gourd-melons and cucumbers have been developed and released for cultivation in subsequent years by both public and private sector research organizations. Hybrids of vegetable crops which carry heat tolerance, for example cabbage, have been developed which enabled off-season cultivation of these vegetables. Watermelons, which were earlier available only during summers in India, are now available in winter months of November- December. Cauliflower is raised during March to May in north-western plains of the country enabling availability in hot season.

1.8.2.2 Fruit Breeding

Breeding of fruits has lagged. This could be attributed to a few factors like:

i) More time requirement

ii) Laborious task

iii) Mode of inheritance of several fruit crops is not yet perfectly known.

iv) Varying degrees of self-incompatibility mechanism is observed among fruit crops.

v) Long juvenile phase

vi) Long time is required to identify promising hybrid combinations.

Despite these hurdles, breeders have made considerable contribution to fruit crops. Regular bearing habit has been incorporated in mango through breeding. For example, Mallika, Amrapalli, Ratna and Sindhu are recently bred regular bearing mango cultivars. In Punjab, a hybrid 'Kinnow' ('King' × 'Willow leaf') has replaced the traditional citrus fruits such as sweet orange and local mandarin. High yielding cultivars of papaya, e.g.,'Honey Dew', 'Pusa Dwarf', 'Pusa Giant', 'Pusa Delicious'; hybrids of grapes such as 'Arka Hans', 'Arka Kanchan', 'Arka Shyam' have been released for cultivation. Hybrids of apple 'Lal Ambri' ('Red Delicious' × 'Ambri') and 'Sunehari' ('Ambri' × 'Golden Delicious') have also been evolved and released for cultivation. Ambri is a local type of apple grown in Jammu and Kashmir.

1.8.2.3 Ornamental Crop Breeding

Plant breeding has also played a significant role in the evolution of novel cultivars of gladiolus, rose, chrysanthemum etc.

1.8.3 Present Scenario

Globally, India is the second largest producer of fruits and vegetables; largest producer of mango, banana, coconut, cashew, papaya and pomegranate; largest producer and exporter of spices; ranks first in productivity of grapes, banana, cassava, peas and papaya.

There is a total accession of 72,974 of genetic resources (germplasm) in horticultural crops maintained in the country including 9,240 accessions in fruits, 25,400 accessions in vegetables and tuber crops, 25,800 accessions in plantation and spice crops, 6,250 accessions in medicinal and aromatic plants, 5,300 accessions in ornamental plants and 984 accessions in mushroom. Molecular characterization of available germplasm has been done in several horticultural crops including mango, banana and citrus.

A total of 1,596 high yielding cultivars and hybrids of horticultural crops including 134 in fruits, 485 in vegetables, 115 in ornamental plants, 467 in plantation and spice crops, 50 in medicinal and aromatic plants and 5 in mushrooms were developed. As a result, the productivity of horticultural crops *viz* banana, grapes, potato, onion, cassava, cardamom, ginger and turmeric has increased significantly. The cultivars of quality standard for export purpose have been developed in apple, mango, grapes, banana, orange, guava, litchi, papaya, pineapple, sapota, onion, potato, tomato, peas and cauliflower. The cultivars for processing purpose and the cultivars tolerant/resistant to various biotic and abiotic stresses have been developed in different fruits, vegetables, and medicinal and aromatic plants. By using biotechnological tools, transgenic in brinjal and tomato has been developed.

1.8.4 Goal Setting for Future

Plant breeding is an ongoing process always striving for better cultivars in as many aspects as feasible. The future goals of horticultural breeding in the country may be as follows:

1. The cultivars of most of the vegetable crops are poor yielder and season-bound; respond poorly to higher doses of nutrients. Their shelf life and resistance to bruises in transportation are not optimum. Breeders should strive to develop cultivars that are high yielding, responsive to nutrients, photo-thermo-insensitive, resistant/tolerant to bruises and having good shelf life. Drought resistant cultivars are required in vegetables to make them important components in rainfed farming. Development of cultivars for non-traditional areas is another important area of vegetable breeding. Breeding for biotic stresses may be assisted by biotechnological tools. Quality breeding must be stressed upon for improvement in quality as per local preferences.

2. If breeding annual agricultural crops takes time, imagine the patience required to breed trees or other perennials. In trees that bear fruit, after planting the first hybrid, a breeder must wait upto 10 years for the hybrid to flower for use in a breeding programme. It is due to the long juvenile phase. This is the phase after germination during which the tree does not yet flower or bear fruit. Because several consecutive crosses are necessary to obtain a new cultivar, breeding a new fruit tree can easily take 30 years. The juvenile phase can be shortened by means of DNA technology. The method is based on the discovery of genes that determine when a plant flowers. Under the influence of specific internal and external signals, these flowering genes are activated, after which the plant prepare itself to flower and bear fruit. Over expression of these flowering genes can be induced in a tree through genetic

engineering, meaning that the seedlings will not wait 5 to10 years to flower for the first time but rather do so in the first year. The total breeding period of fruit tree can be shortened in this way from 30 years to 5 years. Like the method of reverse breeding, the extra flowering genes added can be removed again in the final breeding step leading to an end product that does not differ at all from a classic hybrid. This technique is called '**fast-track breeding**' and used nowadays in combination with marker-assisted breeding to introduce disease resistance to apple trees and plum trees. Fast-track breeding can be adopted in tropical and sub-tropical fruit trees (e.g., mango, guava, citrus, sapota etc.).

3. Novel pattern of flowers and ornamental plants may be bred with longer blooming period, keeping quality and vase life with resistance to diseases and insect pests.
4. Breeding of spices, tubers, aromatic and medicinal plants, plantation crops etc. should be given proper attention with conventional methods assisted by biotechnological tools.

1.9 Undesirable Consequences of Plant Breeding

Plant breeding has several useful applications in the improvement of crop plants. However, it has four undesirable consequences on crop plant *viz* (1) reduction in genetic diversity, (2) narrowing of genetic base, (3) danger of uniformity and (4) undesirable combinations.

1.9.1 Reduction in Genetic Diversity

Modern improved cultivars are more uniform than land races. Thus plant breeding leads to reduction in genetic diversity. The uniform cultivars are more prone to the new races of pathogen than land races which have high genetic diversity.

1.9.2 Narrowing of Genetic Base

Uniform cultivars have narrow genetic base. Such cultivars generally have poor adaptability. On the other hand, diverse genotypes like land races or primitive cultivars have wide adaptability.

1.9.3 Danger of Uniformity

Most of the improved cultivars have some common parents in the pedigree which may cause danger of uniformity. However, it can be avoided through development of multiline cultivars in self-pollinated crops, and synthetic and composite cultivars in cross-pollinated crops.

1.9.4 Undesirable Combinations

Sometimes plant breeding leads to undesirable combination. The examples of man-made crops having undesirable combinations of characters are *Raphanobrassica* and *Pomato.*

1.10 Brief References

Abraham 2017, Acquaah 2012, Breseghello and Coelho 2013, Chaudhari 1971, Chaudhary 2014, Dabholkar 2006, De and Bhattacharjee 2011, Janardhan 2012, Kumar 2018, Peter *et al.* 2015, Poehlman 1987, Rajani and Joshi 2017, Ram H.H. 2012, Ram M. 2014, Sharma J.P. 2009, Simmonds 1979, Singh B.D. 2011, 2018; Singh Phundan 2011, 2013a; Singh and Pawar 2006, VIB 2016, http://www.icar.org.in/en/horticulture.htm

Review Questions

Part A

Answer the following questions true or false.

1. Plant breeding is *only* science and technology. (True/False)
2. Dr. B.P. Pal was a famous flower breeder. (True/False)
3. Dr. D.C. Mohanty released the first turmeric cultivar in India. (True/False)
4. All India Coordinated Vegetable Improvement Project was established in 1970-71. (True/False)
5. Resistance breeding is otherwise known as stress breeding. (True/False)
6. Plant breeding leads to reduction in genetic diversity. (True/False)
7. Fruit plants have short juvenile phase. (True/False)
8. The Indian Central Cotton Committee was established in 1921. (True/False)
9. G.B. Pant University of Agriculture and Technology is the second oldest agricultural university in India. (True/False)
10. Plant breeding is nature-directed evolution. (True/False)
11. 'Amrapalli' is a cultivar of mango. (True/False)
12. 'Sindhu' is not a regular bearing mango cultivar. (True/False)
13. Fruit breeding is not a laborious task. (True/False)
14. Organised agricultural research in India dates back to 1871. (True/False)
15. Evaluation is not an important step in plant breeding. (True/False)

Part B

Answer the following questions.

1. Define horticultural breeding.
2. Plant breeding leads to reduction in __________
3. Activities in plant breeding include__________, __________, __________, and __________.
4. What is transgenic breeding?
5. When was ICAR established?
6. Dr. Pushkarnath developed__________cultivars.
7. Genetic resources is also called__________.
8. Who was the first Director General of ICAR?
9. Name the first tomato hybrid in India.
10. What is a pre-requisite in a plant breeding programme?
11. Name the first capsicum hybrid in India.
12. Who is Dr. M. S. Swaminathan?
13. Longer blooming period is desirable in __________.
14. Define molecular breeding.
15. Write one synonym of genetic engineering.

Part C

Write a brief note on each of the following topics.

1. Fast-track breeding.
2. Goals of horticultural breeding.
3. Branches of horticultural breeding.
4. Limitations of horticultural breeding.
5. Major achievements in horticultural breeding.
6. Activities of plant breeding.
7. Undesirable consequences of plant breeding.
8. Agricultural breeding versus horticultural breeding.
9. Vegetable breeding in India.
10. Fruit breeding in India.
11. Present scenario of horticultural crop breeding.

12. History of plant breeding—Phase III.
13. History of plant breeding—Phase IV.
14. Cold tolerance.
15. Photo-insensitivity

Part D

1. Plant breeding is an art, a science and a technology. Explain it.
2. A plant breeder should have good knowledge of the cognate disciplines. Discuss.
3. Describe the history of horticultural breeding in India.
4. What are the steps of plant breeding?
5. Summarize the history of plant breeding phase-wise.
6. Plant breeding may be defined as the art, multidisciplinary science and technology of developing genetically better cultivars with the application of genetic principles and practices to upgrade the productivity of crop plants per unit area, time and input (energy) coupled with tolerance to biotic and abiotic stresses and adaptability to cropping system and socio-economic reach of the farmers of the region/state. Explain with examples.

2

Plant Biotechnology An Introduction

2.1 Introduction

Dear students, you shall be taught plant biotechnology during the third semester in the course 'Elementary Plant Biotechnology'. But, for understanding of non-conventional breeding methods/techniques you should acquire some knowledge of basic plant biotechnology. Hence, in this chapter plant biotechnology is introduced to you in brief.

2.2 Definition

The term **'biotechnology'** stems from fusion of **'biology'** and **'technology'**. It was coined by Karl Ereky, a Hungarian agricultural engineer, in the beginning of 20th century. According to him, all processes used to yield finished products from raw materials with the help of living organism come under biotechnology. It may be defined in number of ways. However, in simple terms, **biotechnology** may be defined as 'the controlled use of biological agents, such as microorganisms, or cellular components, for beneficial use' (U.S. National Science Foundation).

2.3 Branches of Biotechnology

Biotechnology is divided into five branches based on its application. They are briefly discussed below:

2.3.1 Animal Biotechnology

It deals with development of transgenic animals for improvement of milk or meat production with resistance to various diseases. It also deals with *in vitro* fertilization and transfer of embryo in animals including human being.

2.3.2 Medical Biotechnology

It deals with diagnosis of various diseases; large scale production of various drugs and hormones such human insulin and interferon; vaccines for chicken pox, rabies, polio etc.; and growth hormones such as bovine.

2.3.3 Industrial Biotechnology

It deals with commercial production of various useful organic substances such as acetic acid, citric acid, acetone, glycerine etc. and antibiotics like penicillin, streptomycin, mitomycin using microorganisms especially fungi and bacteria.

2.3.4 Environmental Biotechnology

It deals with detoxification of waste and industrial effluents, treatment of sewage water, and control of plant diseases and insects using biological agents such as viruses, bacteria, fungi etc.

2.3.5 Plant Biotechnology

Plant biotechnology is a combination of tissue culture and genetic engineering. It deals with development of transgenic plants with resistance to biotic and abiotic stresses; development of haploids, embryo rescue, clonal multiplication, cryopreservation etc.

2.4 Plant Biotechnology

Plant biotechnology makes use of two basic approaches/techniques, *viz*., (i) tissue culture and (ii) genetic engineering. They are discussed in brief.

2.4.1 Plant Tissue Culture

Plant tissue culture is the *in vitro* (literally means "under glass") manipulation of plant cells and tissues, which is a keystone in the foundation of plant biotechnology. It is useful for plant propagation and the study of plant hormones, and is generally required to manipulate and regenerate transgenic plants. Whole plants can be regenerated *in vitro* using tissues, cells, or a single cell to form whole plants by culturing them on a nutrient medium in a sterile environment.

2.4.1.1 Important Terms

Important terms such as nutrient medium, explants, totipotency and callus which are frequently used are explained as follows:

1. *Nutrient Medium*: A medium (pl. media) which contains important nutrients such as essential micro- and macro-nutrients, carbohydrates, vitamins and hormones is known as **nutrient medium** (also called **culture medium**, simply **medium**). Murashige and Skoog (1962) first gave the composition of culture medium. As culture medium differs from one species to other, several modifications of the MS medium have been suggested by various workers. Hence, if a culture medium is not standardized for any plant species, a suitable one should be developed.

2. *Explants*: The plant part that is used for regeneration is called **explant.** It may be a cell, a protoplast, a tissue or an organ. The organ includes any part which has a separate identity such as anther, ovule or bud.
3. *Totipotenccy*: The regeneration ability of a plant cell to develop into a whole plant is referred to as **totipotency**.
4. *Callus*: A mass of unorganized regenerated cells in culture medium is called **callus** (pl. calli). The callus is generally of two types, *viz.*, (i) friable callus and (ii) compact callus. The former can be easily manipulated for suspension; the latter is not suitable.

2.4.1.2 Basic Requirements

In tissue culture techniques, there are certain basic requirements, *viz.*, (1) aseptic conditions, (2) control of temperature, (3) proper culture medium, and (4) sub-culturing. They are briefly described as follows:

1. *Aseptic Conditions*: The tissue culture laboratory must have aseptic conditions. It means it should be well-sterilized against pathogens. A pathogen-free environment will help in maintaining good health of the callus, cell, or protoplast cultures resulting in recovery of healthy plants from such cultures. The explants and glassware should be properly sterilized.
2. *Control of Temperature*: The tissue culture laboratory must be air-conditioned. Usually, 18-25° C temperature is maintained. However, this value varies from one species to other. High temperature adversely affects the growth of callus.
3. *Proper Culture Medium*: Culture media have been developed by various workers for different crop species. The medium must be modified as per requirement of species.
4. *Sub-culturing*: Transfer of tissue or callus from old culture media to fresh ones is called **sub-culturing**. It is necessary to maintain good health of the callus or tissues, because after a period some nutrients are depleted in the culture media.

2.4.1.3 Important Steps

Tissue culture technique generally consists of four important steps, *viz.*, (1) isolation of tissues, (2) regeneration and callus formation in culture media, (3) embryogenesis, and (4) organogenesis. They are described as follows:

1. *Isolation of Tissues*: Tissues for regeneration can be isolated with the help of sterilized blade from any plant part, *viz.*, leaf, stem, apical bud, axillary bud etc. The isolated tissues are sterilized and then grown on culture medium. Tissues should be isolated from disease-free portion.

2. *Regeneration and Callus Formation in Culture Media*: Tissues proliferate on the culture medium and give rise to callus which may be friable or compact. The friable callus can be easily manipulated for suspension culture. However, compact one is not suitable for it.
3. *Embryogenesis*: The process of formation of somatic embryos from the callus is called **embryogenesis**. Sometimes, instead of somatic embryos somatic buds are formed which on germination give rise to a plant. Sub-culturing leads to healthy growth of callus and embryogenesis.
4. *Organogenesis*: The process of differentiation of shoot and root from somatic embryos is called **organogenesis**. Sometimes, a complete plant develops directly from the somatic bud. In such case somatic embryos are not formed. The plants thus obtained are transferred to pots with sterilized soil.

2.4.1.4 Types of Plant Tissue Culture

Various plant parts, namely, meristem, embryo, anther, cell and ovule are used for regeneration. Depending on plant parts used as explants, plant tissue culture is of five main types as briefly described below:

1. *Meristem Culture*: Meristem culture refers to regeneration of a whole plant from meristems such as stem tip, root tip or axillary bud. Usually, shoot apical meristem is used for regeneration. This culture has been widely used in vegetatively propagated crops.
2. *Embryo Culture*: Regeneration of a whole plant from an embryo in the culture medium is called **embryo culture**. It is used to make distant (wide) crosses successful.
3. *Anther Culture*: Regeneration of a whole plant from an anther or a pollen in the culture medium is called **anther** or **pollen culture**. It is used to obtain haploid plants.
4. *Cell Culture*: Regeneration of a whole plant from callus and suspension cultures in the culture medium is known as **cell culture**. Protoplasts (cells without cell wall) are also used for regeneration.
5. *Ovule Culture*: Regeneration of a whole plant from an ovule in the culture medium is called **ovule culture**. It is also known as **ovary culture**. This technique is used for embryo rescue in distant crosses. Haploids plants are obtained from culture of unfertilized ovules and diploid plants from the fertilized ones.

2.4.2 Genetic Engineering

Plant genetic engineering is defined as the isolation, introduction and expression of foreign DNA in the plant. In other words, it refers to direct introduction of foreign gene (DNA) into a plant system by micromanipulation at the cellular level. Gene transfer can be achieved by four methods, *viz*., (1) plasmid method, (2) particle bombardment method, (3) direct DNA transfer method and (4) micro-injection method. Each method is described in brief.

2.4.2.1 Plasmid Method

Plasmids are extra-chromosomal elements that are found in the bacterial cell. They replicate independent of chromosomal DNA and are not essential for normal growth and function of a bacterium. They are used as cloning vectors. The soil-borne bacterium, *Agrobacterium tumefaciens* is used for development of transgenic plants. Main features of this method are described below:

1. *Gene Cloning*: A technique that is used to make several identical copies of a gene is called **gene cloning**. Plasmids of *Agrobacterium* contain tumour producing genes (Ti genes or oncogenes). The foreign gene which must be used for genetic transformation is incorporated in place of oncogene for cloning. Thus, several identical copies of the transgene are produced by self-replication of plasmids.
2. *Genetic Transformation*: The process of transfer, integration and expression of **transgene** (foreign gene) in the host cells is known as **genetic transformation**. For genetic transformation, bacteria with cloned gene are mixed (in millions) in the cell or protoplast culture of the host plant for a day or two. Out of several thousands of cells or protoplasts in the culture medium, only a few are transformed. In other words, plasmid DNA (foreign DNA) can enter, integrate and express in a few cells or protoplasts.
3. *Identification of Transformed Cells*: A marker gene is included with the gene under consideration for easy identification of transformed cells. Generally, either antibiotic resistant gene, such as kanamycin gene, or herbicide resistant gene is used for identification of transformed cells. The plasmid (bacteria) mixed suspension or protoplast culture is grown in the culture medium containing lethal concentration of antibiotic (kanamycin) or herbicide. The non-transformed cells will die and only transformed cells or protoplasts will survive. Thus, transformed cells or protoplasts can be identified.
4. *Regeneration of Transformed Cells*: The transformed cells or protoplasts are transferred into culture medium for regeneration into whole plant. The regeneration of cell or protoplast into whole plant is a basic requirement of *Agrobacterium*-mediated gene transfer system.

2.4.2.2 Particle Bombardment Method

In this method, foreign DNA is delivered into plant cells through high velocity metal particle. Main features of this method of DNA delivery are described below:

1. *Plant Material to be Used*: Generally, meristems, immature embryos, embryonic callus and suspension culture have been used for DNA delivery in different crop plants. Isolation of protoplasts is not essential in this method.
2. *Metal Particles used for Bombardment*: Generally, gold or tungsten particles are used for bombardment. Gold particles, being dense, can penetrate deeper cell layers than tungsten particles.
3. *Technique of Delivering DNA*: The bombardment is done with the help of particle gun. The particle gun is of two types, *viz*., (1) gun powder driven device and (2) helium particle inflow gun. The latter is in more common use than the former. Minute particle (1.0-1.5 μm) of gold or tungsten are coated with the DNA of interest. The DNA coated particles are accelerated with such force that they should penetrate the outer cell walls of target tissues. Some of the DNA of those particles enters the nuclei of target cells and is integrated with the DNA of the host cell resulting in transformation.
4. *Identification and Regeneration of Transformed Tissues*: The transformed tissues are identified with polymerase chain reaction (PCR) technique. Out of several tissues, a few are transformed. These transformed tissues are regenerated in culture medium into whole plant, which are further examined for stability of gene expression.

2.4.2.3 Direct DNA Transfer Method

Direct DNA transfer involves electroporation or chemical fusagens such as polyethyl glycol (PEG) with calcium phosphate. This method is used with protoplast. A suspension of protoplast with desired DNA is prepared. Then, a high voltage current is applied through the protoplast-DNA suspension. The electric current leads to formation of small temporary holes in the membrane of the protoplasts through which the DNA can pass. After entry into the cell, the foreign DNA gets incorporated with the host genome, resulting in genetic transformation.

This method can be used only in those crop species in which regeneration from protoplast is possible.

2.4.2.4 Micro-Injection Method

The plasmid DNA can also be delivered into host cells by mechanical means, i.e., microscopic needles (micro-injection). This method does not have host range limitations. However, in this method also regeneration from protoplasts is a basic requirement. Unfortunately, all crops do not regenerate from protoplasts.

Currently, plasmid and particle bombardment methods are widely used for development of transgenic (genetically engineered) individuals. Direct DNA transfer and micro-injection methods are rarely used for obtaining transgenic plants.

2.5 Brief References

Singh BD 1998, 2011, 2018; Singh Phundan 2013b, Stewart 2008, Trehan 1990, Trevan *et al.* 1987.

Review Questions

Part A

Answer the following questions true or false.

1. 'Biotechnology' is coined by Karl Ereky. (True/False)
2. Plasmids are chromosomal elements that are found in the bacterial cell. (True/False)
3. Nutrient medium is different from culture medium. (True/False)
4. Transgenic plants are genetically engineered. (True/False)
5. *In vitro* literally means 'under glass'. (True/False)

Part B

Answer the following questions.

1. What is totipotency?
2. Define gene cloning.
3. The plant part that is used for regeneration is called __________
4. Genetic transformation is the process of __________
5. What do you mean by callus?

Part C

Write a short note on each of the following topics.

1. Plasmid method of gene transfer.
2. Embryogenesis and organogenesis.
3. Types of plant tissue culture.

4. Direct DNA transfer method.
5. Micro-injection method.

Part D

1. Describe the steps of tissue culture.
2. Discuss the plasmid method of genetic engineering.

3

Classical and Quantitative Genetic Basis

3.1 Introduction

Genetics deals with genes and cytogenetics with chromosomes that harbour genes. On the other hand, plant breeding strives to upgrade the genotypes of crop plants. So, the plant breeding methods should be based on genetic principles. Dear students, you should have clear understanding of genetics in broad sense so that you can appreciate the applicability of breeding methods. As I have already pointed out that you have been taught the course 'Principles of Genetics and Cytogenetics' in the first semester, only a brief discussion is followed for your review and better understanding.

3.2 Character

When we call '**character**' or '**trait**' in genetics, we mean any morphological, anatomical, biochemical or behavioural feature of an organism. For instance, flower colour, blooming period, leaf size, size of guard cells in stomata, vitamin C content, disease resistance etc. A single gene or a number of genes may govern the development of a character and, often it is influenced by the environment. Characters may be qualitative or quantitative.

3.2.1 Qualitative Characters

Qualitative characters exhibit discontinuous variation. They may be classified into discrete classes. One or few **oligogenes** (also called **major genes**) govern each character. Each oligogene has a large and easily detectable effect on the trait. Usually, the genetic background and more particularly, the environment have little effect on it. A genetic background denotes all the remaining genes present in the same nucleus as the oligogene under consideration, while environment consists of all the biotic and abiotic factors other than the organism concerned. Such a trait is studied by counting the number of individuals in its various classes.

3.2.2 Quantitative Characters

Quantitative characters (also called '**metric**', '**polygenic**,' '**multifactorial**' or '**complex**' **characters**) show continuous variation. Hence, they cannot be grouped into distinct classes. Each trait is governed by several genes (the number cannot be reliably estimated) with small individual effects that are **cumulative** (also called **additive**) in nature. Such genes are called **polygenes**. Genetic background and environment have marked influence on such traits. These characters are measured not counted.

3.2.3 Threshold Characters

When a character requires a specific environment for its expression, it is called a **threshold character**. For example, some forms of genetic male sterility are expressed only under either a specific temperature range or a particular photoperiod. Similarly, resistance or susceptibility to disease can be observed only when the plants are exposed to pathogen concerned under an appropriate environment favouring disease development. This is true for resistance to any biotic or abiotic stress.

Many important qualitative and quantitative traits are threshold characters. Evaluation and selection of such traits must be done under the respective threshold environment, i.e., the environment that permits their optimum expression. If it is not carefully followed, the whole effort may be futile.

3.3 Gene-Character Relationship

There is a direct relationship between an individual plant character and the respective gene. Mendel, the father of genetics, proved through his genetical studies that characters are governed by some hereditary units to which Johannsen (1909) named genes. Later studies indicated that genes are of two types: (a) qualitative genes and (b) quantitative genes.

3.3.1 Qualitative Genes

Qualitative genes (also called **oligogenes**, **major genes** or **macro-genes**) which in monogenic or in digenic state gives full type effect, i.e., full expression of character(s) and the variation forms discrete classes. They can be determined by using simple Mendelian ratio.

3.3.1.1 Genotype and Phenotype

The external appearance of a character is called **phenotype**. Each character is governed by a gene. The alternative forms of a character are governed by the alleles of the same gene. **Alleles** are alternative form of a gene. Each somatic cell has two copies of a gene. When two copies are identical, it is called **homozygous** condition. But, when two copies are dissimilar, the condition is termed as **heterozygous**. The genetic constitution of a cell/ an individual is called its **genotype**.

3.3.1.2 Dominance

The ability of an allele to express itself in the heterozygotes is called **dominance**, and such an allele is known as dominant allele. In case of **complete dominance**, the phenotype of the heterozygotes is the same as that of homozygotes for the dominant allele. But, in case of partial or **incomplete dominance**, the phenotype of the heterozygote is intermediate between those of the dominant and recessive homozygotes. The allele that is unable to express itself in the heterozygote is termed as recessive. In some cases, both the alleles express themselves in the heterozygotes (**co-dominance**), which in some others the heterozygote phenotype falls outside the range of phenotypes produced by the dominant and recessive homozygotes (**over-dominance**).

3.3.1.3 Principle of Segregation

The resultant F_1 in a cross is heterozygous. At the time of gamete formation in F_1, the two alleles segregate (separate) and pass into different gametes. This is known as **segregation**. Segregation of the two alleles of a single gene produces two types of gametes in equal frequencies. A random union among the male and female gametes yields three types of zygotes that produce two phenotypes in F_2 in the ratio of 3:1. In case of complete dominance the F_2 ratio is 3:1, while it is 1:2:1 in case of incomplete, co-and over-dominance. Segregation produces genetic variation in the progeny. This happens because all the genes that are in heterozygous condition segregate and give rise to the homozygous genotypes in addition to heterozygotes.

3.3.1.4 Principle of Independent Assortment

When two (or more) genes segregate together, their segregation is independent of each other. So, two genes produce four types of gametes in 1:1:1:1 ratio. A random union among these gametes give rise to 9 different genotypes in the zygotes, which produce four different phenotypes in the ratio 9:3:3:1 in case of complete dominance. It generates genetic variation and produces new gene combinations. As a result, genes present in two different cultivars can be easily combined in to a single, new cultivar. To achieve this, one cultivar is crossed with another, and the F_1 generation is selfed. In F_2 and subsequent generations, plants with the desired new gene combinations are selected.

3.3.1.5 Linkage

Some genes show tendency to be inherited together; this is called **linkage**. Linked genes are present in the same chromosome rather close to each another. New combinations of linked genes are produced due to crossing over. The greater the distance between linked genes in the chromosome, the higher is the frequency of crossing over.

Linkage between dominant alleles of two genes is called **coupling phase linkage**. In contrast, in **repulsion phase linkage**, dominant allele of one gene is linked with the recessive allele of another gene. The phase of linkage (coupling or repulsion) alters drastically the frequencies of various phenotypes in F_2 and other segregating generations. However, it does not affect the frequency of recombination between the linked genes.

By using recombination frequency as a measure of distance, genes linked together are mapped in a linear fashion (**linkage map**). Extensive linkage maps have been prepared in the hope that they will be helpful in making decisions in breeding programmes.

The chief effect of linkage is that it reduces the frequency of new gene combinations (**recombinants**). Therefore, for recovery of new gene combinations of linked genes a much larger F_2 and subsequent populations should be raised. Linkage between tightly linked desirable alleles of two genes is quite helpful in selection for them. Selection for one gene selects for the other as well. When a desirable gene is tightly linked with an undesirable gene, it is a nuisance. Linkage is one of the chief reasons for genetic correlation among quantitative traits. In addition, repulsion phase linkage tends to increase the estimate of dominance variance and to reduce that of additive genetic variance.

3.3.1.6 Gene Interaction

Many qualitative characters are determined by two or more oligogenes. The two or more genes governing a single character affect the expression of each other in various ways giving rise to different ratios in the F_2. This is called **gene interaction**. As a result, a dihybrid or a trihybrid ratio is obtained in the F_2 for single qualitative trait. The gene interaction for two genes affecting single character may be complementary (F_2 ratio 9:7), supplementary (F_2 ratio 9:3:4), inhibitory (F_2 ratio 13:3), masking (F_2 ratio 12:3:1), duplicate (F_2 ratio 15:1) and polymeric (F_2 ratio 9:6:1).

3.3.1.7 Pleiotropy

When a single oligogene governs two or more characters, it is known as **pleiotropy** and such a gene is called **pleiotropic gene**. This type of gene action is called as pleiotropic gene action. Several cases of pleiotropy may be due to tight-linked-genes. It is often difficult to separate such cases from those of true pleiotropy. When the pleiotropic effects are favourable, it is quite useful in plant breeding. But when the pleiotropic effect of a gene (effects on characters other than the one that is its main effect) are unfavourable, it may prevent utilization of the gene.

3.3.1.8 Penetrance

Generally, oligogenes express themselves in all the individuals in which they are present in appropriate genotype. It is called **complete penetrance**. But, some oligogenes express themselves only in a proportion of the individuals carrying them in the appropriate genotype. It is termed as **incomplete penetrance**. Thus, penetrance of a gene is the ratio of the number of individuals expressing it to the total number of individuals having it in the appropriate genotype. The appropriate genotype for a dominant allele will be dominant homozygotes and the heterozygotes, while for a recessive allele it will be the recessive homozygotes. Incomplete penetrance of an allele creates confusion during selections, especially when the selection is against such an allele.

3.3.1.9 Expressivity

Generally, oligogenes express themselves fairly uniformly in all the individuals in which they are present in appropriate genotype (**uniform expressivity**). But in cases of some oligogenes, the degree of expression varies from individual to individual. It is called **variable expressivity**. Variable expressivity creates a quantitative variation for a qualitative trait. This variation is not heritable. Both incomplete penetrance and variable expressivity arise because of environmental factors on the expression of the gene concerned. Variable expressivity of gene creates confusion in the selection for the trait because it generates a quantitative variation in a qualitative trait.

3.3.1.10 Modifying Genes

Many genes have no effect of their own on a character. But they increase or decrease the expression of an oligogene affecting the character. Such genes are called **modifying genes**. Each modifying gene has a small effect. The effects of all modifying genes affecting a character are cumulative (additive). Thus, modifying genes produce a heritable quantitative variation in a qualitative trait. Modifying genes are known to affect the degree of expression of self-incompatibility alleles, many genes for disease resistance, some genes for male sterility, dominance relationship among alleles etc. An effort should be made to accumulate favourable modifying genes to increase or decrease the expression of an oligogene. Modifying genes can also minimize the undesirable pleiotropic effects.

3.3.1.11 Maternal Effects

The development of many traits in higher plants, including many crop species, is influenced by the genotype of the female parent, e.g., seed size, protein content in seeds etc. As a result, F_1 progeny from reciprocal crosses show marked differences for such characters. These differences, however, disappear in F_2, and the amount of variation present in F_2 is usually smaller than that observed in F_3. So, maternal effects would tend to reduce the effectiveness of selection in segregating generations for the traits concerned.

3.3.2 Quantitative Genes

The **quantitative genes** (also called **polygenes, minor genes** or **micro-genes**) are the group of genes with small cumulative effect. In other words, they are additive in nature. The effects of the genes are highly influenced (masked) by environmental factors. The characters controlled by these genes such as height, days to maturity, size and shape of seed, protein content, vitamin content, drought and cold tolerance etc show a continuous variation and can be measured in metric terms. For example, length, time, weight or ratios are called quantitative or metric traits. Quantitative genetics is also derived from the Mendelian concepts of inheritance because all genes behave individually in Mendelian fashion. In other words, quantitative genetics is *de facto* Mendelian genetics in which a group of genes is involved to govern a character. All plant breeding programmes involve quantitative inheritance.

3.3.2.1 Nature of Quantitative Gene Effects

According to the nature of gene effect the quantitative genes are of two types:

1. Additive genes
2. Non-additive genes:
    a) Dominant genes
    b) Epistasis genes:
        i) Additive × Additive gene interaction
        ii) Dominance × Dominance gene interaction
        iii) Additive × Dominance gene interaction

These are also known as genetic parameters to determine nature of quantitative gene action. They have been discussed in detail in Chapter 11.

3.4 Gene Mutation

From an evolutionary point of view, mutation has played an important role in the origin of new genes. And almost all genes have arisen from spontaneous mutation in nature. However, favourable gene mutations have occurred spontaneously from time to time in nature and played significant role in plant breeding. For example, dwarfing gene in wheat and rice brought Green Revolution in 1970s in all the spring wheat and rice growing countries in the world.

H.J. Muller in 1927 demonstrated for the first time that the mutation can be artificially induced by treating *Drosophila* with x-rays. The induced mutations did not differ from spontaneous mutations in expression. Similarly, L.J. Stadler in 1928 demonstrated an increase in the rate of mutations due to x-rays in barley and maize. Besides x-rays, γ-rays can also induce mutations. Both x-rays and γ-rays are ionizing radiations and have high penetrance, and therefore can induce mutations in the seeds. Ultraviolet rays, however, are non-ionizing radiations and have a lower penetrating power than x-rays and γ-rays. Therefore, ultraviolet rays are used for treatment of material, where the nuclei to be attacked are not located too far away from the surface. Since in the seeds, the embryo is present several cell layers away from surface of the seed, it is not a suitable material for treatment with ultraviolet rays. However, pollens can be treated since each pollen has a nucleus, which can be altered.

Besides radiations, chemicals can also be used for inducing mutations. C. Auerbach was the first to find that mutations can also be induced by certain chemicals. She made this important discovery during World War II. The chemicals used for inducing mutations were mustard gas, ethyl urethan, phenol, formaldehyde etc. Mustard gas was found to be highly mutagenic, and it was found to have a delayed effect. For instance, if the treatment is given in the first generation, mutation could be induced in second or third generation. A very long list of chemicals, which can induce mutations, is now available. Recently, ethyl methane sulphonate (EMS) is being most extensively used for inducing mutations both in microorganisms, and higher plants and animals.

Since 1927 extensive studies on producing and utilizing induced mutation in crop plants have been taken up by the plant breeders.

3.5 Heteroploidy

The chromosome number normally found in somatic cells of a species is called **somatic chromosome number** and is denoted by 2n. The chromosomes present in somatic cells constitute the **somatic chromosome complement**. In the normal somatic complement of a species, each chromosome has two copies called **homologues**. Therefore, all such (2n) individuals are called **disomic**. In contrast, **diploid** species/individuals/cells have two copies of a single genome. *A **genome** may be defined as the haploid chromosome complement of a diploid species. It is denoted by x.* The different chromosomes of a single genome are distinct from each other in gene content and homology. Therefore, members of a genome do not tend to pair with each other. The **gametic chromosome complement** is the set of chromosomes normally found in the gametes of a species. Similarly, the **gametic** or **haploid chromosome number** is the number of chromosomes normally found in the gametes of a species, and is represented by *n*.

Normally, chromosome numbers of various species are quite stable. But a low frequency of various irregularities occurs during both mitotic and meiotic cell divisions. These irregularities lead to the origin of changes in chromosome number of some individuals. An individual carrying a chromosome number other than the diploid number is called **heteroploid**, and the condition is called **heteroploidy**. **Polyploidy** refers to certain heteroploid situations.

3.5.1 Types of Heteroploidy

Heteroploidy is divided into two major groups: (1) aneuploidy and (2) euploidy.

3.5.1.1 Aneuploidy

A change of one or few chromosomes from the normal somatic chromosome number (disomic condition) is called **aneuploidy**; it may be written as 2n±l or few. As a rule, aneuploid variations are determined with reference to the normal disomic chromosome complement of the species concerned. The number of chromosomes increased or decreased is, as a rule, different from the genomic number (x) of the species. Therefore, the chromosome number of an aneuploid is not an exact multiple of the basic or genomic number. The main types of aneuploids are as follows:

1. *Nullisomic*: In such individuals/cells, one chromosome pair is missing; they are denoted by 2n–2. The situation is called **nullisomy**.
2. *Monosomic*: These individuals/cells have one chromosome missing. The condition is called **monosomy** and is depicted by 2n–l.
3. *Trisomic*: Such an individual/cell has an extra chromosome, i.e., has 2n+l chromosome number. This condition is termed as **trisomy**.
4. *Tetrasomic*: Tetrasomic individuals/cells have one chromosome pair extra, i. e., have 2n+2 chromosomes. This situation is known as **tetrasomy**.

3.5.1.2 Euploidy

When the chromosome number changes (increases or decreases) by the whole genome, the resulting condition is called **euploidy**. Euploid variation is always determined with reference to the diploid chromosome complement of the species. When we deal with naturally occurring polyploid species, e.g., wheat, we refer to their parental diploid species (usually, determined based on various studies). In euploids, the chromosome number is always an exact multiple of the basic or genomic number.

1. *Haploids*: Individuals having the gametic chromosome number (actually, gametic chromosome complement) of the species are called **haploid** (n), and the situation is called **haploidy**. Haploids of diploid species are known as **monoploid** (x).

2. *Polyploids*: These individuals have more than 2 copies of the same genome or 2 copies (usually) of each of two or more distinct genomes. This condition is called **polyploidy**. Polyploids are of two types: (i) autopolyploids and (ii) allopolyploids.

   i) *Autopolyploids*: Such individuals have 3 (autotriploid), 4 (autotetraploid), 5 (autopentaploid), 6 (autohexaploid), 7 (autoheptaploid), 8 (autooctaploid) or more copies of a single genome. This situation is termed as **autopolyploidy**.

   ii) *Allopolyploids*: When an individual has two copies each of two or more distinct genomes, it is called **allopolyploid**, and the condition is known as **allopolyploidy**. Such individuals are also termed as **amphidiploids** as they have two copies of each of the genomes present. They may contain two copies each of 2 (allotetraploid), 3 (allohexaploid), 4 (allooctaploid) or more distinct genomes.

3.6 Cytoplasmic Inheritance

As chromosomes are present in the nucleus, the genes that are located in chromosomes are called **nuclear genes**. Nuclear genes are usually called simply genes. Their characteristic feature is regular Mendelian segregation in F_2. In contrast, genes governing some traits are located in the cytoplasm. They are called **plasmagenes** and located in the circular chromosomes of mitochondria and chloroplasts. They show an inheritance pattern that is different from that of nuclear genes. This is known as **cytoplasmic inheritance** (also called **extranuclear inheritance**). In this case, there is no segregation in F_2 or later generations; F_1 and subsequent generations show reciprocal differences. Rarely, segregation may occur. But segregation ratios are irregular and very different from those of nuclear genes. *Often, segregation of plasmagenes occurs during mitotic divisions. This produces the typical variegation patterns in many* ***ornamental plants***. The plasmagenes in vast majority of crop plants show maternal transmission. Some economically important characters are governed by plasmagenes, e.g., cytoplasmic male sterility.

3.7 Factors Limiting Genetic Improvement of Crops

3.7.1 Narrow Genetic Base: Progress in plant breeding depends on the extent of genetic variability in the population. The variability gets rapidly exhausted owing to selection, genetic drift and genetic erosion.

3.7.2 Plant Types: Plants are primary factories of agriculture/horticulture which manufacture carbohydrate through photosynthesis by converting atmospheric waste CO_2 in presence of water and solar energy.

$6CO_2 + 6H_2O + \text{Light} = C_6H_{12}O_6 + 6O_2$ (photosynthesis)

$C_6H_{12}O_6 + 6O_2 = 6CO_2 + 6H_2O + \text{energy}$ (respiration)

Thus, yield is a conserved form of solar energy. Therefore, a model plant ideotype should be biologically efficient in manufacturing carbohydrate and its conservation in form of seeds or other economic parts. It should be the basic concept of plant breeding.

3.7.3 Unpredictable Environment: Plants and their production depend upon G×E interaction. A sudden rise in temperature due to global warming in February-March is a cause of concern due to enhanced rate of photorespiration.

3.7.4 Genetic Barrier: Inherent genetic barriers, such as negative genotypic correlation of undesirable traits with desirable traits, low heritability and negative maternal inheritance (Texas cytoplasm in maize hindering non-realization of hybrid vigour), restrict the rapid improvement of crop plants.

3.8 Brief References

Acquaah 2012, Allard 1960, 1999; Chahal and Gosal 2002, Chaudhary 2014, Hayes *et al.* 1955, Poehlman 1987, Ram 2014, Roy 2012b; Sharma JR 1994, Simmonds 1979, Singh BD 2011, 2018; Singh and Pawar 2006.

Review Questions

Part A

Answer the following questions true or false.

1. The plant breeding methods may or may not be based on genetic principles. (True/ False)
2. The genes that are located in chromosomes are called nuclear genes. (True/ False)
3. Some genes show tendency to be inherited together; this is called linkage. (True/ False)
4. Cytoplasmic inheritance is different from extranuclear inheritance. (True/ False)

5. Quantitative genes are the group of genes with small cumulative effect. (True/False)
6. Trisomics have a pair of extra chromosomes. (True/False)
7. Qualitative trait is also known as metric trait. (True/False)
8. Oligogenes are also called major genes. (True/False)
9. Plasmagenes are located on chromosomes. (True/False)
10. Polyploids are of three types. (True/False)
11. In euploids, the chromosome number is sometimes an exact multiple of the basic number. (True/False)
12. Haploids of diploid species are known as monoploid (x). (True/False)
13. H.J. Muller in 1930 demonstrated for the first time that the mutation can be artificially induced by treating *Drosophila* with x-rays. (True/False)
14. Non-additive genes may be dominant or epistatic. (True/False)
15. Maternal effects would not tend to reduce the effectiveness of selection in segregating generations for the traits concerned. (True/False)

Part B

Answer the following questions.

1. Who demonstrated for the first time that the mutation can be artificially induced?
2. Haploids of diploid species are known as __________
3. Name the principles or laws of Mendel.
4. What is aneuploidy?
5. Dominance is a law/phenomenon.
6. What is quantitative character?
7. Heteroploidy is divided into two major groups: __________ and __________
8. Tetrasomic individuals/cells have one __________ extra.
9. In case of incomplete dominance, the F_2 ratio is __________
10. Define genome.
11. Ethyl methane sulphonate (EMS) is a physical/chemical mutagen.
12. Whether segregation of plasmagenes occurs during mitotic divisions?
13. The F_2 ratio in case of complementary gene interaction is __________

14. The principle of independent assortment is/ is not universal.
15. Both x-rays and γ-rays can/cannot induce mutations in the seeds.

Part C

Write a short note on each of the following topics.

1. Cytoplasmic inheritance
2. Euploidy
3. Polygenes
4. Penetrance and expressivity
5. Gene mutation
6. Gene interaction
7. Polyploids
8. Trisomy
9. Quantitative characters
10. Modifying Genes
11. Nullisomy
12. Linkage
13. Genotype and phenotype
14. Nature of quantitative gene effects
15. Threshold Characters

Part D

1. Describe and classify the heteroploidy in brief.
2. Discuss the principles of segregation, and independent assortment.
3. Discuss and differentiate between qualitative and quantitative characters.

4

Molecular Genetic Basis I Markers, QTL Mapping and Marker-Assisted Selection*

4.1 Introduction

Many important traits such as yield, quality and some forms of disease resistance are controlled by many genes and are known as **quantitative traits** or **characters** (also called '**metric**','**polygenic**,' '**multifactorial**' or '**complex**' **traits**). The regions within genomes that contain genes associated with a particular quantitative trait are known as **quantitative trait loci (QTLs)**. The identification of QTLs based only on conventional phenotypic evaluation is not possible. A major breakthrough in the characterization of quantitative traits that created opportunities to select for QTLs was initiated by the development of **DNA** (also called **molecular**) **markers** in the 1980s.

One of the main uses of DNA markers in agri-horticultural research has been in the construction of **linkage maps** for diverse crop species. Linkage maps have been utilised for identifying chromosomal regions that contain genes controlling **simple traits (controlled by a single gene)** and **quantitative traits** using QTL analysis. The process of constructing linkage maps and conducting QTL analysis–to identify genomic regions associated with traits–is known as **QTL mapping** (also called '**genetic**,' '**gene**' or '**genome**' **mapping**). DNA markers that are tightly linked to agronomically important genes (called **gene 'tagging'**) may be used as molecular tools for **marker-assisted selection** (**MAS**) in plant breeding. MAS involves using the presence/absence of a marker as a substitute for or to assist in **phenotypic selection**, in a way which may make it more efficient, effective, reliable and cost-effective compared to the more conventional plant breeding methods. *The use of DNA markers in plant (and animal) breeding has opened a new realm in agriculture and horticulture called* '**molecular breeding**'.

---

*Adapted with permission from: Collard BCY, Jahufer MZZ, Brouwer JB and Pang ECK 2005. An introduction to markers, quantitative trait loci (QTL) mapping and marker-assisted selection for crop improvement: The basic concepts. *Euphytica* 142: 169-196.

DNA markers are widely accepted as potentially valuable tools for crop improvement in rice, wheat, maize, barley, tuber crops, pulses, oilseeds, horticultural crops and pasture species. Some studies suggest that DNA markers will play a vital role in enhancing global food production by improving the efficiency of **conventional plant breeding programmes**. Although there has been some concern that the outcomes of DNA marker technology as proposed by initial studies may not be as effective as first thought, many plant breeding institutions have adopted the capacity for marker development and/or MAS. An understanding of the basic concepts and methodology of DNA marker development and MAS, including some of the terminology used by molecular biologists, will enable plant breeders and researchers working in other relevant disciplines to work together towards a common goal—increasing the efficiency of global food production.

Dear students, this chapter requires only a basic knowledge of molecular biology and/or plant genetics. It consists of five sections: genetic markers, construction of linkage maps, QTL analysis, towards marker-assisted selection and marker-assisted selection.

4.2 Genetic Markers

4.2.1 What are genetic markers?

Genetic markers represent genetic differences between individual organisms or species. Generally, they do not represent the target genes themselves but act as '**signs**'or '**flags**'. Genetic markers that are located in close proximity to genes (i.e. tightly linked) may be referred to as **gene 'tags'**. Such markers themselves do not affect the phenotype of the trait of interest because they are located only near or 'linked' to genes controlling the trait. All genetic markers occupy specific genomic positions within chromosomes (like genes) called 'loci'(singular 'locus').

There are three major types of genetic markers: (1) **morphological** (also called **'classical'** or **'visible') markers** which themselves are phenotypic traits or characters; (2) **biochemical markers**, which include allelic variants of enzymes called isozymes; and (3) **DNA** (also called **molecular**) **markers**, which reveal sites of variation in DNA. Morphological markers are usually visually characterized phenotypic characters such as flower colour, seed shape, growth habits or pigmentation. Isozyme markers are differences in enzymes that are detected by electrophoresis and specific staining. The major disadvantages of morphological and biochemical markers are that they may be limited in number and are influenced by environmental factors or the developmental stage of the plant. However, despite these limitations, morphological and biochemical markers have been extremely useful to plant breeders.

4.2.2 DNA Markers

DNA markers are the most widely used type of marker predominantly due to their abundance. They arise from different classes of DNA mutations such as substitution mutations (point mutations), rearrangements (insertions or deletions) or errors in replication of tandemly repeated DNA. These markers are selectively neutral because they are usually located in non-coding regions of DNA. *Unlike morphological and biochemical markers, DNA markers are practically unlimited in number and are not affected by environmental factors and/or the developmental stage of the plant.* Apart from the use of DNA markers in the construction of linkage maps, they have numerous applications in plant breeding such as assessing the level of genetic diversity within germplasm and cultivar identity.

DNA markers may be broadly divided into three classes based on the method of their detection: (1) **hybridization-based**; (2) **polymerase chain reaction (PCR)-based** and (3) **DNA sequence-based**. Essentially, *DNA markers may reveal genetic differences that can be visualised by using a technique called gel electrophoresis and staining with chemicals (ethidium bromide or silver) or detection with radioactive or colourimetric probes.* DNA markers are particularly useful if they reveal differences between individuals of the same or different species. These markers are called **polymorphic markers**, whereas markers that do not discriminate between genotypes are called **monomorphic markers** (Figure 4.1). Polymorphic markers may also be described as **codominant** or **dominant**. This description is based on whether markers can discriminate between homozygotes and heterozygotes (Figure 4.2). *Codominant markers indicate differences in size whereas dominant markers are either present or absent. Strictly speaking, the different forms of a DNA marker (e.g. different sized bands on gels) are called marker 'alleles'.* **Codominant markers may have many different alleles whereas a dominant marker has only two alleles.**

The technical methods of how DNA markers are generated are beyond the scope of this text book (interested student may read Koh *et al.* 2015). However, the advantages and disadvantages of the most commonly used markers are presented in Table 4.1.

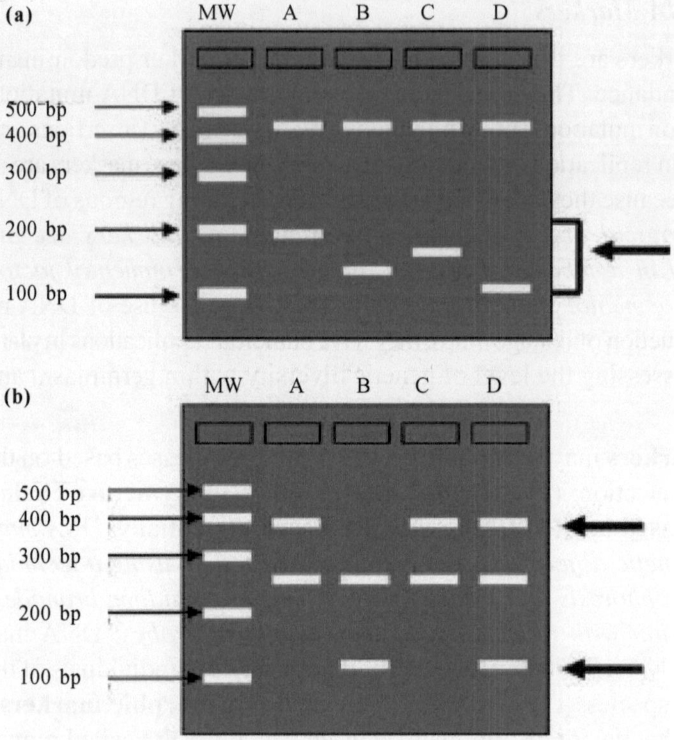

Figure 4.1. Diagram representing hypothetical DNA markers between genotypes, A B, C and D. Polymorphic markers are indicated by arrows. Markers that donot discriminate between genotypes are called monomorphic markers. **a)** Example of SSR markers. The polymorphic marker reveal size differences for the marker alleles of the four genotypes, and represent a single genetic locus, **b)** Examples of markers generated by the RAPD technique, Note that these markers are either presents or absent, Often, the sizes of these markers in nucleotide base pairs (bp) are also provided; these sizes are estimated from a molecular weight (MW) DNA ladder. For both polymorphic markers, there are only two different marker alleles

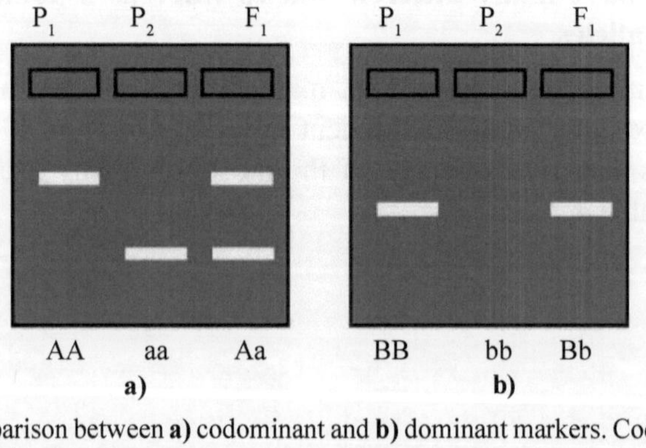

Figure 4.2. Comparison between **a)** codominant and **b)** dominant markers. Codominant markers can clearly discriminate between homozygotes and heterozygotes whereas dominant markers do not. Genotypes at two marker loci (A and B) are indicated below the gel diagrams

Table 4.1. Advantages and disadvantages of most commonly used DNA markers for QTL analysis

Molecular marker	**Codominant (C) or Dominant (D)**	**Advantages**	**Disadvantages**
Restriction fragment length polymorphism (RFLP)	C	• Robust • Reliable • Transferable across populations	• Time-consuming, laborious and expensive • Large amounts of DNA required • Limited polymorphism (especially in related lines)
Random amplified polymorphic DNA (RAPD)	D	• Quick and simple • Inexpensive • Multiple loci from a single primer possible • Small amounts of DNA required	• Problems with reproducibility • Generally not transferable
Simple sequence repeats (SSRs)* or 'microsatellites'	C	• Technically simple • Robust and reliable • Transferable between populations	• Large amounts of time and labour required for production of primers • Usually require polyacrylamide electrophoresis
Amplified fragment length polymorphism (AFLP)	D**/C	• Multiple loci • High levels of polymorphism generated	• Large amounts of DNA required • Complicated methodology
Single Nucleotide Polymorphism (SNP)	C	• Robust • Suitable for high throughput • Different detection methods available • Can be automated • Polymorphisms are identifiable	• Very high development costs • Require sequence information • Can be technically challenging

* SSRs are also known as sequence tagged microsatellite site (STMS) markers

** Dominant, but can also be scored as codominant.

4.3 Construction of Linkage Maps

4.3.1 What are linkage maps?

A linkage map may be thought of as a 'road map' of the chromosomes derived from two different parents. *Linkage maps indicate the position and relative genetic distances between markers along chromosomes, which is analogous to signs or landmarks along a highway.* The most important use for linkage maps is to identify chromosomal locations containing genes and QTLs associated with traits of interest; such maps may then be referred to as **'QTL'** (or **'genetic'**) **maps**. 'QTL mapping' is based on the principle that genes and markers segregate via chromosome recombination (also called crossing-over) during meiosis (i.e., sexual reproduction), thus allowing their analysis in the progeny. Genes or markers that are close together or tightly-linked will be transmitted together from parent to progeny more frequently than genes or markers that are located farther apart (Figure 4.3). In a segregating population, there is a mixture of parental and recombinant genotypes. The frequency of recombinant genotypes can be used to calculate recombination fractions, which may be used to infer the genetic distance between markers. By analysing the segregation of markers, the relative order and distances between markers can be determined–the lower the frequency of recombination between two markers, the closer they are situated on a chromosome (conversely, the higher the frequency of recombination between two markers, the farther away they are situated on a chromosome). *Markers that have a recombination frequency of 50% are described as 'unlinked' and assumed to be located far apart on the same chromosome or on different chromosomes*. For a more detailed explanation of genetic linkage, the student is encouraged to consult basic textbooks on genetics or quantitative genetics. Mapping functions are used to convert recombination fractions into map units called centiMorgans (cM) (discussed later). Linkage maps are constructed from the analysis of many segregating markers. The three main steps of linkage map construction are: (1) production of a mapping population; (2) identification of polymorphism and (3) linkage analysis of markers.

4.3.2 Mapping Populations

The construction of a linkage map requires a segregating plant population (i.e., a population derived from sexual reproduction). The parents selected for the mapping population will differ for one or more traits of interest. Population sizes used in preliminary genetic mapping studies generally range from 50 to 250 individuals, however larger populations are required for high-resolution mapping. If the map will be used for QTL studies (which is usually the case), an important point to note is that the mapping population must be phenotypically evaluated (i.e. trait data must be collected) before subsequent QTL mapping.

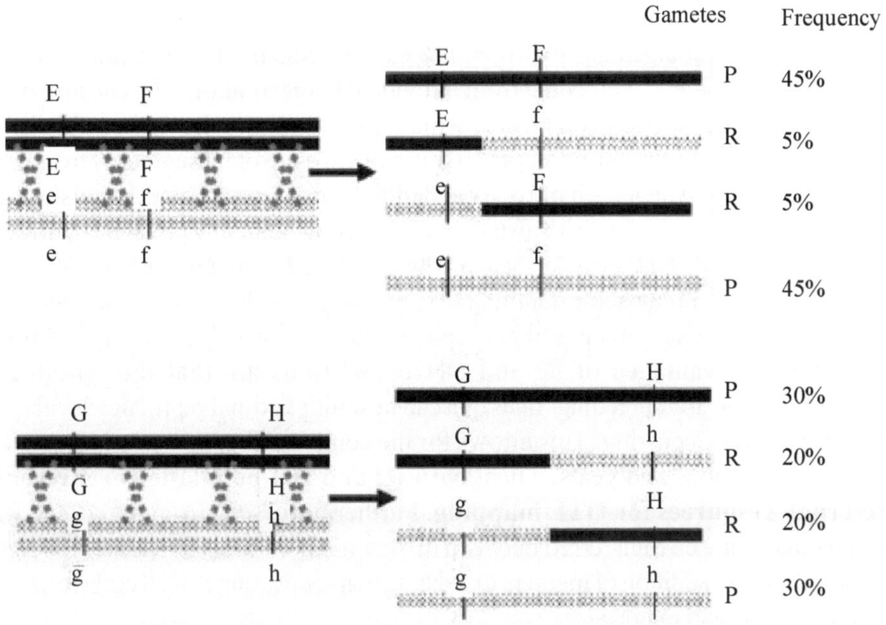

Figure 4.3. Diagram indicating cross-over or recombination events between homologous chromosomes that occur during meiosis. Gametes that are produced after meiosis are either parental (P) or recombinant (R). The smaller the distance between two markers, the smaller the chance of recombination occurring between the two markers. Therefore, recombination between markers G and H should occur more frequently than recombination between markers E and F. This can be observed in a segregating mapping population. By analysing the number of recombinants in a population, it could be determined that markers E and F are closer together compared to G and H.

Generally in self-pollinating species (e.g. naturally self-pollinated annual crops: tomato, pea, cowpea, Indian bean, French bean, cluster bean), mapping populations originate from parents that are both highly homozygous (inbred). In cross pollinating species, the situation is more complicated since most of these species do not tolerate inbreeding. Many cross pollinating plant species are also polyploid (contain several sets of chromosome pairs). Mapping populations used for mapping cross-pollinating species may be derived from a cross between a heterozygous parent and a haploid or homozygous parent. For example, in both the cross pollinating species white clover (*Trifolium repens* L.) and ryegrass (*Lolium perenne* L.), F_1 generation mapping populations were successfully developed by pair crossing heterozygous parental plants that were distinctly different for important traits associated with plant persistence and seed yield.

Several different populations may be utilized for mapping within a given plant species, with each population type possessing advantages and disadvantages (Figure 4.4). F_2 populations, derived from F_1 hybrids, and backcross (BC) populations, derived by crossing the F_1 hybrid to one of the parents, are the simplest types of mapping populations developed for self-pollinating species. Their main advantages are that they are easy to construct and require only a short time to produce. Inbreeding from individual F_2 plants allows the construction of **recombinant inbred (RI) lines**, which consist of a series of homozygous lines, each containing a unique combination of chromosomal segments from the original parents. The length of time needed for producing RI populations is the major disadvantage, because usually six to eight generations are required. Doubled haploid (DH) populations may be produced by regenerating plants by the induction of chromosome doubling from pollen grains, however, the production of DH populations is only possible in species that are amenable to tissue culture. The major advantages of RI and DH populations are that they produce homozygous or 'true-breeding' lines that can be multiplied and reproduced without genetic change occurring. This allows for the conduct of replicated trials across different locations and years. Thus, **both RI and DH populations represent 'eternal' resources for QTL mapping**. Furthermore, seed from individual RI or DH lines may be transferred between different laboratories for further linkage analysis and the addition of markers to existing maps, ensuring that all collaborators examine identical material.

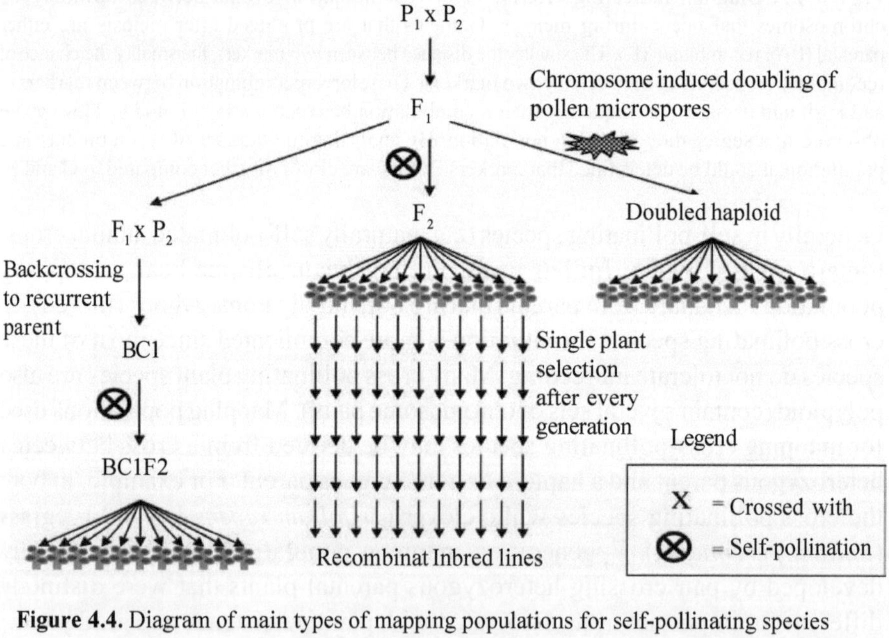

Figure 4.4. Diagram of main types of mapping populations for self-pollinating species

4.3.3 Identification of Polymorphism

The second step in the construction of a linkage map is to identify DNA markers that reveal differences between parents (i.e., **polymorphic markers**). It is critical that sufficient polymorphism exists between parents in order to construct a linkage map. *In general, cross pollinating species possess higher levels of DNA polymorphism compared to inbreeding species; mapping in inbreeding species generally requires the selection of parents that are distantly related.* In many cases, parents that provide adequate polymorphism are selected on the basis of the level of genetic diversity between parents. The choice of DNA markers used for mapping may depend on the availability of characterised markers or the appropriateness of particular markers for a particular species. Once polymorphic markers have been identified, they must be screened across the entire mapping population, including the parents (and F_1 hybrid, if possible). This is known as **marker 'genotyping'** of the population. Therefore, DNA must be extracted from each individual of the mapping population when DNA markers are used. Examples of DNA markers screened across different populations are shown in Figure 4.5. The expected segregation ratios for codominant and dominant markers are presented in Table 4.2. Significant deviations from expected ratios can be analysed using chi-square tests. Generally, markers will segregate in a Mendelian fashion although distorted segregation ratios may be encountered.

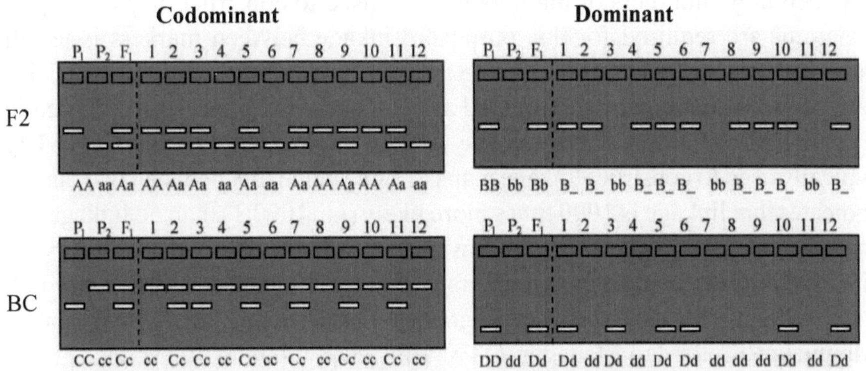

Figure 4.5. Hypothetical gel photos representing segregating codominant markers (left-hand side) and dominant markers (right-hand side) for typical mapping populations. Codominant markers indicate the complete genotype of a plant. Note that dominant markers cannot discriminate between heterozygotes and one homozygote gentoype in F2 populations. The segregation rations of markers can be easily understood by using punnett squares to derive population genotypes

Table 4.2. Expected segregation ratios for markers in different population types

Population type	Codominant markers	Dominant markers
F_2	1: 2:1 (AA:Aa:aa)	3:1 (B- :bb)
Backcross	1:1 (Cc:cc)	1:1 (Dd:dd)
Recombinant inbred or doubled haploid	1:1 (EE: ee)	1:1 (FF:ff)

In some polyploid species such as sugarcane, identifying polymorphic markers is more complicated. The mapping of diploid relatives of polyploid species can be of great benefit in developing maps for polyploid species. However, diploid relatives do not exist for all polyploid species. A general method for the mapping of polyploid species is based on the use of single-dose restriction fragments.

4.3.4 Linkage Analysis of Markers

The final step of the construction of a linkage map involves coding data for each DNA marker on each individual of a population and conducting linkage analysis using computer programs (Figure 4.6). Missing marker data can also be accepted by mapping programs. Although linkage analysis can be performed manually for a few markers, it is not feasible to manually analyze and determine linkages between large numbers of markers that are used to construct maps; computer programs are required for this purpose. Linkage between markers is usually calculated using odds ratios (i.e. the ratio of linkage versus no linkage). This ratio is more conveniently expressed as the logarithm of the ratio, and is called a **logarithm of odds (LOD) value** or **LOD score**. LOD values of >3 are typically used to construct linkage maps. A LOD value of 3 between two markers indicates that linkage is 1000 times more likely (i.e. 1000:1) than no linkage (null hypothesis). LOD values may be lowered in order to detect a greater level of linkage or to place additional markers within maps constructed at higher LOD values. Commonly used software programs for constructing linkage maps include Mapmaker/EXP, Map Manager QTX, Join Map, Q-Gene, ICIM, etc.

A typical output of a linkage map is shown in Figure 4.7. *Linked markers are grouped together into* ***'linkage groups'****, which represent chromosomal segments or entire chromosomes.* **Referring to the road map analogy, linkage groups represent roads and markers represent signs or landmarks**. A difficulty associated with obtaining an equal number of linkage groups and chromosomes is that the polymorphic markers detected are not necessarily evenly distributed over the chromosome, but clustered in some regions and absent in others. In addition to the non-random distribution of markers, the frequency of recombination is not equal along chromosomes.

The accuracy of measuring the genetic distance and determining marker order is directly related to the number of individuals studied in the mapping population.

Ideally, mapping populations should consist of a minimum of 50 individuals for constructing linkage maps.

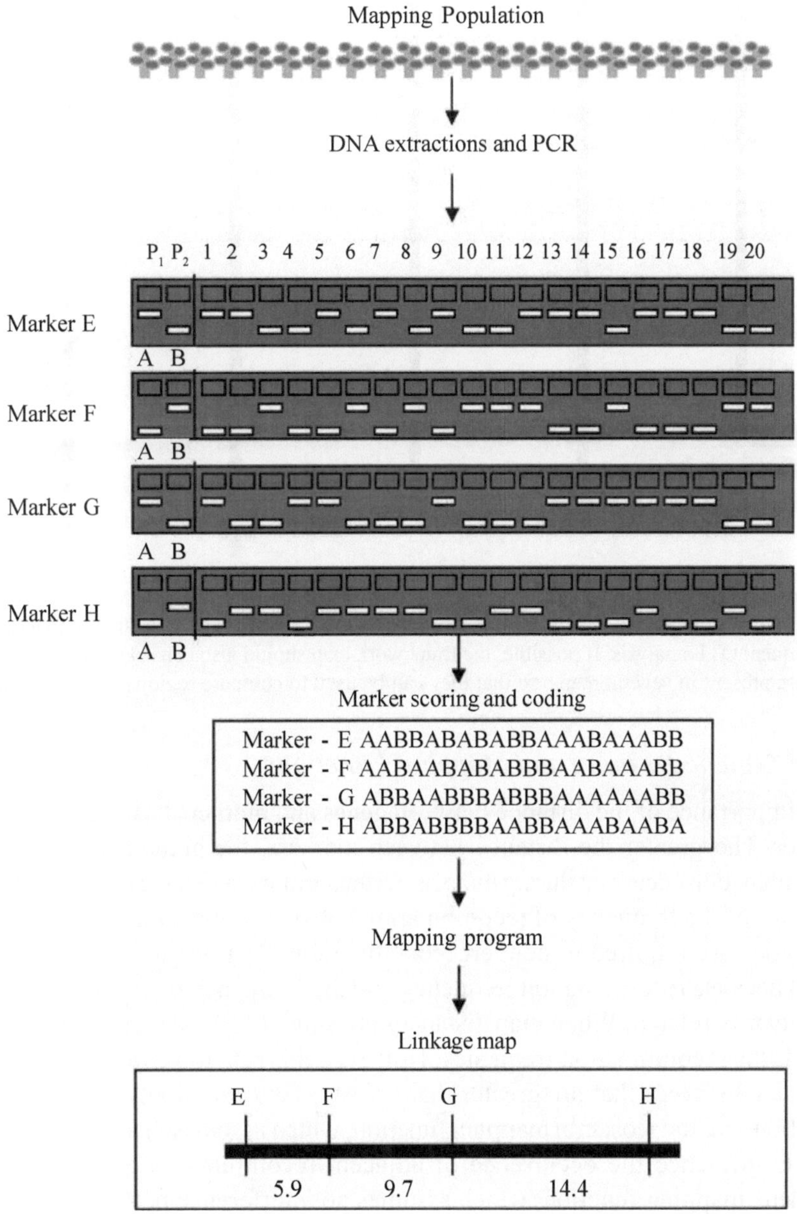

Figure 4.6. Construction of a linkage map based on a small recombinant inbred population (20 individuals). The first parent (P1) is scored as an 'A' whereas the second parent (P2) is scored as a 'B'. Coding of marker data varies depending on the type of population used. This linkage map was constructed using Map Manager QTX (Manly *et al.,* 2001) using the Haldane mapping function.

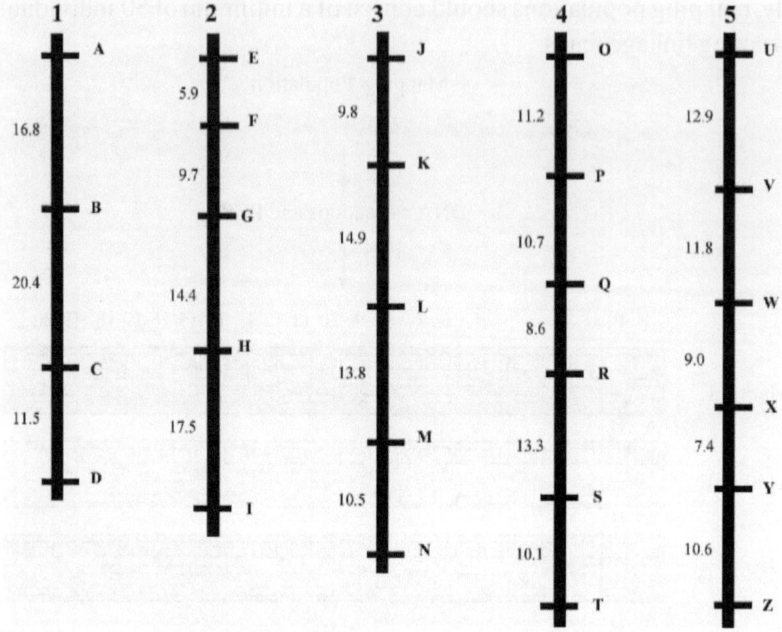

Figure 4.7. Hypothethical 'framework' linkage map of five chromosomes (represented by linkage groups) and 26 markers. Ideally, a framework map should consist of evenly spaced markers for subsequent QTL analysis. If possible, the framework map should also consist of anchor markers that are present in several maps, so that they can be used to compare regions between maps.

4.3.5 Genetic Distance and Mapping Functions

The importance of the distance between genes and markers has been discussed earlier. The greater the distance between markers, the greater the chance of recombination occurring during meiosis. Distance along a linkage map is measured in terms of the frequency of recombination between genetic markers. Mapping functions are required to convert recombination fractions into centiMorgans (cM) because recombination frequency and the frequency of crossing-over are not linearly related. When map distances are small (<10 cM), the map distance equals the recombination frequency. However, this relationship does not apply for map distances that are greater than 10 cM. Two commonly used mapping functions are the Kosambi mapping function, which assumes that recombination events influence the occurrence of adjacent recombination events, and the Haldane mapping function, which assumes no interference between crossover events. It should be noted that distance on a linkage map is not directly related to the physical distance of DNA between genetic markers, but depends on the genome size of the plant species. Furthermore, the relationship between genetic and physical distance varies along a chromosome. For example, there are recombination 'hot spots' and 'cold spots,' which are chromosomal regions in which recombination occurs more frequently or less frequently, respectively.

4.4 QTL Analysis

4.4.1 Principle of QTL Analysis

Identifying a gene or QTL within a plant genome is like finding the proverbial needle in a haystack. However, QTL analysis can be used to divide the haystack in manageable piles and systematically search them. In simple terms, QTL analysis is based on the principle of detecting an association between phenotype and the genotype of markers. Markers are used to partition the mapping population into different genotypic groups based on the presence or absence of a particular marker locus and to determine whether significant differences exist between groups with respect to the trait being measured. A significant difference between phenotypic means of the groups (either 2 or 3), depending on the marker system and type of population, indicates that the marker locus being used to partition the mapping population is linked to a QTL controlling the trait (Figure 4.8).

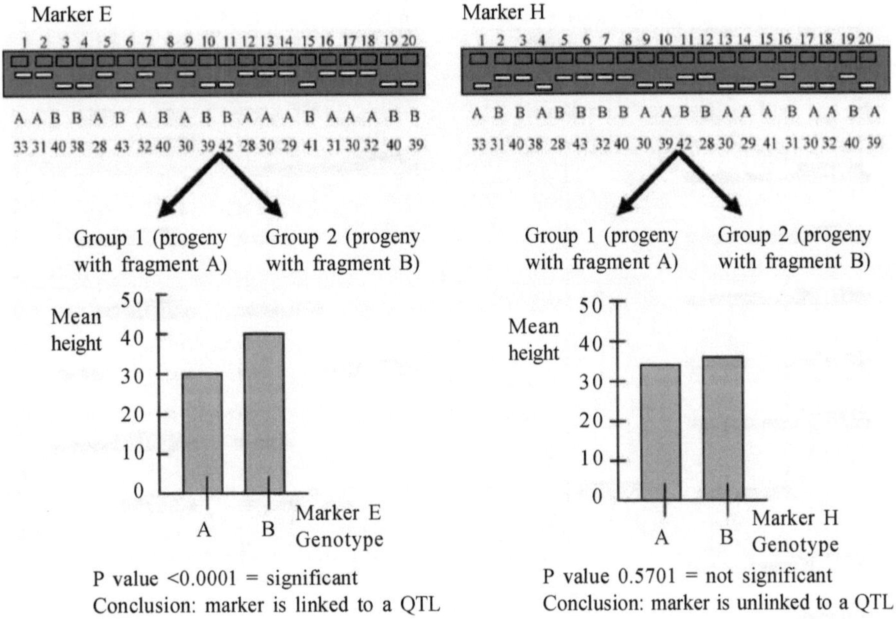

Figure 4.8. Principle of QTL mapping (adapted from Young, 1996). Markers that are linked to a gene or QTL controlling a particular trait (e.g.plant height) will indicate significant differences when the mapping population is partitioned according to the genotype of the marker. Based on the results in this diagram, Marker E is linked to a QTL because there is a significant difference between means. Marker H is unlinked to a QTL because there is no significant difference between means. The closer the marker is to the QTL of interest, the lower the chance for recombination between marker and QTL.

A logical question that may be asked at this point is 'why does a significant P value obtained for differences between mean trait values indicate linkage between marker and QTL?' The answer is due to recombination. The closer a marker is from a QTL, the lower the chance of recombination occurring between marker and QTL. Therefore, the QTL and marker will be usually be inherited together in the progeny, and the mean of the group with the tightly-linked marker will be significantly different ($P < 0.05$) to the mean of the group without the marker (Figure 4.9). When a marker is loosely-linked or unlinked to a QTL, there is independent segregation of the marker and QTL. In this situation, there will be no significant difference between means of the genotype groups based on the presence or absence of the loosely linked marker (Figure 4.9). Unlinked markers located far apart or on different chromosomes to the QTL are randomly inherited with the QTL; therefore, no significant differences between means of the genotype groups will be detected.

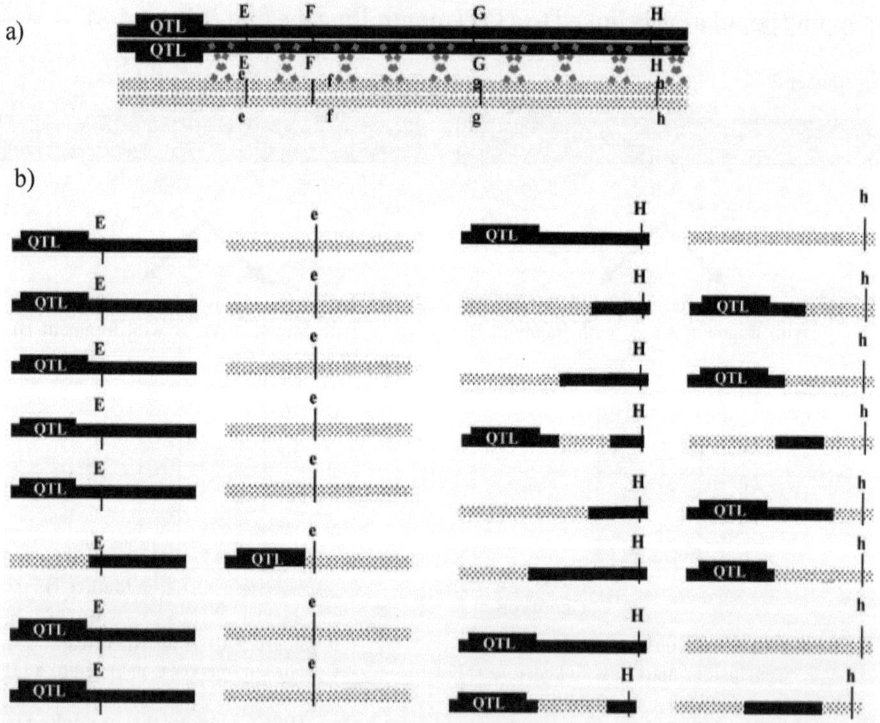

Figure 4.9. Diagram indicating tight and loose linkage between marker and QTL. (a) Recombination events (indicated by crosses) occurs between QTL and marker loci, (b) Gametes in population. Markers that are tightly-linked to a QTL (Marker E) are usually inherited with the QTL in the progeny. Markers that are only loosely-linked to a QTL (Marker H) are randomly inherited together.

4.4.2 Methods to Detect QTLs

Three widely-used methods for detecting QTLs are **single-marker analysis**, **simple interval mapping** and **composite interval mapping**. Single-marker analysis (also called '**single-point analysis**') is the simplest method for detecting QTLs associated with single markers. The statistical methods used for single-marker analysis include t-tests, analysis of variance (ANOVA) and linear regression. Linear regression is most commonly used because the coefficient of determination (R2) from the marker explains the phenotypic variation arising from the QTL linked to the marker. This method does not require a complete linkage map and can be performed with basic statistical software programs. However, the major disadvantage with this method is that the farther a QTL is from a marker, the less likely it will be detected. This is because recombination may occur between the marker and the QTL. This causes the magnitude of the effect of a QTL to be underestimated. The use of a large number of segregating DNA markers covering the entire genome (usually at intervals less than 15 cM) may minimize both problems.

The results from single-marker analysis are usually presented in a table, which indicates the chromosome (if known) or linkage group containing the markers, probability values, and the percentage of phenotypic variation explained by the QTL (R^2) (Table 4.3). Sometimes, the allele size of the marker is also reported. QGene and MapManager QTX are commonly used computer programs to perform single-marker analysis.

Table 4.3. Single-marker analysis of markers associated with QTLs using QGene (Nelson, 1997)

Marker	Chromosome or Linkage group	*p* value	R^1
E	2	<0.0001	91
F	2	0.0001	58
G	2	0.0230	26
H	2	0.5701	2

The simple interval mapping (SIM) method makes use of linkage maps and analyses intervals between adjacent pairs of linked markers along chromosomes simultaneously, instead of analyzing single markers. The use of linked markers for analysis compensates for recombination between the markers and the QTL, and is considered statistically more powerful compared to single-point analysis. Many researchers have used MapMaker/QTL and QGene, to conduct SIM.

More recently, composite interval mapping (CIM) has become popular for mapping QTLs. This method combines interval mapping with linear regression and includes additional genetic markers in the statistical model in addition to an adjacent pair of linked markers for interval mapping. The main advantage of

CIM is that it is more precise and effective at mapping QTLs compared to single-point analysis and interval mapping, especially when linked QTLs are involved. Many researchers have used QTL Cartographer, MapManager QTX and PLABQTL to perform CIM.

4.4.3 Understanding Interval Mapping Results

Interval mapping methods produce a profile of the likely sites for a QTL between adjacent linked markers. In other words, QTLs are located with respect to a linkage map. The results of the test statistic for SIM and CIM are typically presented using a **logarithmic of odds (LOD) score** or **likelihood ratio statistic (LRS)**. There is a direct one-to-one transformation between LOD scores and LRS scores (the conversion can be calculated by: LRS = 4.6 × LOD). These LOD or LRS profiles are used to identify the most likely position for a QTL in relation to the linkage map, which is the position where the highest LOD value is obtained. A typical output from interval mapping is a graph with markers comprising linkage groups on the x axis and the test statistic on the y axis (Figure 4.10).

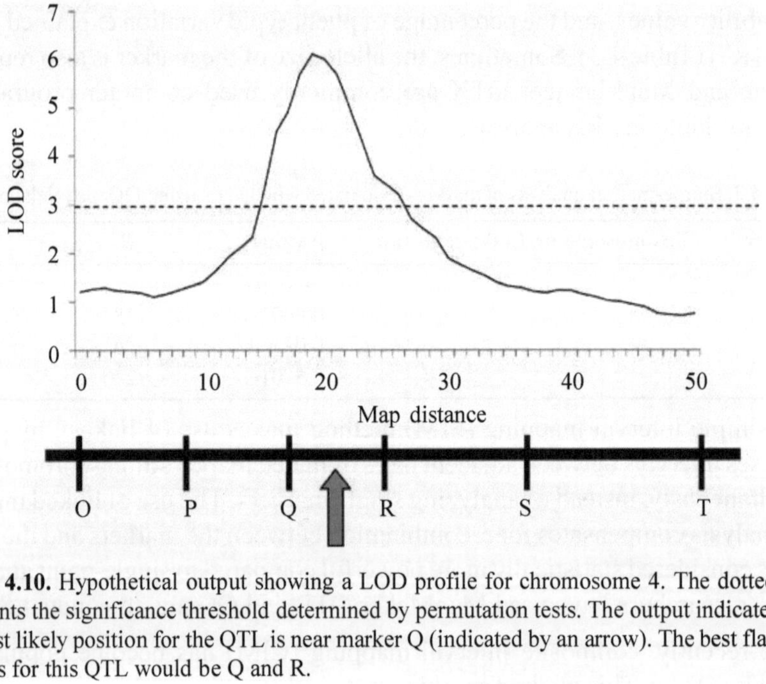

Figure 4.10. Hypothetical output showing a LOD profile for chromosome 4. The dotted line represents the significance threshold determined by permutation tests. The output indicates that the most likely position for the QTL is near marker Q (indicated by an arrow). The best flanking markers for this QTL would be Q and R.

The peak or maximum must also exceed a specified significance level in order for the QTL to be declared as **'real'** (i.e. **statistically significant**). The determination of significance thresholds is most commonly performed using permutation tests. Briefly, the phenotypic values of the population are 'shuffled' whilst the marker genotypic values are held constant (i.e. all marker-trait associations are broken) and QTL analysis is performed to assess the level of false positive marker-trait associations. This process is then repeated (e.g. 500 or 1000 times) and significance levels can then be determined based on the level of false positive marker-trait associations.

Before permutation tests were widely accepted as an appropriate method to determine significance thresholds, a LOD score of between 2.0 to 3.0 (most commonly 3.0) was usually chosen as the significance threshold.

4.4.4 Reporting and Describing QTLs Detected from Interval Mapping

The most common way of reporting QTLs is by indicating the most closely linked markers in a table and/or as bars (or oval shapes or arrows) on linkage maps. The chromosomal regions represented by rectangles are usually the region that exceeds the significance threshold (Figure 4.11). Usually, a pair of markers – the most tightly-linked markers on each side of a QTL – is also reported in a table; these markers are known as **'flanking' markers**. The reason for reporting flanking markers is that selection based on two markers should be more reliable than selection based on a single marker. The reason for the increased reliability is that there is a much lower chance of recombination between two markers and QTL compared to the chance of recombination between a single marker and QTL.

It should also be noted that QTLs can only be detected for traits that segregate between the parents used to construct the mapping population. Therefore, in order to maximize the data obtained from a QTL mapping study, several criteria may be used for phenotypic evaluation of a single trait. QTLs that are detected in common regions (based on different criteria for a single trait) are likely to be important QTLs for controlling the trait.

Mapping populations may also be constructed based on parents that segregate for multiple traits. This is advantageous because QTLs controlling the different traits can be located on a single map. However, for many parental genotypes used to construct mapping populations, this is not always possible, because the parents may only segregate for one trait of interest. Furthermore, the same set of lines of the mapping population used for phenotypic evaluation must be available for marker genotyping, and subsequent QTL analysis, which may be difficult with completely or semi-destructive bioassays (e.g. screening for resistance to necrotrophic fungal pathogens).

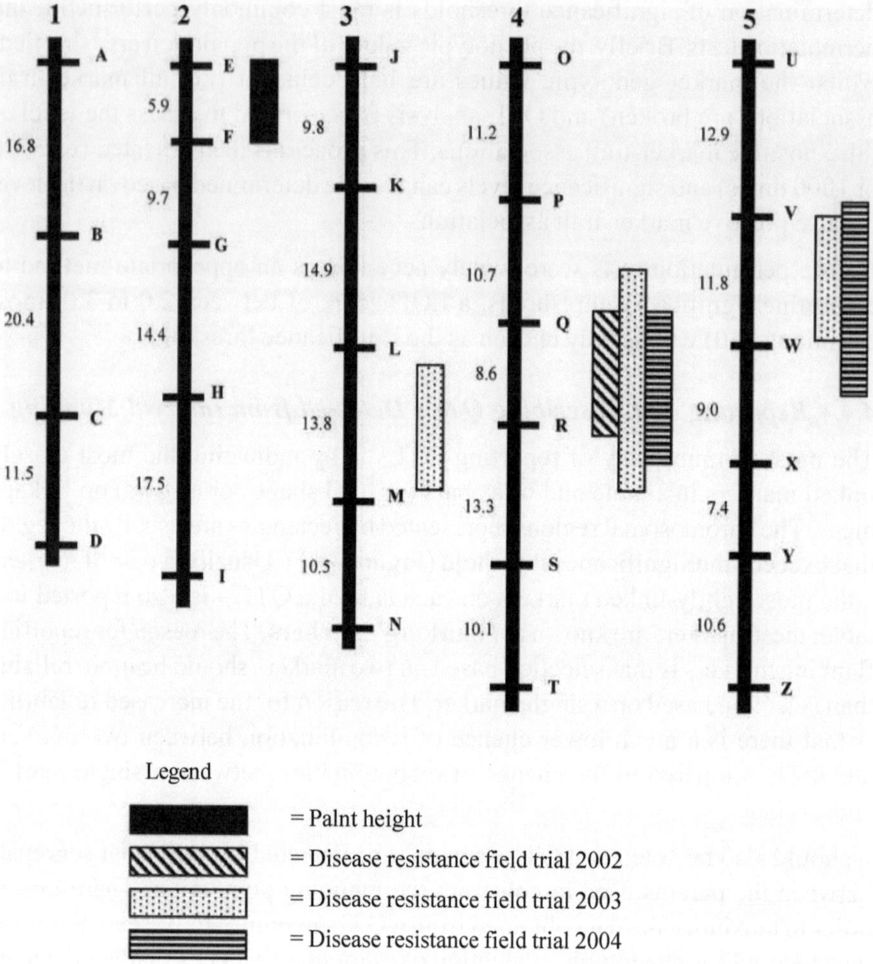

Figure 4.11. QTL mapping of height and disease resistance traits. Hypothetical QTLs were detected on chromosomes 2, 3, 4 and 5. A major QTL for height was detected on chromosome 2. Two major QTLs for disease resistance were detected on chromosomes 4 and 5 whereas a minor QTL was detected on chromosome 3 (adapted from Hartl & Jones, 2001).

In general terms, an individual QTL may also be described as 'major' or 'minor'. This definition is based on the proportion of the phenotypic variation explained by a QTL (based on the R^2 value): **major QTLs** will account for a relatively large amount (e.g. >10%) and **minor QTLs** will usually account for <10%. Sometimes, major QTLs may refer to QTLs that are stable across environments whereas minor QTLs may refer to QTLs that may be environmentally sensitive, especially for QTLs that are associated with disease resistance.

In more formal terms, **QTLs may be classified as: (1) suggestive; (2) significant;** and **(3) highly significant**. Lander and Kruglyak (1995) proposed this classification in order to "avoid a flood of false positive claims" and also ensure that "true hints of linkage" were not missed. Significant and highly significant QTLs were given significance levels of 5 and 0.1%, respectively, whereas a suggestive QTL is one that would be expected to occur once at random in a QTL mapping study (in other words, there is a warning regarding the reliability of suggestive QTLs). The mapping program MapManager QTX reports QTL mapping results with this classification.

4.4.5 Confidence Intervals for QTLs

Although the most likely position of a QTL is the map position at which the highest LOD or LRS score is detected, QTLs actually occur within confidence intervals. There are several ways in which confidence intervals can be calculated. The simplest is the '**one-LOD support interval**', which is determined by finding the region on both sides of a QTL peak that corresponds to a decrease of 1 LOD score. '**Bootstrapping**' – a statistical method for resampling–is another method to determine the confidence interval of QTLs, and can be easily applied within some mapping software programs such as MapManager QTX.

4.4.6 Number of Markers and Marker Spacing

There is no absolute value for the number of DNA markers required for a genetic map, since the number of markers varies with the number and length of chromosomes in the organism. For detection of QTLs, a relatively sparse 'framework' (or 'skeletal' or 'scaffold') map consisting of evenly spaced markers is adequate, and preliminary genetic mapping studies generally contain between 100 and 200 markers. However, this depends on the genome size of the species; more markers are required for mapping in species with large genomes. Darvasi et al. (1993) reported that the power of detecting a QTL was virtually the same for a marker spacing of 10 cM as for an infinite number of markers, and only slightly decreased for marker spacing of 20 or even 50 cM.

4.4.7 Making Comparisons between Maps

All linkage maps are unique and are a product of the mapping population (derived from two specific parents) and the types of markers used. Even if the same set of markers is used to construct linkage maps, there is no guarantee that all of the markers will be polymorphic between different populations. Therefore, in order to correlate information from one map to another, common markers are required. Common markers that are highly polymorphic in mapping populations are called '**anchor**' (also called '**core' markers**). Anchor markers are typically SSRs or RFLPs. Specific groups of anchor markers, that are located in close proximity to each other in specific genomic regions, may be referred to as '**bins**'. Bins are used to integrate maps, and are defined as '10–20 cM regions along chromosomes; the boundaries of each are defined by a set of core RFLP markers'. If common anchor markers have been incorporated into different maps, they can be aligned together to produce 'consensus'. **Consensus maps** are produced by combining or merging different maps, constructed from different genotypes, together. Such consensus maps can be extremely useful for efficiently constructing new maps (with evenly spaced markers) or targeted (or localized) mapping. For example, a consensus map can indicate which markers are located in a specific region containing a QTL, and thus be used to identify more tightly linked markers.

The study of similarities and differences of markers and genes within and between species, genera or higher taxonomic divisions is referred to as **comparative mapping**. It involves analysing the extent of the conservation between maps of the order in which markers occur (i.e. collinear markers); conserved marker order is referred to as '**synteny**'. Comparative mapping may assist in the construction of new linkage maps (or localized maps of specific genomic regions) and in predicting the locations of QTLs in different mapping populations. Previous linkage maps may provide an indication of which markers are polymorphic, as well as provide an indication of linkage groups and the order of markers within linkage groups. Furthermore, *comparative mapping may reveal evolutionary relationships between taxa.*

4.4.8 Factors Influencing the Detection of QTLs

There are many factors that influence the detection of QTLs segregating in a population. The main ones are **genetic properties of QTLs** that control traits, **environmental effects**, **population size** and **experimental error**.

4.4.8.1 Genetic Properties of QTLs

Genetic properties of QTLs controlling traits include the magnitude of the effect of individual QTLs. Only QTLs with sufficiently large phenotypic effects will be detected; QTLs with small effects may fall below the significance threshold of detection. Another genetic property is the distance between linked QTLs. QTLs that are closely-linked (approximately 20 cM or less) will usually be detected as a single QTL in typical population sizes (<500).

4.4.8.2 Environmental Effects

Environmental effects may have a profound influence on the expression of quantitative traits. Experiments that are replicated across sites and over time (e.g., different seasons and years) may enable the researcher to investigate environmental influences on QTLs affecting trait(s) of interest. As discussed previously, RI or DH populations are ideal for these purposes.

4.4.8.3 Population Size

The most important experimental design factor is the size of the population used in the mapping study. The larger the population, the more accurate the mapping study and the more likely it is to allow detection of QTLs with smaller effects. An increase in population size provides gains in statistical power, estimates of gene effects and confidence intervals of the locations of QTLs.

4.4.8.4 Experimental Error

The main sources of experimental error are mistakes in marker genotyping and errors in phenotypic evaluation. Genotyping errors and missing data can affect the order and distance between markers within linkage maps. The accuracy of phenotypic evaluation is of the utmost importance for the accuracy of QTL mapping. A reliable QTL map can only be produced from reliable phenotypic data. Replicated phenotypic measurements or the use of clones (via cuttings) can be used to improve the accuracy of QTL mapping by reducing background 'noise'.

4.4.9 Confirmation of QTL Mapping

Ideally, due to the factors described above, QTL mapping studies should be independently confirmed or verified. Such **confirmation studies** (referred to as '**replication studies**' by Lander & Kruglyak, 1995) may involve independent populations constructed from the same parental genotypes or closely-related genotypes used in the primary QTL mapping study. Sometimes, larger population sizes may be used. Furthermore, some recent studies have proposed that QTL positions and effects should be evaluated in independent populations, because QTL mapping based on typical population sizes results in a low power of QTL detection and a large bias of QTL effects. Unfortunately, due to constraints such as lack of research funding and time, and possibly a lack of understanding of the need to confirm results, QTL mapping studies are rarely confirmed.

Another approach used to confirm QTLs has been to use a specific type of population called **near isogenic lines (NILs)**. NILs are created by crossing a donor parent (e.g. wild parent possessing a specific trait of interest) to a recurrent parent (e.g. an elite cultivar). The F_1 hybrid is then backcrossed to the recurrent parent to produce a first backcross generation (BC_1). The BC_1 is then repeatedly backcrossed to the recurrent parent for a number of generations (e.g. 7). The final BC_7 will contain practically all of the recurrent parent genome except for the small chromosomal region containing a gene or QTL of interest. Homozygous F_2 lines can be obtained by selfing the BC_7 plant. It should be noted that in order to produce a NIL containing a target gene, the gene has to be selected for during each round of backcrossing. By genotyping NILs with important markers, and comparing mean trait values of particular NIL lines with the recurrent parent, the effects of QTLs can be confirmed.

4.4.10 Short Cuts for Gene Tagging

The construction of linkage maps and QTL analysis takes a considerable amount of time and effort, and may be very expensive. Therefore, alternative methods that can save time and money would be extremely useful, especially if resources are limited. Two 'short-cut' methods used to identify markers that tag QTLs are **bulked segregant analysis (BSA)** and **selective genotyping**. Both methods require mapping populations.

4.4.10.1 Bulked Segregant Analysis (BSA)

BSA is a method used to detect markers located in specific chromosomal regions. Briefly, two pools or 'bulks' of DNA samples are combined from 10–20 individual plants from a segregating population; these two bulks should differ for a trait of interest (e.g., resistant vs. susceptible to a particular disease). By making DNA bulks, all loci are randomised, except for the region containing the gene of interest. Markers are screened across the two bulks. Polymorphic markers may represent markers that are linked to a gene or QTL of interest (Figure 4.12). The entire population is then genotyped with these polymorphic markers and a localized linkage map may be generated. This enables QTL analysis to be performed and the position of a QTL to be determined.

BSA is generally used to tag genes controlling simple traits, but the method may also be used to identify markers linked to major QTLs. *'High-throughput' or 'high-volume' marker techniques such as RAPD or AFLP, that can generate multiple markers from a single DNA preparation, are generally preferred for BSA.*

4.4.10.2 Selective Genotyping

Selective genotyping (also known as **'distribution extreme analysis'** or **'trait-based marker analysis'**) involves selecting individuals from a population that represent the phenotypic extremes or tails of the trait being analysed. Linkage map construction and QTL analysis is performed using only the individuals with extreme phenotypes (Figure 4.13). By genotyping a subsample of the population, the costs of a mapping study can be significantly reduced. Selective genotyping is typically used when growing and phenotyping individuals in a mapping population is easier and/or cheaper than genotyping using DNA marker assays. The disadvantages of this method are that it is not efficient in determining the effects of QTLs and that only one trait can be tested at a time (because the individuals selected for extreme phenotypic values will usually not represent extreme phenotypic values for other traits). Furthermore, single-point analysis cannot be used for QTL detection, because the phenotypic effects would be grossly overestimated; interval mapping methods must be used.

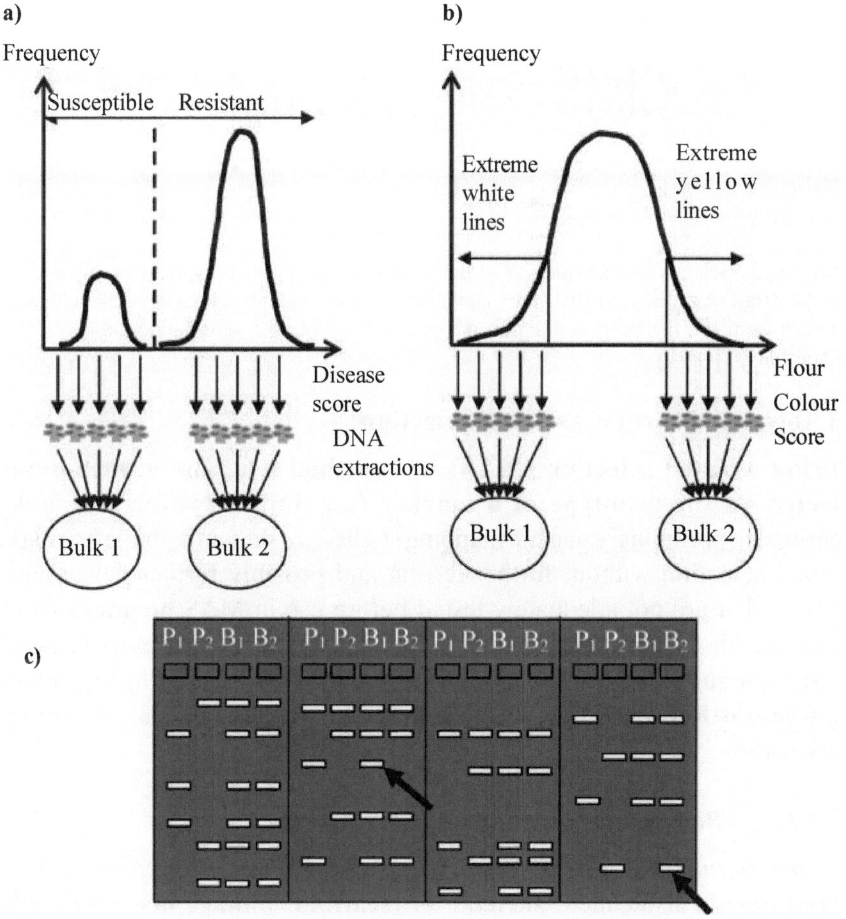

Figure 4.12. The preparation of DNA bulks for a simple disease resistance trait **(a)** and a quantitative quality trait (flour colour) **(b)**. In both cases, two bulks (Bl and B2) are made from individuals displaying extreme phenotypic scores, **(c)** Polymorphic markers (indicated by arrows) that are identified between bulks may represent markers that are linked to genes or QTLs controlling the traits. Such markers are then used to genotype the entire mapping population and QTL analysis performed. (Adapted from Langridge *et al.*, 2001; Tanksley *et al.,* 1995.)

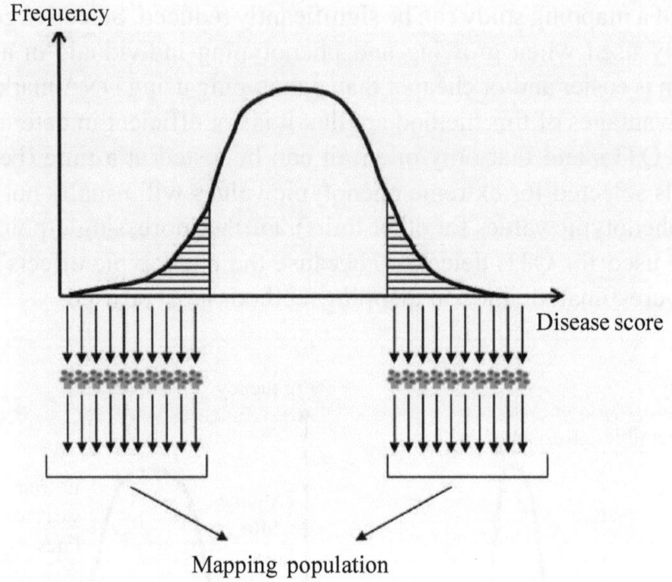

Figure 4.13. Selective genotyping. The entire population is phenotypically evaluated for a particular trait (e.g. disease resistance). However, only individuals representing the phenotypic extremes from the mapping population are selected for marker genotyping, and subsequent linkage and QTL analysis.

4.5 Towards Marker-Assisted Selection

Marker-assisted selection (MAS) is a method whereby a phenotype is selected on the genotype of a marker (see 4.6). However, the markers identified in preliminary genetic mapping studies are seldom suitable for marker-assisted selection without further testing and possibly further development. Markers that are not adequately tested before use in MAS programmes may not be reliable for predicting phenotype, and will therefore be useless. Generally, the steps required for the development of markers for use in MAS includes: **high-resolution mapping**, **validation of markers** and possibly **marker conversion**.

4.5.1 High-Resolution Mapping of QTLs

The preliminary aim of QTL mapping is to produce a comprehensive **'framework'** (also called **'skeletal'** or **'scaffold' linkage map**) *that covers all chromosomes evenly in order to identify markers flanking those QTLs that control traits of interest.* There are several more steps required, because even the closest markers flanking a QTL may not be tightly linked to a gene of interest. This means that recombination can occur between a marker and QTL, thus reducing the reliability and usefulness of the marker. By using larger population sizes and a greater number of markers, more tightly-linked markers can be identified; this process is termed **'high-resolution mapping'** (also called **'fine mapping'**). Therefore, high-resolution mapping of QTLs may be used to develop reliable markers for marker-assisted selection (at least <5 cM but ideally <1 cM away from the gene) and also to discriminate between a single gene or several linked genes.

There is no universal number for the appropriate population size required for high-resolution mapping. However, population sizes that have been used for high-resolution mapping have consisted of >1000 individuals to resolve QTLs to distances between flanking markers of <1 cM.

The mapping of additional markers may saturate framework maps. High-throughput marker techniques that generate multiple loci per primer combination (e.g. AFLP) are usually preferred for increasing marker density (Figure 4.14). Bulked-segregant analysis may also be used to identify additional markers linked to specific chromosomal regions. However, the extent to which framework maps can be saturated depends on the size of the population used to construct the map.

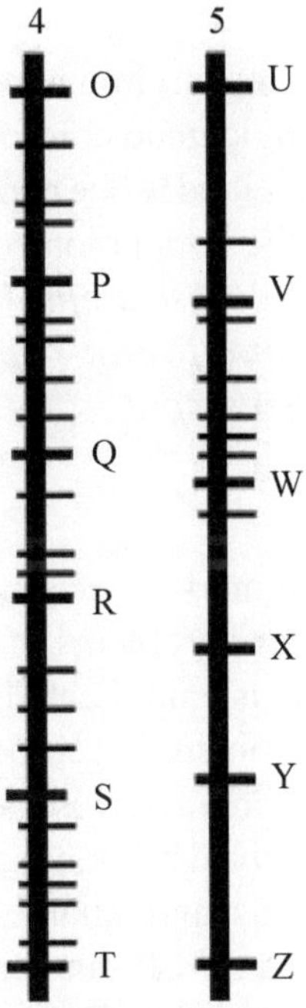

Figure 4.14. High-resolution linkage mapping. Additional markers have been utilised to fill in the gaps' between the anchor markers. The identification of additional markers in the vicinity of QTLs (e.g. between Q and R on chromosome 4) could be useful for MAS. A bulked-segregant analysis approach can be applied to target specific chromosomal regions as shown on chromosome 5 (between V and W).

In many cases, the sizes of segregating populations being used are too small to permit high-resolution mapping, since smaller populations have fewer recombinants than larger populations.

High-resolution maps of specific chromosomal regions may also be constructed by using NILs. Markers that are polymorphic between NILs and the recurrent parent should represent markers that are linked to the target gene and can be incorporated into a high-resolution map.

4.5.2 Validation of Markers

Generally, *markers should be validated by testing their effectiveness in determining the target phenotype in independent populations and different genetic backgrounds, which is referred to as 'marker validation'*. In other words, marker validation involves testing the reliability of markers to predict phenotype. This indicates whether or not a marker could be used in routine screening for MAS.

Markers should also be validated by testing for the presence of the marker on a range of cultivars and other important genotypes. Some studies have warned of the danger of assuming that marker-QTL linkages will remain in different genetic backgrounds or in different testing environments, especially for complex traits such as yield. Even when a single gene controls a particular trait, there is no guarantee that DNA markers identified in one population will be useful in different populations, especially when the populations originate from distantly related germplasm. *For markers to be most useful in breeding programmes, they should reveal polymorphism in different populations derived from a wide range of different parental genotypes.*

4.5.3 Marker Conversion

There are two instances where markers may need to be converted into other types of markers: when there are problems of reproducibility (e.g., RAPDs) and when the marker technique is complicated, time-consuming or expensive (e.g., RFLPs or AFLPs). The problem of reproducibility may be overcome by the development of sequence characterised amplified regions (SCARs) or sequence-tagged sites (STSs) derived by cloning and sequencing specific RAPD markers. SCAR markers are robust and reliable. They detect a single locus and may be codominant. RFLP and AFLP markers may also be converted into SCAR or STS markers. The use of PCR-based markers that are converted from RFLP or AFLP markers is technically simpler, less time-consuming and cheaper. STS markers may also be transferable to related species.

4.6 Marker-Assisted Selection

Selecting plants in a segregating progeny that contain appropriate combinations of genes is a critical component of plant breeding. Moreover, plant breeders typically work with hundreds or even thousands of populations, which often contain large numbers of plants. **'Marker-assisted selection'** (also called **'marker-assisted breeding'** or **'marker-aided selection') may greatly increase the efficiency and effectiveness in plant breeding compared to conventional breeding methods.** Once markers that are tightly linked to genes or QTLs of interest have been identified, prior to field evaluation of large numbers of plants, breeders may use specific DNA marker alleles as a diagnostic tool to identify plants carrying the genes or QTLs (Figure 4.15). The advantages of MAS include:

- Time saving from the substitution of complex field trials (that need to be conducted at particular times of year or at specific locations, or are technically complicated) with molecular tests

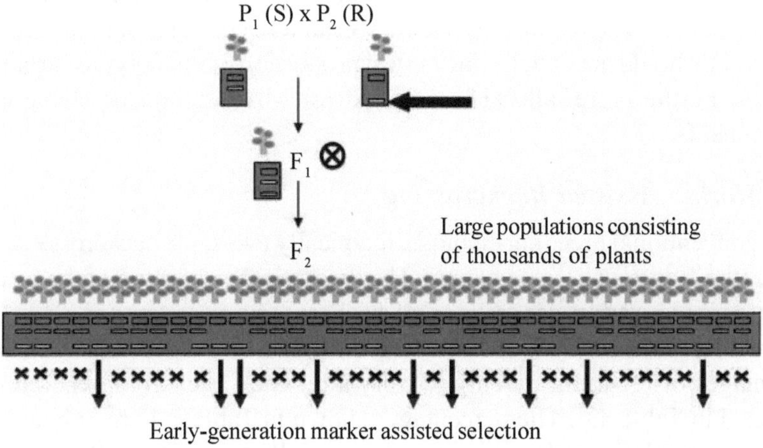

Figure 4.15. MAS scheme for early generation selection in a typical breeding program for disease resistance (adapted from Ribaut & Hoisington, 1998; Ribaut & Betran, 1999). A susceptible (S) parent is crossed with a resistant (R) parent and the Fl plant is self-pollinated to produce a F_2 population. In this diagram, a robust marker has been developed for a major QTL controlling disease resistance (indicated by the arrow). By using a marker to assist selection, plant breeders may substitute large field trials and eliminate many unwanted genotypes (indicated by crosses) and retain only those plants possessing the desirable genotypes (indicated by arrows). Note that 75% of plants may be eliminated after one cycle of MAS. This is important because plant breeders typically use very large populations (e.g. 2000 F_2 plants) derived from a single cross and may use populations derived from hundreds or even thousands of crosses in a single year.

- Elimination of unreliable phenotypic evaluation associated with field trials due to environmental effects

- Selection of genotypes at seedling stage
- Gene 'pyramiding' or combining multiple genes simultaneously
- Avoiding the transfer of undesirable or deleterious genes ('linkage drag'; this is of particular relevance when the introgression of genes from wild species is involved)
- Selecting for traits with low heritability
- Testing for specific traits where phenotypic evaluation is not feasible (e.g. quarantine restrictions may prevent exotic pathogens to be used for screening).

4.6.1 Selection of QTLs for MAS

One commonly asked question is that "since quantitative traits are controlled by at least several QTLs, how many QTLs are typically selected for MAS?" Theoretically, all markers that are tightly linked to QTLs could be used for MAS. However, due to the cost of utilizing several QTLs, only markers that are tightly linked to not more than three QTLs are typically used, although there have been reports of up to 5 QTLs being introgressed into tomato via MAS. Even selecting for a single QTL via MAS can be beneficial in plant breeding; such a QTL should account for the largest proportion of phenotypic variance for the trait. Furthermore, all QTLs selected for MAS should be stable across environments.

4.6.2 Marker-Assisted Backcrossing

Using conventional breeding methods, it typically takes 6–8 backcrosses to fully recover the recurrent parent genome. The theoretical proportion of the recurrent parent genome after n generations of backcrossing is given by: $(2^{n+1}-1)/2^{n+1}$ (where n = number of back-crosses; assuming an infinite population size). The percentages of recurrent parent recovery after each backcross generation are presented in Table 4.4. The percentages shown in Table 4.4 are only achieved with large populations; the percentages are usually lower in smaller population sizes that are typically used in actual plant breeding programmes.

Table 4.4. Percentage of recurrent parent genome after backcrossing

Generation	Recurrent parent genome (%)
BC_1	75.0
BC_2	87.5
BC_3	93.8
BC_4	96.9
BC_5	98.4
BC_6	99.2

Although the average percentage of the recurrent parent genome is 75% (for the entire BC_1 population), some individuals possess more of the recurrent parent genome than others. Therefore, if tightly-linked markers flanking QTLs and evenly spaced markers from other chromosomes (i.e. unlinked to QTLs) of the recurrent parent are used for selection, the introgression of QTLs and recovery of the recurrent parent may be accelerated. This process is called **marker-assisted backcrossing**. The use of additional markers to accelerate cultivar development is sometimes referred to as **'full MAS'** or **'complete line conversion'**.

Simulation studies using PLABSIM–a computer program that simulates recombination during meiosis–indicate that efficiency of recurrent parent recovery using markers is far greater compared to conventional backcrossing (Figure 4.16). Therefore, considerable time savings can be made by using markers compared to conventional backcrossing.

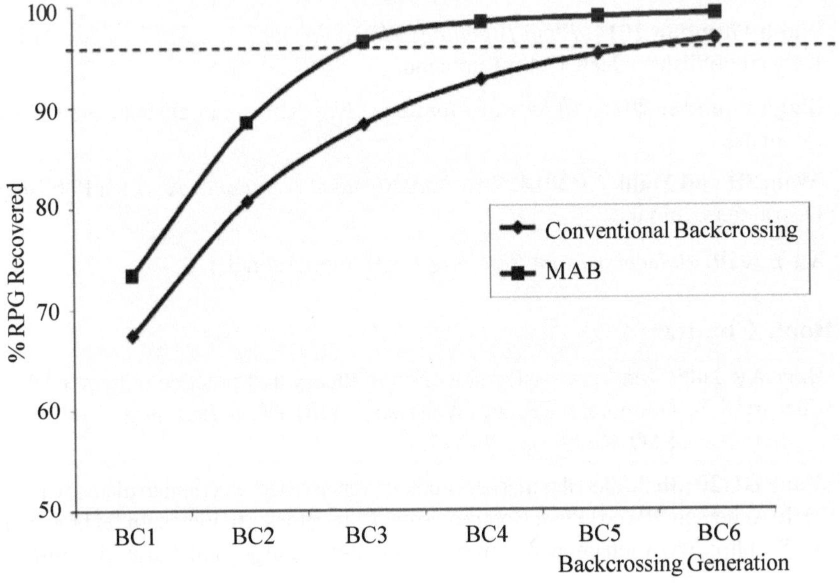

Figure 4.16. Graph of PLABSIM (Frisch *et al.*, 2000) computer simulations of recurrent parent genome (RPG) recovery using marker-assisted backcrossing (MAB) and conventional backcrossing. The use of markers can reduce the number of generations required to achieve the desired proportion of the recurrent parent genome (indicated by the dotted line).

Although the initial cost of marker-assisted backcrossing would be more expensive compared to conventional breeding, the time savings could lead to economic benefits. This is an important consideration for plant breeders because *the accelerated release of an improved cultivar may translate into more rapid profits than by the release of new cultivars in the medium to long-term.*

4.7 Selected Further Readings

A. Books

1. **Acquaah G 2012.** *Principles of Plant Genetics and Breeding*, 2nd ed. Wiley-Blackwell.
2. **Janardhan KV 2012.** *Crop Improvement: A Biotechnological Perspective.* Kalyani Publishers, New Delhi-Ludhiana.
3. **Koh HJ, Kwon SY and Thomson M 2015.** *Current Technologies in Plant Molecular Breeding: A Guide Book of Plant Molecular Breeding for Researchers.* Springer Science+Business Media, Dordrecht.
4. **Newbury HJ 2003.** *Plant Molecular Breeding.* Blackwell Publishing Ltd, UK.
5. **Saxena PK and Balabhaskar KR 2014b.** *Molecular Plant Breeding.* Campus Books International, New Delhi.
6. **Singh BD and Singh AK 2015.** *Marker-Assisted Plant Breeding: Principles and Practices*, 1st South Asian Ed 2017. Springer (India) Pvt Ltd, New Delhi.
7. **Singh Phundan 2014.** *Plant Breeding: Molecular and New Approaches*, 3rd ed. Kalyani Publishers, New Delhi-Ludhiana.
8. **Singh Phundan 2015.** *Molecular Plant Breeding.* Kalyani Publishers, New Delhi-Ludhiana.
9. **Wani SH and Malik CP 2014.** *New Plant Breeding Techniques.* Aaviskar Publishers, Distributers, Jaipur.
10. **Xu Y 2010.** *Molecular Plant Breeding.* CAB International, UK.

B. Book Chapters

1. **Barr AR 2009**. Marker-assisted selection in theory and practice (Chapter 19). In Ceccarelli S, Guimaraes EP and Weltzien E (ed) *Plant Breeding and Farmer Participation.* FAO, Rome, pp 479-517.
2. **Jiang GL 2013b.** Molecular markers and marker-assisted breeding in plants (Chapter 3). In Andersen SB (ed) Plant Breeding from Laboratories to Fields. InTechOpen. Pp 45-83. http://www.intechopen.com/books/plant-breeding-fromlaboratories-to-fields
3. **Kadirvel P, Senthilvel S, Geethanjali *S*, Sujatha M and Varaprasad KS 2015.** Genetic markers, trait mapping and marker-assisted selection in plant breeding (Chapter 4). In Bir Bahadur et al (ed) *Plant Biology and Biotechnology: Volume II: Plant Genomics and Biotechnology*. Springer India, pp 65-88. DOI 10.1007/978-81-322-2283-5_4
4. **Roychowdhury R, Taoutaou A, Hakeem KR, Gawwad MRA and Tah J 2014.** Molecular marker-assisted technologies for crop improvement (Chapter 12). In Roychowdhury R (ed) *Crop Improvement in the Era of Climate Change.* I K International Publishing House Pvt Ltd, pp 248-265.

5. **Thomson MJ, Ismail AM, McCouch SR and Mackill DJ 2010.** Marker assisted breeding (Chapter 20). In Pareek A, Sopory SK, Bohnert HJ and Govindjee (ed) *Abiotic Stress Adaptation in Plants: Physiological, Molecular and Genomic* Foundation. Springer Science + Business Media BV, pp 451–469. DOI 10.1007/978-90-481-3112-9_20

C. Reviews

1. **Babu R, Nair SK, Prasanna BM and Gupta HS 2004**. Integrating marker-assisted selection in crop breeding – prospects and challenges. *Current Science* 87(5): 607-619.
2. ***Collard BCY, Jahufer MZZ, Brouwer JB and Pang ECK 2005.** An introduction to markers, quantitative trait loci (QTL) mapping and marker-assisted selection for crop improvement: The basic concepts *Euphytica* 142: 169-196.
3. **Collard BCY and Mackill DJ 2008.** Marker-assisted selection: an approach for precision plant breeding in the twenty-first century. *Philosophical Transactions of the Royal Society* B 363: 557-572.
4. **Ragimekula N, Varadarajula NN, Mallapuram SP, Gangimeni G, Reddy RK and Kondreddy HR 2013**. Marker assisted selection in disease resistance breeding. *Journal of Plant Breeding and Genetics* 1(2): 90-109.
5. **Schuster I 2011**. Marker-assisted selection for quantitative traits. *Crop Breeding and Applied Biotechnology* S1: 50-55.
6. **Xu Y and Crouch JH 2008**. Marker-assisted selection in plant breeding: from publications to practice. *Crop Science* 48: 391-407.

D. Reports and Bulletins

1. **Brumlop S and Finckh MR 2011**. *Applications and Potentials of MarkerAssisted Selection (MAS) in Plant Breeding,* BfN-Skripten 298, Final report of the F+E project "Applications and Potentials of Smart Breeding"(FKZ 350 889 0020) On behalf of the Federal Agency for Nature Conservation,December 2010, Bundesamt für Naturschutz (BfN),Federal Agency for Nature Conservation, Konstantinstrasse 110, 53179 Bonn,Germany. http://www.bfn.de/0502_skripten.html
2. **Suslow TV, Thomas BR and Bradford KJ 2002.** Biotechnology provides new tools for plant breeding. *Agricultural Biotechnology in California Series Publication No.8043*. University of California, Division of Agriculture and Natural Resources. http://anrcatalog.ucdavis.edu.

Review Questions

Part A

Answer the following questions (Yes/No).

1. Qualititative traits are also called metric traits. (Yes/No)

2. QTLs stands for quantitative trait loci. (Yes/No)
3. DNA markers were developed in the 1990s. (Yes/No)
4. The book 'Marker-Assisted Plant Breeding: Principles and Practices' is written by Dr. Phundan Singh. (Yes/No)
5. The review article 'Collard BCY, Jahufer MZZ, Brouwer JB and Pang ECK 2005. An introduction to markers, quantitative trait loci (QTL) mapping and marker-assisted selection for crop improvement: The basic concepts' was published in 'Genetics'. (Yes/No)
6. Marker-assisted selection and marker-assisted breeding are two different concepts. (Yes/No)
7. There are three major types of genetic markers. (Yes/No)
8. Morphological markers are otherwise known as visible markers. (Yes/No)
9. Marker validation indicates whether or not a marker could be used in routine screening for MAS. (Yes/No)
10. Sample size is not a factor influencing the detection of QTLs. (Yes/No)
11. Selection for traits with low heritability is not possible through MAS. (Yes/No)
12. High-resolution mapping is also called as fine mapping. (Yes/No)
13. Marker-assisted selection (MAS) is a method whereby a phenotype is selected on the genotype of a marker. (Yes/No)
14. Initial cost of marker-assisted backcrossing would be more expensive compared to conventional breeding in the short term. (Yes/No)
15. Anchor is also called core markers. (Yes/No)

Part B

Answer the following questions.

1. What is BSA?
2. RAPD means______
3. Genetic markers are of three types, *viz.*,______, ______and______
4. Full MAS is also known as ______
5. What is LOD?
6. What do you mean by QTL?
7. Metric and polygenic traits are the same / different.
8. What is molecular breeding?

9. RFLP means ______
10. What is AFLP?
11. What is NIL?
12. What do you understand by linkage group?
13. If chromosomes are highways , DNA markers are ______
14. Codominant markers may have many different alleles whereas a dominant marker has only two / four alleles.
15. Differentiate polymorphic and monomorphic markers.

Part C

Write a short note on each of the following topics.

1. Marker-assisted backcrossing
2. Bulked segregant analysis
3. Confidence intervals
4. DNA markers
5. Marker conversion
6. Marker-assisted selection
7. Validation of markers
8. High-resolution mapping
9. Selective genotyping
10. Number of markers and marker spacing
11. Selection of QTLs for MAS
12. Reporting and describing QTLs detected from interval mapping
13. Making comparisons between maps
14. Genetic markers
15. Quantitative traits

Part D

1. Describe the methods to detect QTLs.
2. Discuss the marker-assisted backcrossing.
3. Explain the factors influencing the detection of QTLs.
4. How do you confirm QTL mapping?
5. Discuss the advantages of marker-assisted selection.

5

Molecular Genetic Basis II Association Mapping*

5.1 Introduction

The development of quantitative trait loci (QTL) mapping methods to identify genes or QTLs controlling quantitative traits was a landmark achievement in plant genetics research in the late 1980s. Since then literally thousands of research papers have reported the identification of genes or QTLs for important traits across a diverse range of plant species. This wealth of genetic information was built on the foundation of decades of research in crop molecular genetics and genomics.

The selection and enrichment of QTLs within breeding material was the inevitable next step from QTL discovery to application, and there have been numerous reports of successful tracking of genes and QTLs within breeding programmes. DNA (or molecular) markers have enabled selection of major genes or QTLs for critical or important traits in a process called marker-assisted selection (MAS), which has revolutionized plant breeding. DNA markers are used as tools by breeders to improve the accuracy or efficiency of selection as well cost and time efficiency. Apart from MAS, DNA markers have numerous applications including DNA fingerprinting, genetic diversity analysis and parental characterization.

Recent developments in genotyping platforms and systems to implement molecular breeding schemes offer new tools for modern plant breeders. *Single nucleotide polymorphism (SNP) markers are biallelic, co-dominant markers which are abundant in the genome.* Insertion-deletion (Indel) markers are also prevalent and can be screened using high-throughput genotyping platforms. These platforms are cost-efficient, can handle large sample sizes and provide fast data turn-around time. The availability of genome sequences and genomic resources

---

*Adapted with permission from Verdeprado H, Kretzschmar T, Begum H, Raghavan C, Joyce P, Lakshmanan P, Cobb JN and Collard BCY 2018. Association mapping in rice: basic concepts and perspectives for molecular breeding, *Plant Production Science* DOI: 10.1080/1343943X.2018.1483205

continues to provide a wealth of SNP and Indel markers which will undoubtedly be the marker type of choice for decades to come.

The availability of high-throughput SNP platforms in conjunction with advances in computational methods has also led to the molecular dissection of traits using genome-wide association studies (GWASs) across diversity panels and the development of genomics-based molecular breeding strategies. Association mapping (AM) has now emerged as a standard method for gene and QTL discovery for major crops.

5.2 Approaches to QTL Mapping

Traditional QTL mapping utilizes biparental population (BP) mapping populations from controlled crosses which we refer to as biparental population QTL mapping (BPQM). The selected parents usually differ for traits of interest, and the resulting mapping population should segregate for these traits, although transgressive segregation can be detected even when the parents do not differ for the trait. BP mapping populations include F_2, backcross (BC), recombinant inbred lines (RILs) and near-isogenic lines (NILs). F_2 and BC_1 populations only require two generations to develop. Although they are the simplest population types, these are not homozygous (i.e. 'fixed') and cannot be repeatedly phenotyped. On the other hand, more complex populations, such as NILs and RILs, usually require seven or eight generations to develop, but can be repeatedly phenotyped over multiple years and across environments. Chromosome segment substitution lines (CSSLs) contain chromosome segments from the donor parent in the recurrent parent background. These secondary mapping populations are required to facilitate more comprehensive analysis of target QTLs. More recently in rice, multi-parent populations called multi-parent advanced generation intercross, CSSLs and nested-association mapping populations have been used for QTL mapping.

Constructing linkage maps is necessary to identify chromosomal locations and effects of genes and QTLs associated with traits of interest. A linkage map (or genetic map) shows the relative position and genetic distances between markers or genes along chromosomes. QTL mapping is based on the co-segregation of QTLs and DNA markers. Conceptually, this is based on the principle of chromosome recombination which occurs during meiosis. The principle of QTL detection is based on the association between phenotype and the genotype of markers. The mapping population is partitioned into groups based on the marker each individual carries. A significant difference between the phenotypic means of the groups indicates that the specific marker used is linked to a QTL controlling the trait.

In contrast to ***BPQM, association analysis*** *uses diverse accessions from germplasm collections of varieties, landraces or breeding material referred to as a* **'panel'** (also called **'diversity panel'** or **'association panel'**). Identifying novel QTLs from these panels is the most important use of GWAS for breeding. QTL identification is performed by examining the associations of the markers with the trait that can be explained by the 'linkage disequilibrium' (LD) between markers and polymorphisms across a set of diverse germplasm. QTLs are identified based on historical recombination events between SNPs and QTL at the population level. It is based on the principle that over multiple generations of recombinations, markers that are tightly linked to genes for the trait of interest will generally remain to be associated with the trait.

Population sizes used in preliminary genetic mapping studies generally range from 100 to 250 individuals. A larger population size (approximately >500) is required for the analysis of QTLs having small effects on the target trait. Population sizes used for AM are usually larger than BPQM and, depending on the population structure and diversity, generally require several hundred individuals to identify QTLs.

5.3 Comparison between BPQM and AM

Both BPQM and AM are based on the co-segregation of DNA markers with traits of interest. While the development of a mapping population is required for linkage analysis, AM usually uses diverse populations, or individuals with contrasting geographical origin. Therefore, AM requires less time and resources because phenotypic data are sometimes available for the populations (i.e. panels) that are used for analysis. There is no need to perform controlled crosses to develop mapping populations, although a large number of markers (in the order of thousands or tens of thousands) are required compared to only a few hundred markers for BPQM.

The mapping populations used in BPQM include limited recombination events resulting from fewer generations to establish these populations (i.e. two generations for F_2 population and six to ten generations for RILs). For this reason, QTLs are usually span 10 to 20 cM intervals. In contrast, AM utilizes populations which have undergone many generations of recombination since domestication and therefore in general, only markers that are physically located close to the QTL will be detected as significant. This also explains why linkage analysis has a lower resolution of QTL detection compared to AM. As the size of the interval for localizing the QTL decreases, the number of individuals required to detect at least one recombinant in the region of interest increases, as does the number of molecular markers necessary to detect recombination events.

BPQM uncovers only a small portion of the genetic architecture for a trait because only alleles that differ between the two parental lines will segregate while AM provides an alternative route to identifying QTL that has effects across a broader range of germplasm. On the other hand, BPQM can lead to the discovery of very rare alleles provided the donor accession carries it, while in AM, alleles below a certain minor allele frequency (usually a MAF of < 5%) will be filtered out and hence rare QTLs are usually not detectable. By highlighting the differences between traditional BPQM and AM, it becomes clear that these methods are complementary to each other (Figure 5.1, Table 5.1).

Table 5.1. Comparison of BPQM and AM

Criteria	Traditional QTL mapping (BPQM)	Population-based AM
Number of markers	Relatively few (100–200)	Many (100s–1000s)
Populations	Few parents or grandparents with many offspring (100–500)	Many individuals with unknown or mixed relationships. If pedigreed, family sizes are typically small relative to sampled population (>500)
QTL analysis	Easy. Less sophisticated tools minimize ghost QTL and increase mapping precision	Complex. Sophisticated tools reduce risk of false positives
Detection depends on	QTL segregation in offspring, and marker-trait linkage within-family(s)	QTL segregation in population, and marker-trait LD in mapping population
Mapping precision	Low (5–15 cM) QTL regions may contain many positional candidate genes	Can be high (10–1000 kb). Depends on population LD
Variation detected	Subset (only the portion segregating in sampled pedigrees)	Larger subset (theoretically all variation segregating in targeted regions of genome)
Extrapolation to other families or populations	Poor (other families not segregating QTL, changes in marker phase, etc.)	Good to excellent (although not all QTL will segregate in all population/ pedigree subsamples)
Time spent	More research time in establishing a mapping population	Less research time in establishing an association
Meiosis cycles	Six to seven	Many
Detection power	Low resolution but high statistical power in detecting rare allele	High resolution but low statistical power in detecting rare allele
Population structure	Absent	May be present

RESOURCE or STAGE	TRADITIONAL QLT MAPPING	ASSOCIATION MAPPING
EXPERIMENTAL PLANT MATERIAL	**Bi-parental population** e.g. F2, backcross or RIL derived from A x B	**Panel** e.g. diverse accessions A, B, C, D, E, F, G, H, I, J...Z.....
GENOTYPING	**Linkage map construction** Number of markers"100-400 for framework map	**High marker density** Number of markers generally >10,000 (GWAS) aligned to referencemap position
PHENOTYPING	**Field, greenhouse or lab-based trials** Requires experimental disign and appropriate replication	May need to account for differences in growth and morphology
ANALYSIS	**Interval mapping** e.g. composite interval mapping (CIM) or multiple interval mapping (MIM)	**Association analysis** Generally includers popolation structure analysis and/or marker-based kinship analysis
SIGNIFICANCE TESTING	e.g. Permutation tests	**e.g. Bonferroni Correction or False discovery rate (FDR)**

Figure 5.1. Comparison of QTL mapping methods

5.4 Key Concepts in AM

The main purpose of AM is to dissect complex traits and identify QTLs. QTLs detected using AM (also called **'signals'**, **'peaks'** or **'hits'**) are usually represented using 'Manhattan plots' which show the association of markers with the trait along a chromosome. The y-axis indicates-$\log_{10}$ (P value) for the association plotted against the SNPs along each chromosome on the x-axis, so the map positions of all markers used must be known (Figure 5.2).

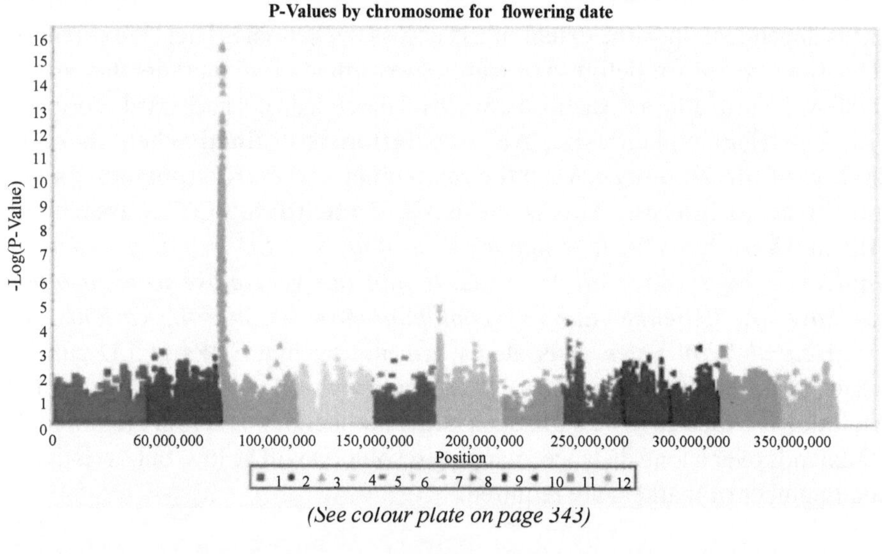

(See colour plate on page 343)

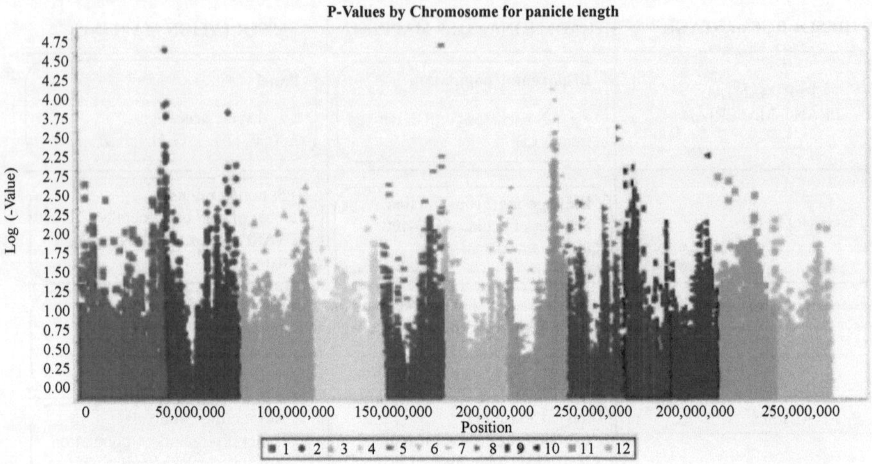

Figure 5.2. AM results for agronomic traits on all 12 chromosomes (2012 dry season data), analysed from Begum et al. (2015). Manhattan plots for flowering date (a) and panicle length (b). Analysis was performed using TASSEL software Version 4.1.34 (Bradbury et al., 2007). Results indicate a major locus for flowering time on chromosome 3 and multiple QTLs for panicle length, typical of quantitative traits. *(See colour plate on page 343)*

5.5 Linkage Disequilibrium (LD)

The fundamental basis of detecting QTLs using AM is due to LD. LD is the 'non-random association of alleles at two or more different loci' in a population. It measures the strength of correlation between markers caused by their shared genetic history. It is a characteristic of pairs of SNP that describes the degree to which an allele of one SNP is inherited or correlated with an allele of another SNP within a population. **The terms LD mapping and AM are often used interchangeably.**

AM is dependent upon the extent of LD across the genome. Thus, the extent of LD should be known before AM can be performed. Two markers that are in LD show non-random association between alleles, but do not necessarily correlate with a particular phenotype. **An association is defined when there is significant correlation between the covariance of a marker polymorphism and a trait of interest. This is the basis of identifying QTLs associated with markers by AM.** *It is important to note that LD is not the same as physical linkage; many allelic variants that are very close to each other may have low LD either due to recombination or because the variants are not at equal frequencies.* It is also worth noting that SNPs in LD can be located on different chromosomes. If LD decays within a short distance, mapping resolution is expected to be high, but a larger number of markers are required. If LD extends over a long distance, mapping resolution will be low, but a relatively small number of markers are required.

In practice, combinations of closely linked, adjacent SNPs or 'haplotypes' are used to characterize a QTL region or allele of a gene rather than a single SNP. Generally, about 5–15 SNP markers per locus are usually sufficient to characterize haplotypes in crops. Haplotypes based on multiple SNPs are generally multi-allelic, in contrast to a single SNP which is biallelic. Variations in SNP haplotypes are of interest to breeders in order to identify genomic regions under selection.

5.6 Analysis Methods for AM

In early reports of AM in crops, relative simple statistical association tests (e.g. general linear models for normally distributed traits or non-parametric tests) were performed, analogous to the initial use of single-marker analysis in BPQM. More advanced methods for AM were subsequently developed, which have become routinely used in crops.

The **effect of population structure** (referred to as **'Q'**) must be accounted for when performing AM. The most commonly used approach to assess the level of population structure is to use marker information to detect sub-groups within the experimental population. This is sometimes called 'structured association' and pieces of information on population structure are considered as fixed effects and used as cofactors in the analysis. Using this method, a set of random markers is used to infer the structure of the population as well as the ancestry of the panel. Common methods used to calculate population structure include: (1) using a computer programme called STRUCTURE or (2) principal component analysis (PCA). One of the main advantages of PCA is that the computational analysis is considerably simpler.

It was subsequently determined that taking into account **the level of genetic relatedness** (called **'kinship'**; referred to as **'K'**) improved the accuracy of AM. Here, **mixed linear models (MLM)** were used to include information on population structure and kinship (i.e. **'Q + K' mixed-model**) that was superior in terms of reducing the false positive rate while maintaining statistical power. This has now become commonly used in crops and algorithms have been streamlined to improve the efficiency (i.e. speed) of the 'data crunching'. The most commonly used approaches in rice include efficient mixed model association, EMMA eXpediated, compressed MLM and population parameters previously determined (P3D). Recently, even more advanced methods have been developed (e.g. Settlement of MLM Under Progressively Exclusive Relationship and multi-locus MLM).

False positive signals are an inherent problem in GWAS, which is often referred to as the problem of 'multiple testing' because a large number of markers are tested and a P value is generated for each individual test. Therefore, the cumulative possibilities of false positive results are large when all of the multiple tests are considered. Among the methods to define significance levels (i.e. that will reduce the chance of false positives but increase the chance of false negatives and vice versa), the Bonferroni correction and the false discovery rate are most commonly used.

5.7 Practical Considerations – Scale of AM

There are two broad categories of AM: (1) candidate gene (CG) analysis and (2) GWAS. While CG analysis is a hypothesis-driven approach based on prior studies about genes involved in the trait of interest, GWAS is a more comprehensive approach which does not require any initial information about the genetic control of a trait of interest. **Due to the increasing use of GWAS over CG, the term is now used interchangeably with AM.**

Genotyping is a critical component of GWAS experiments, and two methods are usually employed. **Genotyping by sequencing (GBS)** is a cheaper way to generate medium to high marker densities (i.e. 10,000–100,000 filtered data points per sample). Although GBS has been optimized and successfully applied in rice, it requires considerable wet-lab and computational expertise for library preparation and bioinformatics data analysis, respectively. **Fixed rice arrays (i.e. 'SNP chips')** are easier to use and several options for different marker densities are available. User-friendly, open-access software to perform AM, such as TASSEL and GAPIT, have been commonly used in rice and allow data formatting, data filtering, data visualization for quality control and advanced analyses.

5.8 Limitations of AM

There are several inherent limitations of AM which researchers must consider when interpreting results, especially for down-stream applications. Some of these limitations are based on the fundamental architecture of complex quantitative traits and are briefly discussed below.

5.8.1 Rare Variants

Previous research has indicated that combinations of rare alleles are usually involved in controlling complex traits. The power to detect rare alleles, however, is low especially since markers used in GWAS are often excluded based on the MAF (i.e. < 0.05), since the statistical methods used for AM are not reliable for very low MAF. Furthermore, these variants may also be in strong/complete LD with non-causative rare variants (e.g. specific SNPs within an individual which have nothing to do with the target trait, but appear to be associated). This is referred to as a **'synthetic association'**. Importantly, indirect associations may be detected due to LD between multiple factors affecting a single trait, especially caused by adaptation. **Thus, conventional QTL mapping approaches still are and will likely remain more effective in discovering rare alleles.** The deliberate inclusion of lines derived from BPs with parents containing rare alleles (if known) to supplement the panel and the use of **'joint-linkage association mapping'** are two approaches to improve the detection for rare variants.

5.8.2 Small-Effect QTLs

Many complex traits are controlled by multiple loci, many of which are small-effect QTLs. Therefore, their presence may simply be too small to be reliably detected, which is a problem when using any QTL mapping method. Detection of these small-effect QTLs can be increased by using a larger population size and accurate phenotypic measurements (i.e. to increase heritability).

5.8.3 Ascertainment Bias

This refers to the fact that SNPs are pre-selected during the SNP development stage prior to their use for association analysis. This is relevant for array-based SNP genotyping platforms, which are usually derived by sequencing a small number of accessions for SNP discovery. Therefore, SNPs that are associated with a trait might not be included in SNP chips, and therefore cannot be detected. Furthermore, fixed arrays might be designed to be maximally informative across certain subgroups (e.g. *indica*), but might detect many monomorphic markers when deployed across panels of a different sub-group (e.g. *japonica*) and hence may not be useful.

5.8.4 Genetic Heterogeneity

This can occur when different genes controlling a single trait under investigation are included in the panel (i.e. the mechanism of genetic control differs within the population and causes confounding results because the ability to detect associations is lower). This can be understood by considering biology – plants originating from different areas may have evolved different mechanisms of adaptation. In rice, genetic heterogeneity is present (i.e. in *O. sativa*). *Indica* and *japonica* varietal groups have evolved through independent domestication process. The consequence of genetic heterogeneity reduces the strength of the association between markers and traits.

5.9 Will AM Always 'Land' on Genes?

One interesting finding from several GWAS experiments in animal and plant species is that the SNPs at the QTL peak are not always within the

functional gene. For example, in *Arabidopsis*, the highest associated SNPs were not the causal SNPs for the vernalization-response gene (FRI); this was attributed to complexities due to LD with specific alleles and population structure. Similar observations have been reported in independent GWASs in rice and may be confusing for researchers who are familiar with BPQM. In the pioneering work done by Yano *et al.* (2016), this issue of undetectable causal SNPs was addressed by using GWAS based on whole genome sequencing, followed by the screening of candidate genes based on the estimated effect of nucleotide polymorphisms.

5.10 From GWAS to Molecular Breeding

The large number of published QTLs or GWAS signals would imply that there are thousands of trait markers (i.e. highly predictive of target traits) available for breeders to use in selection programmes. *Several authors have addressed the issue why only few reports of QTL discovery from GWAS are resulting in the actual use in MAS. Gupta et al. (2014) and Zhang et al. (2016) suggested that the high* ***FDR (false discovery rates)*** *was the most likely reason for a lack of application of GWAS signals in breeding.*

The lack of validation of QTL-marker trait associations is a major factor explaining the lack of applied outcomes from GWASs. It is generally accepted that BPQM research needs to be validated in a range of genetic backgrounds and different environments (i.e. field testing) prior to deployment. Due to **high false discovery rates (FDR) in AM approaches**, there is an even greater need to verify marker-trait associations arising from AM experiments. In other words, the utility of a marker to accurately predict trait phenotype needs to be **verified** (referred to as **'marker-trait validation'**). For example, Breseghello and Sorrells (2006) suggested that this could simply be performed by developing BPs and confirming this at the F_2 or F_3 stage. Readily available breeding populations (e.g. F_2, BC, RILs or elite material) should be used to validate GWAS results in rice. Other methods involving gene expression of candidate genes can be used if the objective is to identify the causal gene for the trait, but this step is not required for breeding.

For large-scale routine screening in breeding populations, GWAS-based markers need to be converted to low-plex high-throughput marker assays (e.g. KASP, Taqman, Fluidigm) and considerable work is required to design and verify these assays. Verification is typically performed for the markers using 48 to 96 samples, which contain known donors and recipients as well as a representative collection of the breeding germplasm. This tests whether the assay is robust in a variety of genetic backgrounds and whether it reliably scores favourable and unfavourable alleles in known backgrounds. Breeders should also be informed by molecular geneticists about details regarding the markers that have been developed (i.e. reliability in terms of selection accuracy for a trait, as well as sensitivity, specificity and effectiveness in different genetic backgrounds). A versatile analysis tool called 'SNP Seek' (http://snp-seek.irri.org) was developed in conjunction with the 3000 rice genomes and provides a valuable resource for the validation of markers located in QTL peaks. Single highly significant SNPs or linked SNP clusters/haplotypes of significant association can be queried, analyzed for allele frequencies on a subpopulation specific basis and put into context with global variation within the region of interest, all of which can greatly aid in the development of low-plex assay.

5.11 Phenotyping: Critical Considerations and New High-Throughput Phenotyping Methods

The importance of accurate trait data for GWAS has been previously emphasized by several researchers and now, the importance of phenotypic data is reiterated for the success of any QTL mapping experiment. Proper experimental design and generation of high-quality phenotypic data are absolutely critical. In practice, accurate phenotyping of panels may be more complicated than molecular geneticists may realize. In rice, for example, ordinal scales are widely used for trait characterization because they are commonly used by rice breeders (for example IRRI standard evaluation system; IRRI, 2014). Ideally, scoring scales should not be based on subjective rating scales. In order to properly characterize quantitative traits, reliable phenotyping methods based on quantitative measurements are required to accurately dissect genetic variation. Broad- or narrow-sense heritability should be calculated for each trait in order to understand the proportion of genetic variance that has been explained by the detected QTLs.

Furthermore, GWAS experiments often do not consider genotype by environment (G x E) interactions. Breeders are well aware that G x E interactions may complicate trait phenotyping. Variation usually exists even within and between controlled environment greenhouse trials and certainly occurs between years, seasons and environments in field trials. Further improvement of phenotyping can be achieved under controlled environment conditions, which is useful for reliable and precise phenotyping of plant responses to abiotic stresses. However, transferability of controlled environment observations to actual field conditions is difficult for some traits such as drought. In these cases, completely new phenotyping methods are required.

Plant phenomics for accurate high-throughput phenotyping has undergone rapid development in the last decade. Both ground-based proximal sensing and aerial remote sensing systems are now routinely used for field phenomics. Rapid, GPS-guided, high-throughput semi- or fully automated phenomics systems have been developed for quantitative measurement of above-ground biomass, stem and canopy attributes, photosynthesis and pigment content, flowering, abiotic stress responses, pathogenesis, leaf traits and agronomic traits. Advancements in aerial vehicle engineering, automation, sensor-based imaging, software capability, data storage and analytical capacity underpin this phenomics revolution. Ground-based platforms have the advantage of generating high-resolution data, but they cannot screen large populations simultaneously, which is critical for many applications.

Remote sensing platforms such as unmanned aerial vehicles (UAVs) and satellites generate data with different spatial resolution and spectral coverage. Manned aircrafts are also used for aerial remote sensing. Most of the commercially available phenomics systems can be deployed directly for many crops though crop-specific features may demand considerable system optimization. UAV-based aerial phenomics platforms are the most commonly used systems for germplasm development and crop research. A range of RGB (red, green, blue), multi-spectral, hyperspectral, thermal-fluorescence, 3D, LIDAR and SONAR sensors are available for plant phenotyping, and many of them can be integrated with UAVs depending on their payload capacity. Rapid progress in automated sensor-based imaging and image-processing software solutions became the cornerstones of field phenomics.

The success of high-throughput phenomics depends on the accuracy and precision of data collected. Accuracy can be improved with the use of known standards, while the stability of the object sensing system determines the precision, which is important to reduce error variance. Unfortunately increasing the throughput may compromise accuracy and precision and this necessitates crop-specific system optimization. Despite the best efforts to maximize accuracy and precision of primary sensory data, which are collected as images over a relatively long period of time, they carry a significant level of genetic, environmental and experimental noise. Developing algorithms for converting the primary sensory data to useful phenotypic data for performance prediction of breeding lines thus can be quite demanding.

Measuring phenotype during the entire crop cycle requires the collection of a very large volume of data. This makes storage, processing and management of phenomics data challenging. Successful high-throughput field phenotyping is contingent on several factors. These include the type of the phenomics platform used, sensor systems with the required spectral coverage, computational and analytical capacity to convert sensory data to useful phenotypic data and database management capability. Identifying the appropriate phenotypic parameters that can be quantified accurately, rapidly & cost-effectively using a carefully designed experiment established with an appropriate genetic population is also vital.

5.12 Closer Integration with Breeding is Needed

Over the last decade, GWAS results have been poorly utilized in breeding programmes which have created an ***'application gap'****. Molecular geneticists, who typically perform most of the gene discovery work and who are main users of GWAS, are often not well aligned with breeding programmes*. The objectives of molecular geneticists are to elucidate and begin to unravel the physiological basis of phenotypic variation at the gene or molecular level. While these are valid and noble endeavours, most of the knowledge obtained from these efforts is not needed in actual breeding programmes.

In order to align gene discovery programmes utilizing GWAS and breeding programmes, discovery efforts need to be more closely integrated with actual breeding activities. Traits that gene discovery groups work on should be the same as those used to set breeding priorities. Reviewing the literature, this is seldom the case and gene discovery work and breeding are rarely connected in the public sector. Often morphological or physiological traits are selected for gene discovery because they are easily measured in controlled environments with high accuracy, rather than their relevance for breeding. These secondary traits may be utilized, provided they are more heritable and highly correlated with the trait of interest like yield. Some notable exceptions were studies to specifically identify loci associated with field blast resistance. The challenge is that high priority complex traits for breeders (e.g. yield) often defy reductionist descriptions using component traits due to complex interactions involving genetics, physiology and the physical environment. When attempts are made to measure complex traits, often the phenotyping methods employed by molecular geneticists are limited to controlled environments like growth chambers and greenhouses, where the correlation with field conditions is either low or unknown. Loci discovered in this way may be highly reliable for a specific assay (e.g. greenhouse-based disease or abiotic stress tolerance test) but there would be little value using markers developed for these loci if the assay was not an effective predictor of field performance. In private sector breeding organizations, gene discovery and breeding teams are structured in a unified product development pipeline, to efficiently evaluate germplasm and identify new alleles of large effect that can be easily deployed in a breeding programme. In rice and other cereals, however, there are traits (notably disease resistance, abiotic stress tolerance and quality traits) for which single genes of large effect suitable to MAS are well known. Any breeding programme interested in doing MAS for these traits will need to prioritize the loci of interest. Particularly for traits where well-developed trait markers still need to be developed, GWAS using breeding germplasm is suited to inform marker development priorities by quickly identifying which loci are both polymorphic and at high frequency in the breeding germplasm and thus would have high utility rates for MAS. In this context, GWAS results can provide useful information by identifying loci (i.e. due to the lack of a QTL peak where one might be expected) that are already fixed in the breeding programme. If fixed for the favourable allele due to selection, the expenditure of marker development resources can be avoided for a locus that has no utility within the breeding programme. For example, one important finding by Begum *et al.* (2015), who used a panel comprised of elite breeding lines, was that major genes (e.g. for flowering time) were often not segregating in elite *indica* rice breeding germplasm, presumably because fixation had already occurred. This information sheds light on the genetic basis of flowering time among the breeding lines they are using, and can thus help direct future strategy for further modifications of flowering time within the breeding programme.

Likewise, if a new disease or specific trait suddenly becomes important to a breeding programme, GWAS cannot only help to determine if the existing breeding lines contain loci easily amenable to MAS, but also to screen panels of exotic germplasm in hope of finding large effect genes/QTLs (provided they are present at reasonable allele frequencies). However, the utility of GWAS to a pre-breeding programme is of limited use when the genetic variants of large effect for a trait are already well known. In these cases, gene discovery efforts are focused on searching for rare alleles of large effect which GWAS is unable to detect. In these situations, a more effective strategy is to screen a large number of exotic lines to identify donors and create BPs with each donor. Such populations can then be targeted for fine mapping and marker development such that NILs in elite backgrounds can be deployed in the breeding programme using marker-assisted forward breeding.

There has been some research using GWAS results as covariates for inclusion in genomic selection (GS) models in a breeding programme and potentially increase prediction accuracies). GS is a new molecular breeding method that was originally developed in animals and has recently been implemented in crops. The method is based on using high-density marker data and advanced computational methods to calculate predict trait performance (in contrast to MAS, which is based on using a single or small number of target alleles). This area of research urgently needs more attention and could represent a critical application of GWAS to breeding programmes.

5.13 Conclusion and Future Perspective

AM has become an increasingly popular approach in crop genetics to understand the architecture of quantitative traits and to identify QTLs controlling important traits. This method complements traditional linkage-based mapping methods based on BPs. It is likely that future progress using AM approaches will be based on two major factors: (1) greater integration with functional analysis or gene annotation data (i.e. 'post-GWAS research'); and (2) improvements in statistical and computational methods (e.g. methods using Bayesian methods, haplotypes and SNP imputation). Further advances in rice genome re-sequencing (3,000 rice genomes project, 2014) and developments in mammalian and other model species will surely have a major impact on GWAS. Current research in AM includes methods to combine GWAS results from multiple studies (e.g. meta-analysis), strategies to account for G x E and tests for epistatic interactions. User-friendly analytical tools and genomics resources will always need to be improved and further developed.

In the meantime, researchers engaged in applied AM activities need to focus on the fundamental factors such as population size, marker density and accounting for population structure. Accurate and relevant phenotypic data will always remain to be one of the critical factors for success in terms of integrating results with breeding. Ultimately, a holistic and integrated approach (i.e. **'systems genetics'**) for GWAS would be the ideal configuration.

5.14 Selected Further Readings

Koh H-J, Kwon S-Y and Thomson M 2015. *Current Technologies in Plant Molecular Breeding: A Guide Book of Plant Molecular Breeding for Researchers.* Springer Science+Business Media, Dordrecht.

Singh BD and Shekhawat NS 2018. *Molecular Plant Breeding.* Scientific Publishers, Jodhpur.

Singh BD and Singh AK 2015. *Marker-Assisted Plant Breeding: Principles and Practices*, 1st South Asian Ed 2017. Springer (India) Pvt Ltd, New Delhi.

Verdeprado H, Kretzschmar T, Begum H, Raghavan C, Joyce P, Lakshmanan P, Cobb JN and Collard BCY 2018. Association mapping in rice: basic concepts and perspectives for molecular breeding, *Plant Production Science* DOI: 10.1080/1343943X.2018.1483205.

Review Questions

Part A

Answer the following questions Yes/No.

1. SNP markers are biallelic and dominant. Yes/No
2. Panel is also called diversity panel. Yes/No
3. AM is based on the co-segregation of DNA markers with traits of interest. Yes/No

4. The main purpose of AM is to dissect simple traits. Yes/No
5. Traditional QTL mapping does not utilize biparental population. Yes/No

Part B

Answer the following questions

1. GWAS stands for ____________.
2. What is FDR?
3. What is the difference between genomics and phenomics?
4. LD means ____________.
5. What is marker-trait validation?

Part C

Write a short note on each of the following topics.

1. Practical considerations-scale of AM.
2. Closer integration of AM with breeding.
3. Key concepts in AM.
4. High false discovery rates (FDR) in AM approaches.
5. Signals.

Part D

1. Compare biparental population QTL mapping (BPQM) and association mapping (AM).
2. Discuss the limitations of AM.
3. Elaborate the analysis methods for AM.
4. Throw light on new high-throughput phenotyping methods.
5. What is the role of linkage disequilibrium (LD) in association mapping?

6

Mode of Reproduction in Relation to Horticultural Crop Breeding

6.1 Introduction

The process by which living beings give rise to the progeny of similar kind is called **reproduction**. Reproduction allows both multiplication and maintenance of a genotype/line/cultivar/species. So, the process of reproduction is also called **propagation**, and the plant part used for this purpose is called **propagule**. The way progeny originate is known as **mode of reproduction** (also called **breeding system** or **reproductive system**). The mode of reproduction of a crop determines its genetic composition which, in turn, is the deciding factor to develop suitable breeding methods. Knowledge of mode of reproduction is also essential for its artificial manipulation to breed improved types. The methods to be adopted for multiplication and maintenance of improved cultivars are also governed by the mode of reproduction. Especially in undertaking breeding investigations with new plants the nature of reproduction must be known before efficient breeding plans can be developed. The mode of reproduction has been well established for most of our important crop species.

6.2 Mode of Reproduction

Mode reproduction in crop plants may be **sexual** or **asexual.** In **sexual reproduction** (also called **amphimixis**), specialized reproductive cells called gametes are formed by the process of gametogenesis. Fusion of the male and female gametes leads to the development of an embryo and eventually the seed. In **asexual reproduction**, fusion of the male and female gametes does not occur, and new plants arise from unfertilized gametes (but not from zygote), or from specialized vegetative organs such as rhizome, tuber, bulb, corm, runner, sucker, stolon, bulbil etc., or by various artificial means of propagation such as rooting of stem and root cuttings, layering, grafting, gootee, etc., or tissue culturing.

6.2.1 Sexual Reproduction

In conventional breeding methods, genetic variability is created and utilized in plants by the plant breeder through sexual reproduction. The importance of the reproductive process to him is so great that it is reviewed in detail including floral parts, types of flowers, sporogenesis, gametogenesis, anthesis, pollination, fertilization and embryogenesis.

6.2.1.1 Floral Parts

A flower is a modified shoot. It contains the reproductive structures of the plant (Fig. 6.1). A typical flower consists of four whorls: **calyx**, **corolla**, **androecium** and **gynoecium** (also called **pistil**). The androecium and gynoecium are upper **essential** or **reproductive whorls**, and the calyx and corolla are lower **non-essential** or **accessory whorls**. The **calyx** and **corolla** consist of a few **sepals** and **petals** respectively. They may be free (**poly-sepalous** or **-petalous**) or united (**gamo-sepalous** or **-petalous**). Sometimes, when calyx and corolla are not differentiated, they are called **perianth** (e.g. banana, onion, garlic, ginger, turmeric, mango ginger, lily, tuberose, etc.) and its members **tepals**.

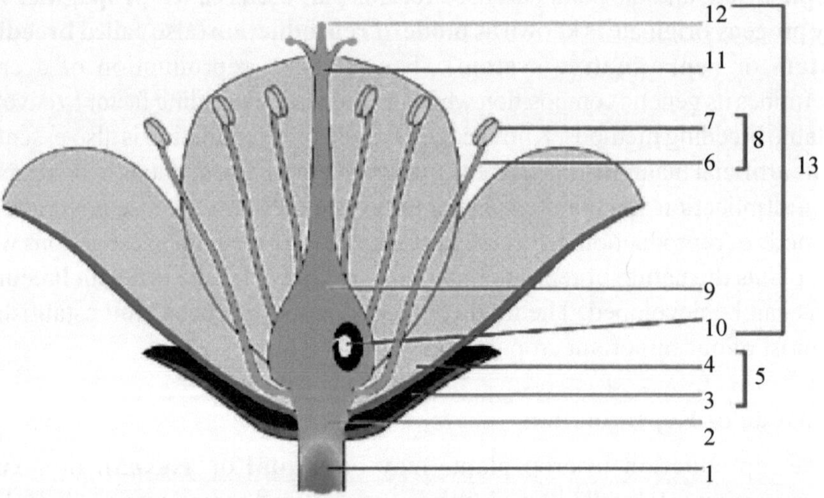

Generalized Flower

Figure 6.1. 1. Pedicel, 2. Receptacle, 3. Calyx, 4. Corolla, 5. Accessory or non-essential whorls, 6. Filament, 7. Anther, 8. Stamen, 9. Ovary, 10. Ovule, 11. Style, 12. Stigma, 13. Reproductive or essential whorls (*Courtesy*: http://courses.botany.wisc.edu) *(See colour plate on page 344)*

The **stamens** (also called **microsporophylls**) are individual units of **androecium** (male organ). They have two parts: (i) **filament**, and (ii) **anther** within which **pollen grains** (also called **pollens**) develop. They may be free (e.g. mango, guava, custard apple, onion, rose, lotus, cashewnut, etc.) or united (coherent). When only filaments are united anthers being free, the **cohesion** (union of similar units) is called **adelphous** (**monadelphous**, e.g., lady's finger (okra),

China rose, hollyhock, etc.; **diadelphous**, e.g., pea, bean, cowpea, etc.; or **polyadelphous,** e.g., lemon). When only anthers are united filaments being free, the cohesion is called **syngenesious**, e.g., lettuce, sunflower, zinnia, cosmos, dahlia, daisy, *Calendula*, aster, chrysanthemum, gerbera, marigold, etc. Again, when both filaments and anthers (whole stamens) are united, it is called **synandrous** cohesion, e.g., ash gourd, taro (*Colocasia*), etc. The **adhesion** (union of dissimilar units) **of stamens with petals** wholly or partially by their filaments is called **epipetalous**, e.g., potato, brinjal, chilli, tomato, petunia, *Ixora*, sunflower, zinnia, cosmos, dahlia, daisy, *Calendula*, aster, chrysanthemum, gerbera, marigold, etc. The **adhesion of stamens with perianth** is called **epiphyllous**, e.g., lily. When **the stamens are united with the carpels** either wholly or by their anthers only, the adhesion is called **gynandrous**, e.g., orchids. Both cohesion and adhesion of stamens except gynandrous condition are helpful for easy emasculation of bisexual flower.

The **gynoecium** (female organ) consists of one or more **carpels** (also called **megasporophylls**) that bear **female spores** (also called **megaspores** or **embryo-sacs**). It may be **simple** (made of one carpel), e.g., pea, bean, cowpea, etc., or **compound** (made of two or more carpels). In a compound pistil, either the carpels may be free (**apocarpous**), e.g., rose, lotus, *Michelia*, *Artabotrys*, etc., or all the carpels may be united together (**syncarpous**), e.g., tomato, onion, etc. The gynoecium has three parts: (i) **ovary**, (ii) **style** and (iii) **stigma**. The **cohesion of carpels** (called **syncarpy**) is of several types. The carpels may be united either throughout their whole length, as in most syncarpous pistils; or, they may be united in the region of ovary alone, styles and stigmas remaining free, as in *Dianthus*, *Linum*, etc.; or, in the region of the ovary and the style, stigmas remaining free, as in hollyhock, China rose, etc.; or, in the region the style and the stigma, ovaries only remaining free, as in *Vinca*, *Nerium*, etc.; or, in the region of the stigma (and the style partly), as in *Calotropis*. The **ovary** may be **superior** (e,g. brinjal, China rose, etc.), **inferior** (e.g. sunflower, guava, gourd, cucumber, apple, pear, etc.) or **half-inferior** (e.g. rose, plum, peach, *Lagerstroemia*, gold mohur, etc.). Consequently, the **flower** may be **hypogynous**, **epigynous** or **perigynous** respectively. The stigma is receptive to pollen deposited on it. Within the ovary are found the **ovules**, which, after **embryo** formation within, develop into mature **seeds**. The number of ovules within an ovary may vary from one, as in mango or cashewnut, to many, as in brinjal or guava.

6.2.1.2 Types of Flowers

Different types of flowers are found in horticultural crops. They have been described in brief:

1. **Complete** *versus* **Incomplete Flower**: If all the four whorls are present in a flower, it is called **complete flower**, e.g., pea, cowpea, potato, tomato, betel, hollyhock, rose, guava, etc. whereas, **incomplete flower** (e.g. male flowers of mango and cashewnut, *Colocasia*, *Caladium*, *Anthurium*, lemon grass, citronella, vetiver, etc.) lacks one or more whorls.
2. **Perfect** *versus* **Imperfect Flower**: If both the essential whorls, androecium and gynoecium, are present in a flower, it is called **perfect flower** and if either androecium or gynoecium is present, it is known as **imperfect flower**. In other words, perfect flowers are **bisexual** (also called **hermaphrodite**), but they may be **complete** (e.g., pea, cowpea, potato, tomato, brinjal, chilli, betel, hollyhock, rose, guava, etc.) or, **incomplete** (e.g. lemon grass, citronella, vetiver, etc.), and imperfect flowers are **unisexual** and, consequently, **incomplete** (e.g. male and female flowers of cucurbits, male flowers of mango and cashewnut, etc.).
3. **Unisexual Flower**: The **unisexual** (imperfect) flower is **male** (also called **staminate**) when it bears only androecium (stamens) but no gynoecium or pistil (carpels), and it is **female** (also called **pistillate**) when it bears only pistil (gynoecium) but no androecium (stamens).
4. **Monoecy** *versus* **Dioecy**: When both staminate and pistillate flowers are borne on the same plant either in the same inflorescence or separate inflorescences, it is called **monoecy** (also called **monoecism**) and if they are borne on different plants, it is known as **dioecy** (also called **dioecism**). Consequently, the crop plants are called **monoecious** (e.g. *Casuarina equisetifolia*, pine, *Begonia*, *Petunia*, etc.) and **dioecious** respectively. The **typical dioecious plant** may be either **male** (also called **androecious**) or **female** (also called **gynoecious**), i.e., unisexual plants, e.g., papaya, date palm, palm, pointed gourd (*parwal*), ivy gourd or little gourd (*kundru*), *Juniperus*, *Polygonum cuspidatum*, etc. There are variants of monoecious and dioecious plants which are described in brief:

a) Variants of Monoecy

1. *Andromonoecious*: Staminate and bisexual flowers are present on the same plant but at different places.
2. *Gynomonoecious*: Pistillate and bisexual flowers are present on the same plant but at different places.
3. *Trimonoecious*: Staminate, pistillate and bisexual flowers are present on the same plant but at different places.

b) Variants of Dioecy

1. *Androdioecious*: Male (androecious) and bisexual plants are separately found.
2. *Gynodioecious*: Female (gynoecious) and bisexual plants are separately found. Development of gynodioecious cultivars ensures fruiting on each plant, e.g., papaya.
3. *Tridioecious*: Male (androecious), female (gynoecious) and bisexual plants of the same species are found separately.

c) Variants of Monoe-Dioecy

Male (androecious) and monoecious, female (gynoecious) and monoecious or male (androecious), female (gynoecious) and monoecious plants of the same species occur separately. All these variants of monoecoius and dioecious plants are found in case of papaya.

It appears that evolution in respect of sex in plants began from **hermaphroditism (bisexuality)** and **typical dioecious species** are the most recent ones.

6.2.1.3 Sporogenesis

Spores are haploid (n) cells produced by meiosis in androecium and gynoecium. They produce male and female gametes. Production of spores is known as **sporogenesis** (Fig 6.2). Sporogenesis is of two types: (1) microsporogenesis and (2) megasporogenesis.

1. *Microsporogenesis*: Each anther has four **pollen sacs (**also called **microsporangia)** that contain number of **microspore mother cells** (also called **pollen mother cells**). Each microspore mother cell undergoes meiosis to produce four haploid **microspores**. Each microspore matures into a **pollen grain** mainly by thickening of its cell wall and accumulation of starch in the cytoplasm.
2. *Megasporogenesis*: A single cell in each ovule differentiates into a **megaspore mother cell**. The megaspore mother cell undergoes meiosis to produce four haploid **megaspores**. Three of them degenerate one remaining as a **functional megaspore** per ovule.

6.2.1.4 Gametogenesis

The production of male and female gametes is known as **gametogenesis** (Fig 6.2). It occurs within microspores and megaspores. It involves only mitosis. It is of two types: (1) microgametogenesis and (2) megagametogenesis.

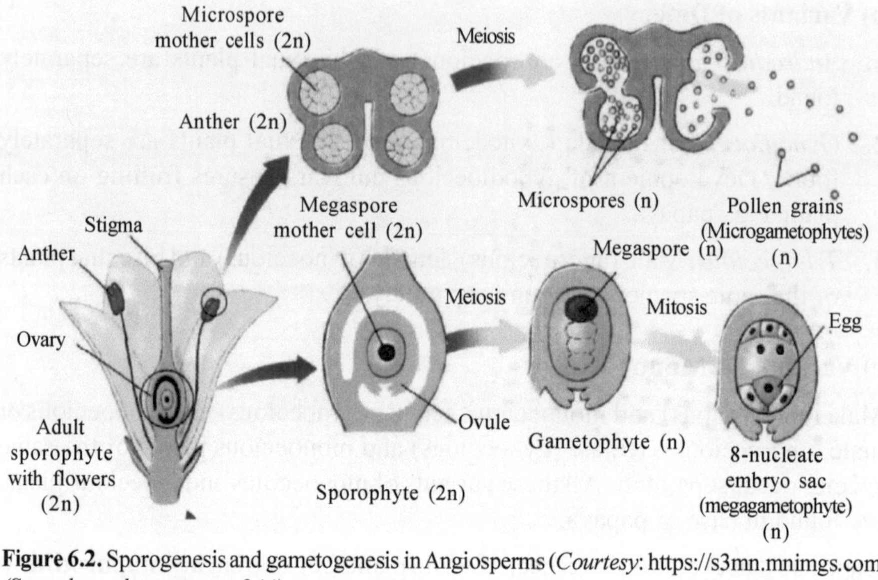

Figure 6.2. Sporogenesis and gametogenesis in Angiosperms (*Courtesy*: https://s3mn.mnimgs.com) *(See colour plate on page 344)*

1. *Microgametogenesis*: Production of **male germ cells** or **gametes** (also called **sperms** or **sperms cells**) through mitosis of microspore is called **microgametogenesis**. The nucleus of the microspore divides mitotically to produce a **generative nucleus** and a **tube nucleus** (also called **vegetative nucleus**). Such pollen grains are called **binucleate**; pollination usually occurs in this state. After pollination, pollen germinates and the **pollen tube** enters stigma and passes through the style. The generative nucleus migrates to and remains near the tip of the pollen tube; it divides mitotically to produce **two sperms** (also called **sperm cells**). The pollen together with pollen tube constitutes the **microgametophyte**.

2. *Megagametogenesis*: Production of **female germ cell** or **gamete** (also called **egg cell**, **egg**, **ovum** or **oosphere**) through mitosis of megaspore is called **megagametogenesis**. Generally, the nucleus of megaspore divides three times to produce eight haploid nuclei. Three of them move to one pole forming **egg-apparatus**; the central nucleus develops in to an **egg cell**, while the two nuclei on its either side give rise to **synergids**. Another three nuclei move to the opposite pole giving rise to **antipodal cells** (also called **antipodals**). The remaining two nuclei move to the centre; they are called **polar nuclei**, and ultimately fuse to form a 2n **secondary nucleus** (also called **definitive nucleus** or **fusion nucleus**). This elaborate cell housing the egg-apparatus (egg cell and two synergids), three antipodals and a secondary nucleus is known as **embryosac** (also called **megagametophyte**).

6.2.1.5 Anthesis

The first opening of a flower is called **anthesis**. It usually happens in the morning, & is influenced by environmental factors such as temperature, humidity, etc. At the time of anthesis the stigma is completely receptive. Pollen grains are shed at this time from the anthers. *The relative timing of opening of a flower and pollen shedding from anthers affects the mode of pollination of a crop species.*

6.2.1.6 Pollination

The transference of pollen grains from the anther to the stigma is called **pollination**. The pollens should be live and the stigma receptive. Generally, pollen grains are in the binucleate stage at the time of pollination. They germinate and pollen tubes enter the stigma (Fig 6.3). A pollen tube passes through the style and finally enters the ovule through the micropyle. Eventually, it releases the two sperms into the embryosac. Pollination is discussed in detail in 6.2.1.8.

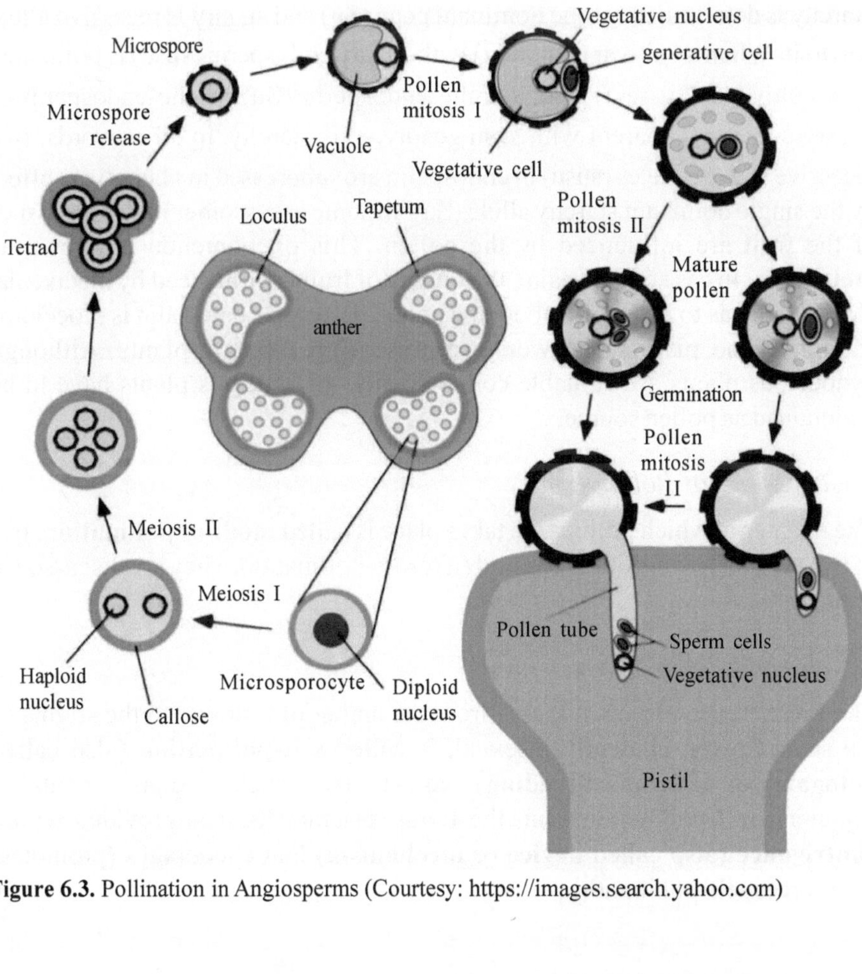

Figure 6.3. Pollination in Angiosperms (Courtesy: https://images.search.yahoo.com)

6.2.1.7 Fertilization and Embryogenesis

Fusion of one male gamete (sperm) with the egg cell is called **fertilization** (also called **syngamy**). It gives rise to a zygote (2n). The other sperm unites with the secondary nucleus (2n) yielding a **primary endosperm nucleus** (3n). It is named as **triple fusion**. The fertilization and triple fusion are combinedly called **double fertilization**. The double fertilization must occur in the embryosac to enable the ovule to develop into a seed.

The zygote divides mitotically to produce an **embryo**. The primary endosperm nucleus divides repeatedly to produce **endosperm**. Eventually, the ovule develops into a **seed** and the ovary into a **fruit**. The two integuments develop into two **seed coats**—the outer **testa** and the inner **tegmen**. Embryo is the *only* part in seed which serves to continue the hereditary stream of plant characteristics of species. Endosperm, when present, may show influence traceable to the pollen. The effect of pollen on embryo and endosperm is called **xenia**. Example of xenia can be seen in maize (corn). The endosperm may be sugary or starchy. Starchy is determined by one dominant gene (Su) and sugary is recessive allele (su) to it. When a cultivar (♀ parent) with sugary endosperm (susu) is pollinated by a cultivar (♂ parent) with starchy endosperm (SuSu), the endosperm of sugary cultivar (♀ parent with susu genotype) is starchy. In other words, two recessive sugary alleles (susu) in endosperm are suppressed in phenotypic effect by the single dominant starchy allele (Su). In some cases, other maternal tissues of the fruit are influenced by the pollen. This phenomenon is known as **metaxenia**. In case of date palm, the quality of fruit is influenced by metaxenia. Hence, one has to be careful about the pollen parent. As date palm is dioecious, staminate and pistillate flowers are present on separate plants. Although gynoecious plants are valuable commercially, androecious plants have to be maintained as pollen source.

6.2.1.8 Mode of Pollination

The manner in which pollination takes place is called **mode of pollination**. It is of two types–(i) self-pollination and (ii) cross-pollination. They are discussed in brief:

6.2.1.8.1 Self-Pollination

The transference of pollen grains from the anther of a flower to the stigma of the same flower, evidently bisexual, is called **self-pollination** (also called **autogamy** or **natural inbreeding**). No external agencies are instrumental in achieving self-pollination. But, the floral structure itself is provided with a **contrivance** (also called **device** or **mechanism**) that encourages (promotes) or enforces self-pollination.

Contrivances Encouraging/Enforcing Self-Pollination

There are several **contrivances (mechanisms)** which encourage/enforce self-pollination in crop plants. They are briefly discussed below:

i) *Bisexuality*: Presence of androecium and gynoecium in the same flower is known as **bisexuality**. The presence of bisexual flowers is a must for self-pollination. All the self-pollinated plants have bisexual (hermaphrodite) flowers.

ii) *Homogamy*: Maturation of anthers and stigma of a flower at the same time is known as **homogamy**. As a rule, homogamy is essential for self-pollination, e.g., *Lathyrus odoratus*, lupin, aster, calendula, etc.

iii) *Cleistogamy*: The situation where flowers never open is called **cleistogamy** and hence, pollen grains may only pollinate the stigma of the same flower. Cleistogamy is found in *Commelina bengalensis*, some species of pansy (*Viola*), balsam (*Impatiens*), wood-sorrel (*Oxalis*), sage (*Salvia*), etc.

iv) *Chasmogamy*: Opening of flowers only after the completion of pollination is known as **Chasmogamy**. It also promotes self-pollination and found in crops like lettuce, pansy, *Clitorea ternatea*, etc. In such cases, some cross-pollination may occur.

v) *Position of Anthers*: In some species, stigmas are surrounded by anthers in such a way that self-pollination is ensured. Such situation is found in tomato, brinjal, potato, etc.

   In some legumes, the anthers and stigma are enclosed in keel (carina) ensuring self-pollination, e.g., pea, French bean, etc.

vi) *Sticky Stigma*: The stigma becomes sticky as in okra, hollyhock, China rose, etc. It promotes self-pollination.

vii) In some species, stigmas become receptive and elongate through the staminal column. This results in predominant self-pollination.

6.2.1.8.2 Cross-Pollination

The transference of pollen grains from one flower to another borne on the same or different plant of the same or allied species, or even allied genera, regardless of whether the flower is bisexual or unisexual, is called **cross-pollination** (also called **allogamy**). It is of three types-(a) **xenogamy** when pollination takes place between flowers borne by two different plants of the same species; (b) **geitonogamy** when it takes place between two flowers borne by the same plant; and (c) **hybridism** when it takes place between two flowers borne by two different plants of allied species or even allied genera. In **autogamy** and **geitonogamy** only one parent plant is concerned in producing the offspring, while in **xenogamy (cross-pollination in strict sense)** two parent plants are concerned and, therefore, a mingling of two sets of parental characters takes place resulting in healthier offspring. Both the methods are wide spread in nature.

Cross-pollination is brought about by external agents such as **insects** (bees, flies, moths, etc.), **animals** (birds, snails, etc.), **wind** and **water**. It is the rule in unisexual flowers, while in bisexual flowers it is of general occurrence. Nature favours cross-pollination and, therefore, adaptations in flowers to achieve it through external agents are many and varied.

1. *Entomophily*: Pollination by insects is of very general occurrence. **Entomophilous** (also called **insect-loving**) **flowers** have various adaptations by which they attract insects and use them as conveyors of pollen grains for the purpose of pollination. Main adaptations are **colour**, **nectar** and **scent**.

   a) *Colour*: One of the most important adaptations is the colour of the petals. The brighter the colour and the more irregular the shape of the flower the greater is the attraction. Sometimes, when the flowers are not conspicuous, other parts may become coloured and showy to attract insects, e.g., one sepal modified into a large white or coloured leafy structure in *Mussaenda*, bracts highly coloured and attractive in *Bougainvillea*, spathes brightly coloured in banana and aroids.

   b) *Nectar*: Nearly all gamopetalous flowers secrete nectar that attracts the cleverer insects like bees. Nectary, a special gland, contains nectar. Sometimes spur, a special structure, also contains nectar. When the bees collect the nectar from the nectary or the spur, they bring about pollination.

   c) *Scent*: Most of the nocturnal flowers are insect-loving. They emit at night a sweet scent that attracts insects from a distance. At night when colour fails, the scent is particularly useful in directing the insects to the flowers. Thus, nocturnal flowers are mostly sweet-smelling. Common examples are night jasmine (*Nyctanthes*), queen of the night (*Cestrum*), jasmines, etc. Sometimes the smell is offensive and nauseating to human beings. But, it is very much liked by certain small insects, e.g., the appendix of the mature inflorescence of *Amorphophallus* and *Arum*.

   The pollens of insect-loving flowers are either sticky or provided with spinous outgrowths. The stigma is also sticky. Pollens and nectar sometimes afford excellent food for certain insects. Sometimes insects visit the flowers in search of shelter from sun and rain, incidentally bring about pollination.

d) *Special Adaptations*: In sunflower, marigold, *Cosmos*, *Anthocephalus*, etc., the individual flowers are small and inconspicuous. They are massed into a dense inflorescence which evidently becomes much showier and more attractive. Dense inflorescence has another advantage. As the flowers are close together, they have every chance of being pollinated.

In snapdragon (*Antirrhinum*) and toad-flax (*Linaria*), the corolla is personate. So, only the insects of particular size and weight can open the mouth of the corolla. Again, long-tongued insects are only useful in the case of the flowers with long corolla tube.

2. *Anemophily*: In some cases, pollination is brought about by wind. **Anemophilous** (also called **wind-loving) flowers** are small and inconspicuous. They are never coloured or showy. They do not emit any smell nor do they secrete any nectar. The anthers produce a large quantity of pollens as the wastage during transit from one flower to another is considerable. They are also light and dry. Sometimes, as in pine, they are provided with wings for facility of distribution by wind. Stigmas are comparatively large and protruding, sometimes branched and often feathery. Examples are grasses, bamboos, pines, palms, etc.
3. *Hydrophily*: Pollination may also be brought about in some aquatic plants, particularly the submerged ones, through the agency of water, e.g., *Vallisneria*, *Hydrilla*, etc. Those aquatic plants that lift their flowers above the water level normally achieve pollination through insects or wind. **Hydrophilous** (also called **water-loving) flowers** are as a rule small and inconspicuous.
4. *Zoophily*: Birds, squirrels, bats, snails, etc., also act as useful agents of pollination; for example, birds and sometimes squirrels bring about pollination in coral tree (*Erythrina*), silk cotton tree (*Bombax*), rose-apple (*Syzygium*), *Bignonia*, etc.; bats in *Anthocephalus*; snails in certain large varieties of aroids and in snake plant (*Arisaema*). (Insects are also instrumental in bringing about pollination in them).

There are some **contrivances** (also called **devices**, **mechanisms**) either encouraging or enforcing cross-pollination in the flower. They are discussed in brief below:

A. *Contrivances Encouraging Cross-Pollination*

There are certain contrivances that have little or no effect on the genetic differentiation among individuals, and hence cause no alteration of the composition of population. In fact, such contrivances encourage cross-pollination, but they also do not prevent self-pollination from occurring. So, controlling pollination in crops with such contrivances requires considerable precaution. However, inbreds can be generated with no difficulty. These contrivances are listed below:

i) *Dichogamy*: In many bisexual flowers the anther and the stigma often mature at different times. This condition is known as **dichogamy**. Because the anther and the stigma mature at different times, dichogamy often prevents self-pollination. There are two types of dichogamy:

   a) *Protogyny*: It occurs when the gynoecium matures earlier than the anthers of the same flower. Here the stigma receives the pollen grains brought from another flower. Common examples are fig, four o'clock plant (*Mirabilis*), custard-apple (*Annona*), some palms, etc.

   b) *Protandry*: It occurs when the anthers mature (burst and discharge their pollens) earlier than the stigma of the same flower. Here the pollen grains are carried over to the stigma of another flower. Common examples are China rose, lady's finger (okra), sunflower, marigold, coriander, rose, carrots, raspberries, beets, etc.

ii) *Monoecism*: Monoecism, male and female flowers borne separately on the same inflorescence or on the same plant, favours cross-pollination much more than self-pollination. Common examples are mango, banana, coconut, cucurbits, walnut, chestnut, strawberries, rubber, grapes, cassava, etc.

iii) *Herkogamy*: There may be some sort of barrier standing between stamens and pistil of the same flower. The keel in alfalfa (lucerne) and a hood covering the stigma in pansy, iris, etc. are the common forms of such natural barriers. These barriers may be broken down by insects which can take some pollen grains to other plants also, thus causing cross-pollination.

B. *Contrivances Enforcing Cross-Pollination*

By contrast to these operating at individual level, some contrivances bring about genetic differentiation among individuals, and change in the gene-frequency. Hence, they tend to alter the population structure. These systems are so efficient in their functioning that they enforce perfect or almost perfect cross-pollination in their carriers. Therefore, they are of great advantage in plant breeding through hybridization. These contrivances include:

i) *Male Sterility*: Due to irregular microsporogenesis or inadequate nourishment, spores (pollen grains) are rendered non-functional, or, due to non-dehiscence of anthers, pollen grains are not released in some crop plants. So, the normal stigma needs to be pollinated by foreign pollens only. Male sterility is genetically or cytoplasmically controlled. This system has been greatly exploited in crops such as pearlmillet, maize, onion, etc. (see next chapter).

ii) *Self-Incompatibility*: In a large number of crops, although pollen grains and stigma both are normally functional, pollen grains from the flowers of a plant have no fertilizing effect on the stigmas of the same flower or different flowers borne on the same plant. So, for producing the next generation, cross-pollination is required. This situation is called **self-incompatibility** (also called **self-sterility**). It is most frequent in several species of *Brassica* (mustard, rai, cauliflower, etc.) radish, rye & many grasses (see next chapter).

iii) *Dioecism*: In dioecy, male and female flowers are borne on different plants. Hence, *cross-pollination is indispensable* for the production of seeds. Common examples are papaya, date palm, palm, mulberry, spinach, pointed gourd, little gourd, spine gourd, asparagus, etc. In dioecy, there is considerable wastage of gametes in the absence of suitable pollinating agent.

6.2.1.8.3 Determination of Mode of Pollination

For breeding of any crop, the mode of pollination must be known before undertaking any breeding programme. We know the mode of pollination of most of the important crops. *But, for a new crop species we have to determine*. There are some natural contrivances (favouring either self-pollination or cross-pollination) that are of great use while determining the mode of pollination of the new crop. However, some other methods are also available that can be appropriately applied along with natural devices for this purpose. The mode of pollination can be broadly determined by:

a) *Floral Structure*: Bisexuality and homogamy are essential for self-pollination. Position of anthers, sticky stigma, etc. promote self-pollination. Cleistogamy and chasmogamy permit only self-pollination. Dichogamy (protogyny and protandry), monoecism and herkogamy favour cross-pollination while male sterility, self-incompatibility and dioecism enforce cross-pollination.

b) *Inbreeding Depression*: If selfing of plants leads to considerable loss of vigour, cross-pollination is anticipated. Conversely, little or no loss of vigour would indicate self-pollination. However, it is not a fool-proof method because many cross-pollinated crops, e.g. pearl millet and cucumber, can tolerate inbreeding to the extent that many self-pollinated crops do.

c) *Bagging and Seed-set*: Cross-pollination need not prevent self-pollination from occurring and hence, from normal seed-setting under controlled selfing, i.e., bagging (as in maize and pearl millet). When there is seed-setting in controlled selfing, the crop may be either self-pollinated or cross-pollinated. So, bagging can detect cross-pollination if there is no seed-setting. However, bagging may create unfavourable environment for pollination and seed-setting. Therefore, the (space) isolation as detailed below is preferred to.

d) *Isolation*: If a single plant is raised in isolation, normal seed-setting will indicate self-pollination, while erratic or no seed-set means flowers need pollens from other sources. Thus, cross-pollination can easily be identified. But, self-pollination is more difficult to be established by isolation planting.

The same objective can also be achieved on emasculating a few flowers and leaving rest of the flowers intact. Then, seed-set in emasculated flowers suggests cross-pollination and no seed-set self-pollination.

e) *Use of Recessive Markers*: A strain with a simply inherited recessive marker gene may be inter-planted in isolation along with the normal strain (possessing dominant allele). The marker genes may preferably be of seed/seedling traits (characters). After harvest of seeds from marker strain only, the seeds are examined for the absence of seed marker trait. Or, after raising of seedlings from those seeds, they are examined for the absence of seedling marker trait. Marker trait, if dominated by normal trait, indicates the occurrence of cross-pollination. If no such crossing has occurred, self-pollination may be indicated.

A dominant marker strain can also be used. But, in that case the seeds of the marker strain are not to be harvested; and seeds only from recessive strain should be collected. Further testing is based on the presence of marker traits in contrast to their absence in the former case.

6.2.1.8.4 Determination of the Amount of Cross-Pollination

The extent of cross-pollination (out-crossing) is determined by growing two strains of the crop. One of them is homozygous dominant for a seed/seedling trait, while the other one is recessive for that trait. Both the strains are sown/planted in such a manner that plants of the dominant strain surround each plant of the recessive strain to facilitate abundant pollens. Utmost care should be taken for synchrony of flowering of both the strains. At maturity, seeds are harvested from the plants of the recessive strain only. These seeds (or the seedlings developed from them) are scored and classified on the basis of the dominant and recessive forms of the trait concerned. The percentage of seeds (or seedlings) with the dominant form of the trait concerned is the percentage of cross-pollination (out-crossing) in the crop. The percentage of cross-pollination in a crop varies greatly with cultivar, weather conditions and location. So, such study should include several cultivars of the crop and it should be conducted at several locations for two or more years. In view of this, information on the amount of cross-pollination in many crops may be regarded as incomplete.

6.2.1.8.5 Classification of Crops Based on Pollination

On the basis of mode of pollination, the horticultural crops are grouped into four categories, *viz.*, (i) naturally self-pollinated crops, (ii) often cross-pollinated crops, (iii) naturally cross-pollinated crops, and (iv) dioecious cross-pollinated crops (Table 6.1). They are discussed in brief below:

i) *Naturally (Normally) Self-Pollinated Crops*: Many crops show more than 95% natural self-pollination and they are called naturally (normally) **self-pollinated** (also called **self-fertilized, selfers, inbreeders, autogamous**) (Table 6.1). In such crops, the amount of natural out-crossing (cross-pollination) is less than 5% and affected by the cultivar, temperature, humidity, location, etc. Self-pollinated crops are believed to have originated from cross-pollinated ancestors.

ii) *Often Cross-Pollinated Crops*: When the amount of natural cross-pollination is more than 5% and may be as high as 30%, the crops are called **often cross-pollinated** (also called **often cross-fertilized**, **intermediates**) (Table 6.1). The genetic make-up of such crops is intermediate between that of self-pollinated and cross-pollinated crops. Hence, breeding methods applicable to both the groups can be applied to them. In general, hybrids are superior to purelines and used wherever feasible.

iii) *Naturally (Normally) Cross-Pollinated Crops*: Many crops exhibit natural cross-pollination and are called naturally (normally) **cross-pollinated** (also called **cross-fertilized, outbreeders, allogamous**) (Table 6.1). In many of such crops, a small amount of natural self-pollination (upto 5-10%) may occur. Such crops are either monoecious or have bisexual flowers and are not self-incompatible. Wherever self-incompatibility occurs, the crops will be *completely cross-pollinated*. Cross-pollinated crops are highly heterozygous, show high inbreeding depression and considerable heterosis, and follow Hardy-Weinberg Law. They differ completely in their genetic make-up from self-pollinated crops. So, the breeding methods and their end products (= cultivar) differ drastically in the two groups of crops.

iv) *Dioecious Cross-Pollinated Crops*: As in dioecy male and female flowers are borne on separate plants, cross-pollination is indispensable and 100% in these crops (Table 6.1). But, in this case both male and female plants with the characters desired must be selected and their progeny is tested to determine the breeding value of parents. By this means, cultivars of superior type may be synthesized.

Table 6.1. A partial list of horticultural crops based on pollination and habit

S.No.	Groups	Crops
1.	Naturally Self-Pollinated Crops	Annual: tomato, potato, pea, cowpea, Indian bean, French bean, cluster bean Perennial: apricot, nectarine, citrus, peach
2.	Often Cross-Pollinated Crops	Annual: okra, brinjal, chilli
3.	Naturally Cross-Pollinated Crops	Annual: onion, garlic, coriander, cucumber, muskmelon, watermelon, pumpkin, summer squash, winter squash, other cucurbits, marigold, *Petunia*, *Begonia* *Biennial: cauliflower, cabbage, knol-khol, broccoli, carrot, turnip, radish, beets Perennial: sweet potato, apple, avocado, mango, pear, blackberries, raspberries, walnut, chestnut, hazelnut, banana, cherry, fig, coconut, grapes, plum, loquat, strawberries, almond, *Casuarina equisetifolia*, pine, roses
4.	Dioecious Cross-Pollinated Crops	Annual: pointed gourd, little gourd, spine gourd, spinach Perennial: asparagus, papaya, date palm, palm, *Juniperus*, *Polygonum cuspidatum*

*Biennial but cultivated as annual.

6.2.2 Asexual Reproduction

In asexual reproduction, new plants arise from unfertilized gametes, or from specialized vegetative organs, or by various artificial means of propagation, or tissue culturing. In horticultural crop plants, asexual reproduction is of paramount importance. So, it is discussed in some details including apomixis, natural vegetative reproduction and artificial vegetative reproduction (conventional and biotechnological).

6.2.2.1 Apomixis

When embryos and seeds develop without fertilization, it is called as apomixis. Winkler (1908) *defined apomixis as "the substitution for sexual reproduction of another asexual reproductive process that does not involve nuclear or cellular division (i.e. fertilization)"*. It is prerequisite for apoximis that fertilization does not take place at all in any form. Thus, there is no normal fertilization of the egg cell, hence no normal development of embryo from the egg cell. However, embryo may develop from an unfertilized egg cell or from a cell other than the egg cell within the embryo sac.

Seedling plants produced by apomixis are known as **apomicts**. Apomicts are two types: **obligate**, producing only apomictic embryos, and **facultative**, producing both apomictic and normal (zygotic) embryos. A simplified classification of apomixis is as follows: (1) adventive embryony, (2) apospory, and (3) diplospory.

6.2.2.1.1 Adventive Embryony

In this case, embryos arise directly from a vegetative cell or a group of vegetative cells of the ovule such as nucellus, integuments and chalaza, and there is no embryo sac formation. Adventive embryos are generally developed in addition to the zygotic embryos of seeds; this situation is called **polyembryony**, e.g., in mango, *Citrus*, orchids, roses, *Opuntia*, etc.

6.2.2.1.2 Apospory

Embryos arise from unreduced egg cells or some other cells of embryo sacs derived from vegetative cells of ovules, e.g., in *Malus*, *Crepis*, etc.

6.2.2.1.3 Diplospory

Embryo sacs are produced from megaspores, which are generally diploid, but may be haploid. It is of two types: (i) parthenogenesis and (ii) apogamy.

i) *Parthenogenesis*: Embryos develop from egg cells, which may be haploid (**haploid parthenogenesis**) or diploid (**diploid parthenogenesis**). Diploid parthenogenesis occurs naturally in many grasses. But haploid parthenogenesis occurs accidentally in low frequencies in many species, e.g., in maize, tobacco, barley, etc.

ii) *Apogamy*: Embryos arise from synergids or antipodals of the embryo sacs, and they may be diploid or haploid. Diploid apogamy occurs in *Antennaria*, *Alchemilla*, *Allium*, etc. Haploid apogamy can be induced in many species by culture of unfertilized ovaries.

6.2.2.1.4 Detection of Apomixis

The occurrence of apomixis in a crop may be detected by crossing a recessive strain to a dominant one. The former is used as female while the latter as male. It is convenient to consider seed/seedling traits. While emasculating the former, utmost precaution must be taken to avoid chance selfing. The seeds are harvested from the recessive female parent and examined for the seed/seedling character (trait) considered. If a sufficiently high frequency of recessive offspring is obtained in the progeny, the crop is most likely apomictic. Apomixis is indicated by several other features also, that are enumerated below:

1. In case of **obligate apomixis**, all the progeny from a cross resemble the female parent.

2. In case of **facultative apomixis**, the percentage of progeny that are identical to the female parent gives the degree of apomixis.
3. In a selfed generation, presence of plants showing no inbreeding depression is indicative of apomixis.
4. Presence of multiple seedlings (from single seeds) indicates apomixis (**polyembryony**).
5. Embryological studies of young ovules confirm apomictic behaviour. The presence of more than one embryo sac in an ovule is an important indicator of apomixis. But progeny tests are essential to determine the degree of apomixis.

6.2.2.2 Vegetative Reproduction

When a new individual arises from a somatic cell without formation of a seed, it is called **vegetative reproduction**. Vegetative reproduction occurs naturally in many crops, and profitably exploited in horticulture using either conventional or biotechnological techniques.

6.2.2.2.1 Natural Vegetative Reproduction

In nature, vegetative reproduction occurs through underground stems, sub-aerial stems, bulbils, leaves and even roots.

i) *Underground Stems*: Rhizome (e.g., ginger, turmeric, arrowroot, *Canna*, water lily, lotus, banana, etc.), stem tuber (e.g., potato), bulb (e.g. onion, garlic, tuberose, most lilies, tulip, etc.) and corm (e.g., *Colocasia*, *Amorphophallus*, *Gladiolus*, saffron (*Crocus*), water chestnut, etc.) are storage organs and contain many buds giving rise to new plants.

ii) *Sub-aerial Stems*: Runner (e.g., strawberry), stolon (e.g. mint, wild strawberry), suckers (banana, pineapple, *Chrysanthemum*, date-palm, pear, guava, pomegranate, red-raspberry, etc.), etc. are used in propagation.

iii) *Bulbils*: Bulbils are modified vegetative or floral buds. They give rise to new plants without fertilization and seed formation. They are seen in *Dioscorea bulbifera*, *Agave americana*, onion, garlic, *Lilium bulbiferum*, etc.

iv) *Leaves*: Foliar buds develop on leaves in sprout leaf plant (*Bryophyllum*) and elephant ear plant (*Begonia*). New plants originate from these buds.

v) *Roots*: The roots of sweet potato, *Aegle*, *Trichosanthes*, *Dalbergia*, lemon and ipecac bear buds. Such plants are sometimes propagated by root cuttings.

6.2.2.2.2 Artificial Vegetative Reproduction

This includes vegetative propagation of plants by man using either conventional or biotechnological techniques.

a) *Conventional Artificial Vegetative Reproduction*: These techniques are based on sections of modified stems (e.g., potato, ginger, turmeric, etc.), stem cuttings (grapes, roses, sweet potato, etc.), root cuttings (e.g., ipecac), grafting (e.g., mango, peach, loquat, sapota, etc.), budding (e.g., roses, *Ziziphus*, etc.), layering (e.g., guava, litchi, kagzi lime, pomegranate, fig, loquat, etc.), suckering (e.g., banana, etc.), etc. They have been widely used for commercial propagation of fruit trees and ornamental shrubs.

b) *Biotechnological Artificial Vegetative Reproduction*: Through tissue culture large number of plantlets can be produced from small pieces of tissues in a very short time. It is called **micropropagation**. New plants may arise from pre-existing shoot buds, adventitious shoot buds, adventitious buds, etc. and somatic embryos. They are of considerable commercial value.

6.3 Groups of Horticultural Crops

The horticultural crops may be asexual/sexual (naturally self-pollinated/often cross-pollinated /naturally cross-pollinated/dioecious cross-pollinated) and annual or perennial. Based on modes of reproduction and pollination, and habit, they can be classified as:

A. Annual Horticultural Crops

1. Naturally (Normally) Self-Pollinated Crops (e.g., tomato, cowpea)
2. Often Cross-Pollinated Crops (e.g., okra, chilli)
3. Naturally (Normally) Cross-Pollinated Crops (e.g., radish, pumpkin)

B. Perennial Horticultural Crops

1. Crops exclusively propagated by vegetative means (e.g., ginger, turmeric)
2. Crops propagated by both vegetative and sexual means (e.g., mango, guava)
3. Crops exclusively propagated by sexual means (e.g., coconut, arecanut)

C. Dioecious Horticultural Crops

1. Annual Crops (e.g., pointed gourd, little gourd, spine gourd, spinach)
2. Perennial Crops (e.g., asparagus, papaya, date palm, palm)

Common breeding methods can be applied to the naturally self-pollinated and often cross-pollinated annual crops. However, the parents chosen for breeding under often cross-pollinated crop need to be selfed for few generations to select desirable, homozygous lines. Again, common breeding methods can be applied to naturally cross-pollinated and dioecious cross-pollinated crops. But, in latter case both male and female plants with the characters desired must be selected and their progeny is tested to determine the breeding value of particular parents. By this means, cultivars of superior type may be synthesized.

6.4 Implications of Reproductive Systems on Population Structure

The implications (consequence) of reproductive systems (modes of reproduction) on population structure of a crop species are discussed below in cases of sexual and asexual reproduction:

6.4.1 Implications of Sexual Reproduction

Self-pollination leads to homozygosity. So, self-pollinated populations are normally highly homozygous. Self-pollinated crops do not exhibit inbreeding depression but may show considerable heterosis. Self-pollination maintains intact all homozygous genotypes. The progeny of a single homozygous plant has the same genotype as the parent. They constitute a **pureline**. Cultivars of self-pollinated crops are ordinarily purelines.

Cross-pollination preserves and encourages heterozygosity in a population. So, cross-pollinated crops are highly heterozygous. They show severe to high inbreeding depression and high heterosis. So, cultivars of such crops should be as heterozygous as possible, e.g., hybrids. Such populations are random mating. Hence, their gene and genotype frequencies follow the Hardy-Weinberg Law. This makes the consequences of selection in populations very different from those obtained in self-pollinated populations.

6.4.2 Implications of Asexual Reproduction

Asexual reproduction (including apomixis producing diploid embryos) produces progeny that are genetically identical to the parent. The progeny of a single plant propagated asexually is called a **clone**. Clones are used as cultivars in many horticultural crops. Both homozygous and heterozygous genotypes can be maintained indefinitely. Hence any desirable plant/mutant can be used as a cultivar. Diploid apomixis can be utilized to multiply/maintain hybrid cultivars. Haploid apomixis produces haploid plants through tissue culture of ovary and anther. Chromosome doubling of such plants allows rapid isolation of homozygous lines from heterozygous plants/hybrids. The major disadvantage of asexual reproduction system is that it does not enable genes present in two different individuals/clones to be combined into a single clone. Apomixis creates confusion during hybridization.

6.5 Relation of Mode of Reproduction to Horticultural Breeding

The modes of reproduction and pollination determine the following aspects of a crop species:

6.5.1 Genetic Composition

The genetic composition of a crop species depends on its modes of reproduction and pollination. Both cross-pollinated and asexually propagated crops are naturally heterozygous. They also show severe to high inbreeding depression. On the other hand, self-pollinated crops are naturally homozygous. They exhibit little or no inbreeding depression.

6.5.2 Breeding Methods and Cultivars

The breeding methods are designed to restrict inbreeding to the minimum possible in cross-pollinated and asexually propagated crops. On the other hand, in self-pollinated crops they are designed to isolate homozygous purelines. The cultivars of cross-pollinated crops are highly heterozygous (hybrids, composites, synthetics and open-pollinated). In self-pollinated crops, purelines are the most widely used cultivars. The cultivars of asexually propagated crops are clones. So, any desirable plant can be asexually propagated to give a cultivar.

6.5.3 Controlled Pollination

Most of the breeding methods involve crossing between selected strains/ genotypes. In self-pollinated crops, selfing occurs naturally. But, in cross-pollinated crops it must be ensured by suitable techniques. In asexually propagated crops, sexual progeny has to be generated for making crosses. The ease in making crosses depends on the floral structure and the mode of pollination of the crop.

6.5.4 Gene Transfer and Genetic Variation

The gene transfer and creation of genetic variation are based on sexual reproduction. They are possible only in those asexually propagated crops in which sexual reproduction also takes place. Special techniques, e.g., somatic hybridization and genetic engineering, must be adopted if sexual reproduction is totally absent.

6.5.5 Stability of Cultivars

Since the cultivars of self-pollinated crops are purelines, they remain genetically pure for a long period. The farmer can use his own produce as seed. Similarly, the clones of asexually propagated crops are stable so that the produce is reusable as seed. In cross-pollinated crops, fresh seed of hybrids must be purchased every year. But synthetic, composite and open-pollinated cultivars can be maintained for several years if inbreeding is avoided.

6.6 Brief References

Acquaah 2012, Allard 1960,1999; Chahal and Gosal 2002, Chaudhari 1971, Chaudhary 2014, Dabholkar 2006, De and Bhattacharjee 2011, Dutta 1974, Hayes *et al.* 1955, Khan 2013, Kumar 2018, Kute and Aher 2013, Mohanan 2010, Poehlman 1987, Ram M 2014, Roy 2012a, Sharma JP 2009, Sharma JR 1994, Singh BD 2011,2018; Singh NP 2015, Singh Phundan 2011,2013a; Singh and Pawar 2006, Tyagi 2015, Yadav and Krishna 2012.

Review Questions

Part A

Answer the following questions true or false.

1. Sexual reproduction and amphimixis are different. (True/False)
2. In obligate apomixis, the progeny from a cross resemble the male parent. (True/False)
3. The adhesion of stamens with petals wholly or partially by their filaments is called epipetalous. (True/False)
4. Self-pollination is also known as allogamy. (True/False)
5. In asexual reproduction, fusion of the male and female gametes occurs. (True/False)
6. The individual units of perianth are called sepals. (True/False)
7. In protogyny androecium matures earlier than gynoecium. (True/False)
8. Mode of reproduction is also called reproductive system. (True/False)
9. The first opening of a flower is called pollination. (True/False)
10. Self-incompatibility is also called self-sterility. (True/False)
11. Self-pollination leads to heterozygosity. (True/False)
12. The process of reproduction is also called propagation. (True/False)
13. The production of male and female gametes is known as sporogenesis. (True/False)
14. A complete flower contains two whorls. (True/False)
15. Secondary nucleus is also called definitive nucleus. (True/False)

Part B

Answer the following questions.

1. What is self-pollination?
2. The egg-apparatus consists of __________ and __________.
3. What do you mean by apomixis?
4. What is triple fusion?
5. Define megasporophylls.
6. Tomato is a __________ pollinated crop.
7. On the basis of mode of pollination, the horticultural crops are grouped into __________categories.
8. What is megagametogenesis?
9. Purelines are found in self-/cross-pollinated crops.
10. What is apomixis?
11. Production of spores is known as __________.
12. The progeny of a single plant propagated asexually is called a __________.
13. Micropropagation means __________.
14. What is double fertilization?
15. __________is an example of polyembryony.

Part C

Write a brief note on each of the following topics.

1. Implications of asexual reproduction on population structure.
2. Groups of horticultural crops.
3. Detection of apomixis.
4. Classification of crops based on pollination.
5. Diplospory.
6. Megagametogenesis.
7. Artificial vegetative reproduction.
8. Calyx and corolla.
9. Fertilization.
10. Cross-pollinated population.
11. Stability of cultivars.

12. Dioecious horticultural crops.
13. Reproductive whorls of a flower.
14. Natural vegetative reproduction.
15. Implications of sexual reproduction on population structure.

Part D

1. Define cross-pollination. Discuss the contrivances encouraging or enforcing cross-pollination.
2. How can you determine mode of pollination?
3. Elaborate the relevance of mode of reproduction in horticultural breeding.
4. What are the implications of reproductive systems on population structure?
5. Discuss gametogenesis with diagram.

7

Pollination Control Mechanism

7.1 Introduction

Breeding plans are based on **controlled mating** (i.e., mating between selected strains) and selfing of selected strains or a combination of both. Continued selfing is practised to isolate purelines and inbreds. On the other hand, crossing (hybridization) is necessary to produce hybrids and synthetics. In crossing, self-pollination is prevented in the female parent. **Male sterility** and **self-incompatibility** easily exclude self-pollination, and so they are used in hybrid seed production.

7.2 Male Sterility

Koelreuter (1763) for the first time reported male sterility in flowering plants. **Male sterility is a reproductive deficiency of some plants in which male organs in bisexual flowers become defunct.** *It relates particularly to non-viable pollen grains* that are formed through a chain of vital processes during microsporogenesis. These processes are so delicately balanced under genetic control of many loci that mutation of any one locus may lead to the formation of non-functional microspores, and hence male sterility. In male sterility, pollen grains are non-functional (inactive); but the **female gametes (egg cells)** are normally functional.

7.2.1 Significance of Male Sterility

Self-incompatibility occurs frequently in crop plants. But, **male sterile** (also called **androsterile) plants** appear rarely. In fact, male sterility is a **genetic abnormality (defect)** in a natural population. Hence, male sterile plants are promptly eliminated by natural selection forces. However, if properly maintained under domestication, they would be quite helpful as a natural and effective means for genetic emasculation of plants. *So, male sterility greatly simplifies the commercial hybrid seed production.*

7.2.2 Morphological Features of Male Sterility

Due to mutation/chromosomal aberrations, cytoplasmic factors or interaction of cytoplasmic and genetic factors, the morphological features of flowers change basing on which male sterile plants can be identified. They are discussed in brief below:

1. Viable pollens are not formed (**pollen sterility**). They can be identified by the transparent nature of anthers or by very light to light acetocarmine stain of pollen grains. Most of the crops exhibit such behaviour, e.g., onion.
2. Viable pollens are enclosed in the anthers. But the anthers are non-dehiscent probably due to hard outer layer surrounding the **microsporangium** (also called **pollen sac**). So, the pollens are not released (**structural sterility**).
3. The androecium may abort before pollens are formed.
4. The androecium may be severely malformed. So, there is no possibility of formation of pollen grains (**staminal sterility**), e.g., phyllody.

7.2.3 Mechanism of Male Sterility

The disturbances in normal microsporogenesis lead to the formation non-viable microspores and so male sterility. They may be due to several factors. The explanations put forth by different scientists are discussed below:

7.2.3.1 Mitochondrial Mutation

Male sterility may arise by mutation in mitochondrial DNA (mt DNA) of a normal fertile plant. Mitochondria are the respiratory sites in the cell. Mutation in mitochondria may reduce the respiratory capacity of the cell and hence result in loss of fertility. Thus, male sterility is associated with cytoplasmic spherical particles and recognizable types of double stranded RNA. It is also associated with mtDNA and polypeptides of mitochondrial origin.

7.2.3.2 Barrier of Tapetal Layer

Male sterility in flax (linseed) occurs due to delayed degeneration of tapetal cells that block the availability of nutrients to microspores. Thus, nourishment of pollens is adversely affected leading to abortive pollens.

7.2.3.3 Improper Timing of Callase Activity

It was demonstrated in *Petunia* that proper timing of the activity of the enzyme, callase, during microsporogenesis may be critical for normal development of pollen. Early or delayed activity of callase in different plasma-types leads to male sterility. Further research indicated that pH within the anther locule regulates the callase activity. Differences in the free amino acid balance in fertile and sterile anthers which are reflected at early meiotic stages, are also responsible for male sterility. Thus, alternate cytoplasmic and chromosomal heredity elements exist for male sterility.

7.2.3.4 Based on Operon Concept of Gene Action

Cytoplasmic male sterility may arise from operon-type of control rather than due to independent cytoplasmic factors. A specific condition guided by the nuclear gene(s) is propagated in the cytoplasm that imposes a permanent repression upon the genetic system that determines the pollen producing capacity of that plant. Male sterility is the consequence. It perpetuates over generations since the cytoplasm of the progeny is contributed by the female parent. However, this hypothesis fails to provide an explanation for fertility restoration in the male sterility system.

7.2.4 Types of Male Sterility

Male sterility may be of four types, *viz*., genetic, cytoplasmic, cytoplasmic-genetic and chemically-induced male sterility. They have been discussed in brief below:

7.2.4.1 Genetic Male Sterility (GMS)

Genetic male sterility (GMS), *also called* ***nuclear male sterility,*** *is governed by nuclear genes.* Here, the cytoplasm is not involved. GMS occurs widely in plants, e.g., watermelon, muskmelon, pumpkin, tomato, peas, sugarbeet, etc. Male sterility alleles are generally monogenic recessive, rarely dominant e.g. safflower. Male sterility allele is usually designated by 'ms' and the dominant allele for male fertility by 'Ms'. Male sterility occurs in plants due to mutation of the fertility locus situated on a chromosome within the nucleus. There could be three possible genotypes* for this locus and one of them is male (♂) sterile:

1. [MsMs] : ♂ fertile
2. [Msms] : ♂ fertile
3. [msms] : ♂ stertile

* Letters inside the bracket represent nuclear genes and those outside cytoplasmic genes.

In a plant species, several different ms genes act monogenically to produce male sterility, e.g., 64 different ms genes are reported in tomato, 54 in peas, etc.

Genetic male sterility (GMS) is divided into (1) *environment-insensitive* and (2) *environment-sensitive*. In the former case, ms gene expression is little influenced by the environment. But, in the latter it is influenced by the photoperiod and/or temperature, e.g., rice, tomato, wheat, etc. Environment-sensitive GMS is classified into *photoperiod-sensitive genetic male sterility* (PGMS) and *temperature-sensitive genetic male sterility* (TGMS). Besides, genetic engineering has produced a novel type of dominant genetic male sterility known as *transgenic genetic male sterility* (TrGMS).

7.2.4.1.1 Photoperiod-Sensitive Genetic Male Sterility (PGMS)

*In **PGMS** the ms gene expression is highly influenced by the photoperiod (day-length) if the ambient temperature is within a critical range.* For instance, in rice, long day (photoperiod longer than 13 hr 45 min) produces male sterility, while short day (photoperiod shorter than 13 hr 45 min) gives rise to male fertility if the ambient temperature is between 23 and 29° C. But at temperatures above 29° C the plants are male sterile, while below 23° C, they are male fertile irrespective of the photoperiod. PGMS is being used to develop hybrid rice in China.

7.2.4.1.2 Temperature-Sensitive Genetic Male Sterility (TGMS)

*In **TGMS** the ms gene expression is mainly influenced by the ambient temperature.* When the temperature is higher than a critical point, the ms gene expresses and male sterility is obtained. But, when it is lower than the critical point, normal male fertility results. For example, in rice, the critical point ranges from 23 to 29° C for different male sterile lines. TGMS is being used to produce hybrid rice in China.

7.2.4.1.3 Transgenic Genetic Male Sterility (TrGMS)

The **transgene** is also used for induction of male sterility. *The male sterility which is induced by genetic engineering is called **transgenic genetic male sterility (TrGMS)** and it is dominant.* The gene responsible for inducing male sterility generally comes from microorganisms. TrGMS is induced in tobacco and rapeseed by transferring a gene from *Bacillus amyloliquefaciens*. In this system two types of transgenes are used. One transgene called *barnase* that encodes an RNAase causes male sterility. Another transgene namely *barstar* suppresses the male sterility transgene *barnase* by producing a specific inhibitor of the *barnase* encoded RNAase, and leads to restoration of fertility. This is an expensive method of obtaining male sterility. Moreover, maintenance of the male sterility is a problem. This method is quite promising for commercial hybrid seed production of various crops in the near future.

7.2.4.2 Cytoplasmic Male Sterility (CMS)

***Cytoplasmic male sterility (CMS)** is governed by **cytoplasmic genes** (also called **plasmagenes**).* Nuclear genes are not involved. In most cases, the plasmagenes determining male sterility are located in mtDNA (mitochondrial

DNA). CMS occurs due to the mutation of mitochondria or to other cytoplasmic factors outside the nucleus, resulting in transformation of male fertile cytoplasm into a male sterile one. With two types of cytoplasm (i.e. male sterile and male fertile), at the most only two kinds of genotypes are possible. One of them is male sterile, and the other, male fertile. Let us designate male fertile cytoplasm by F and male sterile one by S. Then two genotypes are:

1. [ ] F : ♂ fertile
2. [ ] S : ♂ sterile

CMS has been reported in different plants.

7.2.4.3 Cytoplasmic-Genetic Male Sterility (CGMS)

Cytoplasmic-genetic male sterility (CGMS)*, also called* ***nucleoplasmatic male sterility,*** *is governed by both* **cytoplasmic** *and* **nuclear genes.** It was first discovered by Jones and Davis (1944) in onion. Now, it has been reported in several crops such as tomato, sugarbeet, carrot, etc. In CGMS, male sterility is determined by the cytoplasm, and the plasmagenes producing male sterility are generally located in mtDNA like CMS. But a nuclear gene, called **restorer gene**, restores male fertility and, thereby, eliminates the effects of male sterile cytoplasm. The restorer gene is generally dominant (R), and is found in some cultivars of the same species or in a related species. In fact, CGMS is essentially a CMS with a provision for restoration of fertility (which is not possible in CMS). *In brief, the male fertility is restored by the nuclear dominant gene(s); the combination of nuclear and cytoplasmic genes determines the male-fertility or -sterility in respective plants.* Based on these combinations there can be a maximum of six kinds of genotypes; only one of them is male sterile. Let us designate restorer nuclear dominant gene by R and its recessive allele by r (non-restorer), male fertile cytoplasm by F and male sterile cytoplasm by S. Then the six genotypes are:

1. [RR] F : ♂ fertile
2. [RR] S : ♂ fertile
3. [Rr] F : ♂ fertile
4. [Rr] S : ♂ fertile
5. [rr] F : ♂ fertile
6. [rr] S : ♂ sterile

7.2.4.4 Chemically-Induced Male Sterility

Moore (1950) and Naylor (1950) first reported chemically induced male sterility in maize using maleic hydrazide. Male sterility can be induced by some chemicals known **male gametocides** (though there are other synonyms, this is better one). However, one of its synonyms namely **chemical hybridising agent** (CHA) proposed by McRae (1985) may be preferred to from standpoint of hybrid seed production. A chemical hybridising agent is defined as a chemical that induces artificial nongenetic male sterility in plants so that they can be effectively used as female parents in hybrid seed production. Some examples of male gametocides are ethrel, GA_3, maleic hydrazide (MH), naphthalene acetic acid (NAA), sodium methyl arsenate, zinc methyl arsenate, etc. Unfortunately, none of them is an ideal male gametocide. For instance, they often produce incomplete male sterility, repeated applications are required, female fertility is also reduced and often undesirable side effects are produced. But, they have the unique advantage of application with any line/cultivar of a crop whenever desired. Some effective male gametocides are NAA in cucurbits, gibberellins in onion and lettuce, MH in tomato, cucurbits and onion, etc. The main difference between mutagen-induced male sterility and gametocide-induced male sterility is that the former is **heritable (genetic)** and the latter is **non-heritable (non-genetic)**.

7.2.5 Inheritance of Male Sterility

The pattern of inheritance differs for GMS, CMS and CGMS.

(i) GMS is usually controlled by recessive gene ms. A cross between ♂ sterile and ♂ fertile parents produces ♂ fertile F_1, and on selfing in F_2, a phenotypic ratio of 3 ♂ fertile : 1 ♂ sterile is obtained (Figure 7.1). When F_1 is backcrossed with ♂ sterile parent, a ratio of 1 ♂ sterile : 1 ♂ fertile is obtained. The ♂ fertile plants are heterozygous.

Parents [msms] × [MsMs]
♂ sterile ↓ ♂ fertile

F1 [Msms]
♂ fertile
↓ Selfing

F2 1 [MsMs] : 2 [Msms] : 1 [msms]
└♂ fertile┘ ♂ sterile

Phenotypic ratio 3:1

Figure 7.1. Inheritance of GMS due to recessive gene ms

ii) CMS shows cytoplasmic inheritance. When a cross is made between ♂ sterile and ♂ fertile plants, the F_1 is always ♂ sterile, i.e., '[ ] **S**' as it receives cytoplasm from ♀ parent.

iii) In CGMS the F_1 would be ♂ sterile, only when the nuclear gene is recessive (rr) and the cytoplasm is sterile (S), i.e., '**[rr] S**'.

7.2.6 Transfer of Male Sterility

The male sterility can be easily transferred from one genotype to another by backcross method. Generally, 6-7 backcrosses are adequate to transfer male sterility gene into the background of a popular cultivar.

i) For transfer of GMS, the adapted cultivar is used as ♂ parent and male sterile genotype as ♀ parent. The F_1 is ♂ fertile which on selfing segregates in F_2 with a phenotypic ratio of 3 ♂ fertile : 1 ♂ sterile. The ♂ sterile plants are again backcrossed with the adapted cultivar to which male sterility is to be transferred.

ii) For transfer of CMS, a cross is made between a ♂ sterile line and an adapted cultivar. The F_1 is ♂ sterile which is repeatedly backcrossed with the adapted cultivar for 6-7 generations. Thus, an adapted cultivar can be converted to a cytoplasmic male sterile line.

iii) For transfer of CGMS, a cross is made between a ♂ sterile line and an adapted cultivar with recessive nuclear genes (rr). The F_1 is ♂ sterile which is repeatedly backcrossed with the adapted cultivar for 6-7 generations. Thus, an adapted cultivar can be converted to a cytoplasmic-genetic male sterile line.

7.2.7 Maintenance of Male Sterility

The term **m.s. line** (also called **A-line**) is used to represent a male sterile line belonging to any of the male sterility systems, namely, GMS, CMS and CGMS. By nature, A-lines are always used as female/seed parents. **B-line** (also called **maintainer**) is used to maintain the male sterility of A-line. The maintenance of male sterility implies that by crossing A and B lines, viable and normal F_1 seeds are formed; but the F_1 plants developed from them do not produce viable pollens, and so remain male sterile. These sterile F_1 progenies will again produce hybrid seeds as above on crossing with B-line, and so forth. Maintenance of male sterility also implies that the characteristic features of A-line must remain unaltered generation after generation on crossing with B-line. For this, B-line should be an **isogenic line** of A-line *which means identical for all characters except male fertility status*.

i) GMS: By crossing A and B lines, ♂-sterile and -fertile progenies are produced in equal proportions as follows:

[msms] × [Msms]
♂ sterile ♂ fertile
(A-line) ↓ (B-line)

[Msms] [msms]
♂ fertile ♂ sterile
1 : 1

The ♂ fertile plants must be removed before pollen shedding, thus leaving only ♂ sterile progenies. The removal of ♂ fertile plants may become easier if fertility is linked with some marker gene(s), expressed preferably at seedling stage.

ii) CMS: As there are only two types of genotypes, the ♂ fertile ones serve as B-line. Then maintenance of male sterility is straight-forward as under:

[ ] S × [ ] F
♂ sterile ♂ fertile
(A-line) ↓ (B-line)

[ ] S
All ♂ sterile progenies

(since they inherit ♂ sterile cytoplasm from female parent)

iii) CGMS: According to their genetic composition and cytoplasm, only [rr] F genotypes can maintain the male sterility of A-line as follows:

[rr] S × [rr] F
♂ sterile ♂ fertile
(A-line) ↓ (B-line)

[rr] S
All ♂ sterile progenies

(since the ♂ sterile cytoplasm of the female parent is inherited by the progenies and nuclear genes are ♂ sterile already)

7.2.8 Sources of Male Sterility

There are three important sources of male sterility, namely, (i) ***spontaneous mutations***, (ii) ***induced mutations***, and (iii) ***interspecific crosses***. They are briefly discussed below:

7.2.8.1 Spontaneous Mutations

Both GMS and CMS have been reported to occur by spontaneous mutation in different crops. For instance, spontaneous GMS has been isolated in pea, sweet pea, tomato, brinjal, lettuce, etc. and CGMS in maize, pearlmillet, sunflower, etc.

7.2.8.2 Induced Mutations

Male sterility can also be induced artificially by mutagens. GMS has been induced by X-rays in *Petunia*. In *Tagetes erecta*, an apetalous and antherless recessive mutant has been induced by X-rays. It is being used in hybrid seed production. GMS has also been induced by gamma-rays in tomato and watermelon; by chemical mutagens in pepper and pea. Ethidium bromide has been used for induction of CMS in *Petunia*, pearlmillet, barley, etc.

7.2.8.3 Interspecific Crosses

Wild species of crop plants are important sources of male sterile cytoplasm. Male sterile cytoplasm from wild species is transferred to cultivated one through interspecific hybridization and backcrossing. The combination of cytoplasm of one species with nucleus of another one leads to induction of CGMS, e.g., cotton, tobacco, pepper, linseed, barley, bread wheat, durum wheat, etc.

7.2.9 Detection of Male Sterility System

Whether a particular male sterile genotype belongs to GMS, CMS or CGMS systems can be detected only by its progeny performance on crossing with a few male fertile genotypes. In subsequent generations, any one of the three patterns in crossed progenies of each genotypes (grown in separate rows) may be found:

Trend-I: Some rows may consist of all ♂ fertile plants, and in some rows ♂ -fertile and - sterile plants may occur in 1:1 ratio.

Trend-II: All the progenies in all the rows may be ♂ sterile.

Trend-III: Some rows may have all ♂ fertile plants, some all ♂ sterile plants, and yet others have ♂ -fertile and -sterile plants in 1:1 ratio.

Now, based on all the possible types of genotypes for ♂ fertility/sterility that can exist in a population, we can readily identify that Trend-I is GMS, Trend -II CMS and Trend-III CGMS.

7.2.10 Applications of Male Sterility

Male sterility has important application in the development of hybrids. All the three types of male sterility are used in hybrid seed production.

i) GMS: It can be used for commercial hybrid seed production in both seed-propagated and vegetatively propagated crops. GMS lines have been used for the development of commercial hybrids in watermelon, cotton, wheat, rice, castor, tomato, pigeonpea, etc.

ii) CMS: It has limited applications and is used for hybrid seed production in vegetatively propagated crops, forage and ornamental plants where vegetative part is of economic value. It is not applicable to seed-propagated crops because the F_1 is male sterile.

iii) CGMS: **Restorer line** (also called **R-line**) is deployed to restore the male fertility (functional pollen production) in resultant (A × R) progenies. Restoration means that when A-line is crossed with R-line, the viable and normal hybrid seeds are formed. This is the hybrid seeds which are of commercial value. These hybrid seeds give rise to F_1 progenies which produce viable pollen grains and normal (F_2) seeds in farmer's fields. However, when F_2 seeds are resown, the male sterile plants will reappear because of segregation. This makes them useless for further cultivation in the farmer's fields. *Since A × R progenies (hybrid seeds) are of great commercial significance, the choice of R-lines should be such that they may produce highly heterotic and disease-free hybrids. Hence, under the appropriate choice of restorer lies the spectacular gain of hybrid vigour.* The CGMS has wide applications and can be successfully utilized for the development of commercial hybrids in both seed propagated and vegetatively propagated crops. CGMS has been effectively used for the commercial production of hybrids in maize, sorghum, onion, sugarbeet, pearlmillet, cotton, rice, sunflower, etc.

7.2.11 Limitations of Male Sterility

Male sterility is useful in hybrid seed production. However it has some limitations.

i) GMS:

   a) The line used as female parent in hybrid seed production is the progeny from a [msms] × [Msms] cross. So, the female parent has 50% male fertile [Msms] plants which must be removed before flowering. This is quite difficult and costly to achieve. This problem is not faced when TGMS or PGMS is used.

   b) Pollen dispersal from the male parent is generally poor.

   Both the above features reduce the quantity of hybrid seed produced from a given area, and increase the cost. This has limited a wider use of the GMS system.

ii) CMS: It cannot be used in such crops where seed is the economic part because F_1 hybrids will be sterile.

iii) CGMS:

a) Male sterile cytoplasm generally produce undesirable side effects.

b) In many cases, fertility restoration is not satisfactory.

c) In many cases, pollination during hybrid seed production is not satisfactory, e.g., *Capsicum* spp. This is particularly so in self-pollinated crops.

d) Modifying genes (nuclear) may reduce the effectiveness of CMS.

e) Male sterility may not be complete under certain environments, e.g., maize, bajra, jowar, etc.

7.3 Self-Incompatibility

Koelreuter, in the middle of 18th century, first reported self-incompatibility in *Verbascum phoeniceum* plants. Later on numerous cases of SI in flowering plants were reported. However, the term self-incompatibility was coined by Stout in 1917. In SI, pollen grains may fail to germinate or enter the stigma, pollen tube may grow too slowly in the style to effect fertilization before the flowers drop, or the embryos (if fertilization does take place) degenerate.

7.3.1 Significance of Self-Incompatibility

SI prevents self-pollination. So, in self-incompatible fruit trees, it is desirable to develop self-fertile (self-compatible) types. Otherwise, two or more cross-compatible cultivars must be inter-planted, and fruit set may be reduced by adverse weather conditions that hamper cross-pollination. When inbreeding is desired, e.g., development of inbreds, techniques that temporarily suppress SI reaction must be used. SI is used for hybrid seed production in cabbage, Brussels sprouts and sunflower. A simple scheme for this is to interplant to such self-incompatible lines that are cross-compatible with each other. Seeds produced on both the lines would be hybrid. This use of SI has been successful mostly with sporophytic system.

7.3.2 Mechanisms of Self-Incompatibility

The mechanism of SI is not clearly understood. It involves the following:

7.3.2.1 Pollen-Stigma Interaction

This interaction occurs just after a pollen lands on the stigma and generally prevents pollen germination. In sporophytic system, the stigma is covered with a hydrated layer of proteins known as pellicle. Pellicle appears to play a role in SI reactions. Within few minutes of landing on the stigma, pollen releases an exine exudates. This exudate induces immediate callose deposition in the stigma cells that are in direct contact of the exudate releasing pollen. This occurs only if the pollen and the stigma cells have the same SI reaction (incompatible mating). Callose may also be deposited on the young pollen tubes; this effectively prevents pollen germination. If somehow the pollen tube crosses stigma and enters the style, there is usually no further inhibition of pollen tube growth.

7.3.2.2 Pollen Tube-Style Interaction

In most cases of gametophytic system, pollen grains germinate, and pollen tubes enter the stigma. But, in incompatible combinations (i.e., when pollen tube and style have the same SI reaction), pollen tube growth is retarded within the stigma or in the style. In the latter case, protein and polysaccharide syntheses are inhibited, which ultimately leads to the busting of pollen tubes.

7.3.2.3 Pollen Tube-Ovule Interaction

In some cases, e.g., *Theobrama cacao*, pollen tubes reach ovules and fertilization takes place. But, in incompatible combinations, embryos degenerate at an early stage. Generally, species showing this type of interaction have hollow styles.

7.3.3 Types of Self-Incompatibility

Lewis (1954) has suggested a simple classification of SI such as (1) heteromorphic and (2) homomorphic systems. The homomorphic system is again divided into gametophytic and sporophytic systems.

7.3.3.1 Heteromorphic System

In heteromorphic system, flowers of different incompatibility groups have different morphology, particularly about the length of styles and stamens. It is of two kinds: (1) distyly, and (2) tristyly.

1. Distyly: In case of distyly, the style may be long or short. Flowers with long style have short stamens. They are called **'pin'** type. On the other hand, flowers with short style have long stamens. They are called **'thrum'** type. Both pin and thrum types belong to different incompatibility groups. Again, pin and thrum flowers are borne on different plants. It does not set seed when flowers of the same incompatibility group are cross-pollinated. So, seeds are produced only when pin flowers are pollinated by thrum flowers and vice-versa. This system has limited occurrence in crop plants, e.g., sweet potato, buck wheat, *Primula,* etc.

Both flower morphology and self-incompatibility are governed by a single locus S with two alleles (S and s). The control of SI reaction is sporophytic. Plants with thrum flowers have Ss genotype, while those with pin flowers have ss genotype. Plants with thrum flowers produce pollen grains of two genotypes, S and s, but all of them exhibit the same SI reaction (denoted by S). The plants with pin flowers produce pollens of only s genotype, which show a different SI reaction (designated by s). Mating between plants with pin and thrum flowers produces thrum (Ss) and pin (ss) progeny in1:1 ratio. It is now known that the locus S is a complex one and it has at least six functional genes (g, s, l_1, l_2, p and a). Plants with pin flowers are homozygous recessive for all of them and those with thrum flowers are heterozygous for all of them.

2. Tristyly: In case of tristyly, styles are of three different length, namely, short, long and medium. Plants with flowers of the same style length are of the same incompatibility group. Tristyly in *Lythrum salicarea* is controlled by two genes S and M.

7.3.3.2 Homomorphic System

In homomorphic system, the floral morphology is same in different incompatibility groups. It is far more common than the heteromorphic system. It is of two types: (1) *Gametophytic System* and (2) *Sporophytic System.*

7.3.3.2.1 Gametophytic System

When SI is controlled by the genotypes of gametes, it is known as gametophytic SI system. It was first reported in 1925 by East and Mangelsdorf in *Nicotiana sanderae*. It is found in rye, sugarbeet, pineapple, diploid clover, alfalfa (lucerne), potato, tomato, etc. In this system:

1. The SI reaction of pollen grains is determined by their own genotypes.
2. In the style, the two S alleles show co-dominance.
3. In many species such as *Nicotiana* sp., *Lycopersicon* sp., *Solanum* sp., *Petunia* sp., etc. the SI reaction is specified by a single gene S with multiple alleles, e.g., S_1, S_2, S_3, etc.
4. In some species, e.g., *Phalaris* sp., *Physalis* sp., etc., two loci (S and Z) govern SI, while in some others like sugarbeet, *Papaver* sp. etc., three loci are involved.
5. A mutant allele Sf of the S locus allows the pollen grains carrying it to effect fertilization of flowers with S genotype.

If a pollen grain lands on a stigma, which has the same S allele as the pollen (e.g., S_1 pollen on S_1S_2, S_1S_3, S_1S_4, etc. stigma), it germinates normally. But the pollen tube growth is very slow, and fertilization does not take place. On the other hand, if the stigma has different S allele, pollen tube grows normally, and fertilization is effected. So, depending on the S locus genotype, mating between different plants is of three types, namely, (1) fully compatible (e.g., $S_1S_2 \times S_3S_4$), (2) partially compatible (e.g. $S_1S_2 \times S_1S_3$) and (3) fully incompatible (e.g. $S_1S_2 \times S_1S_2$).

7.3.3.2.2 Sporophytic System

When the SI is governed by the genotype of pollen-producing plant (sporophyte), it is called sporophytic system. The sporophytic SI system was reported by Hughes and Babcock (1950) in *Crepis foetida*, and Gerstel (1950) in *Parthenium argentatum*. It occurs in radish, diploid *Brassica* sp., sunflower, *Sinapis*, cabbage, cauliflower, cosmos, etc. In this system:

1. SI reaction of a pollen grain is determined by the genotype at the S locus of the plant.
2. The SI reaction is governed by a single locus S with multiple alleles.
3. S alleles show dominance in the manner: $S_1 > S_2$ (S_1 is dominant over S_2)> $S_3 > S_4 > S_5$, etc. Therefore, a plant with genotype S_1S_2 will produce pollen grains of a single SI reaction S_1 (although genotypically the pollen grains will be of two types: S_1 and S_2); the stigmas of its flowers will also have the SI reaction S_1. On self-pollination, pollen of S_1 SI reaction will interact with stigmatic cells of the same (S_1) SI reaction. This interaction generally prevents pollen germination or, the entry or growth of pollen tubes in the stigma. But, when an S_1S_2 plant is crossed (as ♀) with an S_2S_3 plant (used as ♂), pollen grains with S_2 SI reaction are placed on stigma of a different, namely, S_1 SI reaction. As a result, pollen germination and pollen tube growth are not affected, and normal fertilization takes place.
4. S alleles may show co-dominance.
5. The stigma surface is dry.
6. Usually, pollination occurs when pollen grains are trinucleate.
7. Incompatibility reaction of pollen grains appears to be due to some compounds of tapetum deposited on pollen exine.

7.3.4 Elimination of Self-Incompatibility

SI can be eliminated, and self-fertile (self-compatible) forms obtained in the following ways:

1. Doubling the chromosome number is effective in case of single locus gametophytic system.
2. Isolation of self-fertile (S_f) mutants. For this, flower buds are generally irradiated at the PMC stage, and pollen from these buds is used to pollinate flowers having known S allele. It is more successful in the gametophytic system than in the sporophytic system.
3. Transfer of self-compatibility alleles from related species.

7.3.5 Measures to Overcome Self-Incompatibility

There are several ways and means to overcome self-incompatibility in crop plants. The methods used for this purpose include: (1) bud pollination, (2) delayed pollination, (3) late season pollination, (4) irradiation, (5) high temperature, (6) mutilation of style, and (7) in- vitro fertilization. They are briefly discussed below:

1. *Bud Pollination*: Pollination of immature buds with mature pollens has been successful in the production of large quantity of selfed seeds both in gametophytic and sporophytic systems in *Brassica*, *Nicotiana* and *Petunia*. This method is now used for the development of inbreds for hybrid seed production in *Brassica* spp. The bud pollination should be done 2-4 days before opening of flower. Generally, in young buds the stigmas lack adhesivity which can be achieved if the stigma is smeared with exudate from an open flower.
2. *Delayed Pollination*: Delayed pollination also leads to self-fertilization in some self-incompatibility species. Pollination of aged pistils several days after maturity with normal incompatible pollen resulted in some degree self-fertilization in *Brassica* and *Linum*.
3. *Late Season Pollination*: Self-pollination at the end of flowering season also leads to seed set in some species like *Nicotiana*, *Petunia* and *Abutilon*.
4. *Irradiation*: Irradiation of style with X-rays immediately before selfing resulted in breakdown of SI in *Petunia*. Gamma irradiation significantly increased seed setting upon selfing in *Lycopersicon peruvianum* and *Nicotiana alata*. However, these effects were observed in one generation and lost in the next generation.
5. *High Temperature*: Treatment of style at temperature 30-60° C leads to break down of SI in many species, e.g., *Malus, Pyrus, Prunus, Trifolium, Lycopersicon* and rye having gametophytic single locus system of SI. Heat treatment inactivates the substance or enzyme that causes SI.

6. *Mutilation of Style*: Removal of stigma in species having stigmatic SI results in certain amount of seed production in some crops. This method was found to be successful in *Brassica oleracea*.
7. *In-Vitro Fertilization*: Placing of pollen grains in direct contact with ovules resulted in breakdown of SI in many crops. This indicates that restriction of interference from the stigma, the style or the ovary leads to breakdown of SI. Selfed seeds were obtained in *Petunia auxiliaris* by invitro fertilization.

7.3.6 Self-Incompatibility in Relation to Plant Breeding

SI can prove to be bane or boon in plant breeding depending upon the mode of reproduction and the desirability of seed or vegetative part of the crop carrying S-locus.

7.3.6.1 Advantages of Self-Incompatibility

It renders more than one benefit as under:

1. Where male sterility does not exist, SI can alternatively be used for the production of F_1 hybrids in seed crops, forage, grasses, and in ornamental plants to exploit hybrid vigour. A good deal of success has been achieved in breeding for hybrids employing SI in crops, e.g., crucifers and forage grasses. It requires production of S-homozygotes (e.g., S_1S_1, S_2S_2, etc.) and then inter-planting the two S-homozygotes to produce S_1S_2 hybrids.
2. Similarly, double cross hybrids and/or synthetics can also be produced with SI. For example, S_1S_2 and S_3S_4 can be crossed to produce double cross hybrid ($S_1S_2 \times S_3S_4$).
3. If self-fertility is induced temporarily or permanently by mutation of S-allele to S_f through artificial irradiation in clonally propagated orchard species, e.g., cherries and apples, the uniformity of the cropping and crop product (fruit) can readily be ensured. In fact, such a fertile cultivar (with S_f allele) will permit uniform planting as against mixed planting forced-adapted by gardeners (due to the presence of self-sterility) leading to uneven quality. In this way, the gardener can also afford to dispense with the pollinating agents (bees) which are reared artificially at considerable cost.
4. Seedless cultivars where desirable, such as in pineapple, grapes etc., can be evolved if SI is present. This becomes a better proposition in crops where incompatibility is insensitive to polyploidy (such as sporophytic system). Then seedless triploids can be produced simply by inter-planting diploids and tetraploids in alternate rows.

7.3.6.2 Disadvantages of Self-Incompatibility

SI suffers from some drawbacks. It is undesirable in seed crops due to the following reasons:

1. Variation in seed-set due to poor fertility.
2. Poor preservation of genetic purity of improved cultivars since cross-pollination is non-restricted.
3. Difficulties in development and maintenance of homozygous lines (inbreds) which can be utilized for hybridization.
4. Uneven quality of fruits in horticultural crops because of mixed planting of different cultivars based on their cross-compatibility.

7.3.7 Applications of Self-Incompatibility

SI is an important pollination control device that prevents self-pollination and promotes cross-pollination. There are two main applications of SI, *viz.*, (i) production of hybrids and (ii) combining desirable genes from different sources.

7.3.7.1 Production of Hybrids

SI provides means for hybrid seed production without emasculation and without use of genetic or cytoplasmic male sterility. It has been used for production of commercial hybrids in *Brassica* (cabbage and Brussels sprouts) and sunflower. Two self-incompatible lines are planted in alternate rows for hybrid seed production. The harvest from both the lines would be hybrid seed.

7.3.7.2 Combining Desirable Genes

SI system permits combining desirable genes from two or more different sources into a single genotype through natural cross-pollination. Moreover, knowledge of SI especially in fruit crops, helps fruit growers increase the yield of fruits by providing suitable pollinators.

7.3.8 Limitations of Self-Incompatibility

There are some limitations of SI. The main ones are presented below:

1. It is very difficult to produce and maintain homozygous inbred lines in a self-incompatible species. Bud pollination must be made to maintain the parental lines.
2. Continued selfing is required for production and maintenance of inbreds. But, it reduces SI reaction and may select for self-fertility.

3. SI is affected by environmental factors such as temperature and humidity. High temperature and high humidity reduce or even totally suppress SI reaction. Consequently, the hybrid seed produced under this type of unfavourable conditions using SI may contain up to 30% or more selfed seed which is not desirable.
4. Bees often tend to stay within one inbred line, especially when the inbreds differ morphologically. This reduces seed-set and increases selfing. For example, in Brussels sprouts, when two distinct morphological parental lines are used in hybrid seed production programme, bees have preference in visiting only one of them. This also poses problems in the use of SI in hybrid seed production programmes.
5. Transfer of S-alleles from different cultivars/species is problematic
6. Inbreeding may give rise to new SI reactions in the gametophytic system. This limits the use of inbreds as parents of hybrids.

7.4 Pistillate Condition

In case of monoecious crops such as castor, cucurbits, etc., some mutants produce only pistillate (female) flowers in the place of both male and female flowers. It is known as **pistillate condition** and it provides opportunities for its use in hybrid seed production.

7.5 Brief References

Allard 1960, Chahal and Gosal 2002, Chaudhary 2014, Dabholkar 2006, Ram M 2014, Sharma JR 1994, Singh BD 2011,2018; Singh Phundan 2013a, Singh and Pawar 2006.

Review Questions

Part A

Answer the following questions true or false.

1. Koelreuter (1763) for the first time reported self-incompatibility. (True/False)
2. GMS is usually controlled by a recessive gene. (True/False)
3. Self-incompatibility cannot be eliminated. (True/False)
4. Ethidium bromide has been used for induction of CMS in *Petunia*. (True/False)
5. Pistillate condition is always found in cucurbits. (True/False)
6. Self-incompatibility occurs rarely in crop plants. (True/False)
7. Production of hybrids is possible through self-incompatibility. (True/False)

8. Bud pollination is useful to overcome SI. (True/False)
9. Self-incompatibility prevents cross-pollination. (True/False)
10. Tristyly in *Lythrum salicarea* is controlled by two genes. (True/False)
11. Bees often tend to stay within one inbred line, especially when the inbreds differ morphologically. (True/False)
12. CMS is governed by nuclear genes. (True/False)
13. Seedless cultivars where desirable can be evolved if SI is present. (True/False)
14. In male sterility, pollen grains are non-functional (True/False)
15. The sporophytic SI system was reported by Hughes and Babcock (1950) in *Parthenium argentatum*. (True/False)

Part B

Answer the following questions.

1. What is male sterility?
2. What do you mean by controlled mating?
3. SI stands for ______________ in botany/plant breeding.
4. Homomorphic SI are of 2/3 types.
5. S alleles may show dominance/co-dominance.
6. Cite an example of sporophytic SI.
7. Define gametophytic system of self-incompatibility.
8. PGMS stands for__________________.
9. What is male gametocide?
10. Cauliflower shows __________________ system of SI.
11. What did East and Mangelsdorf report in 1925?
12. What do you understand by TGMS?
13. Expand TrGMS.
14. ___________ is an example of gametophytic system of self-incompatibility.
15. Where male sterility does not exist, ______________ can alternatively be used for the production of F_1 hybrids in seed crops, forage, grasses, and in ornamental plants to exploit hybrid vigour.

Part C

Write a brief note on each of the following topics.

1. Limitations of self-incompatibility.
2. Applications of self-incompatibility.
3. Morphological features of male sterility.
4. Advantages of self-incompatibility.
5. Mechanism of male sterility.
6. Chemically-induced male sterility.
7. Cytoplasmic male sterility.
8. Genetic male sterility.
9. Applications of male sterility.
10. Cytoplasmic-genetic male sterility.

Part D

1. Define male sterility. Discuss the different types of male sterility.
2. How can you maintain male sterility?
3. Discuss self-incompatibility in relation to plant breeding.
4. Describe the measures to overcome self-incompatibility.
5. Discuss the inheritance and transfer of male sterility.

Section II
Breeding Principles

8

Plant Domestication Bringing Plants under Cultivation

8.1 Introduction

In pre-historic times, people lived as hunters and gatherers. Then, gradually they began to produce the crops they ate on piece of land, and farming was born. Wild plants have certain traits that keep them protected and productive without human intervention. However, most of them are not desired for food or other use by humans. By spontaneous mutagenesis and hybridization, occasionally variant plants were found that lacked some of the undesired traits, such as bitter taste, thorns, or had particularly attractive features, such as larger fruits and more regular tuber shape. Consequently, as time passes such variation was picked up by humans by selection of such variants, keeping and advancing them. Thus, **plant domestication is the process of bringing wild species under human management**. In other words, *plant domestication is the process by which genetic changes (shifts) in wild plants are brought about through a selection process imposed by humans.* ***It is a method of plant breeding in the sense that, when successful, it provides domestic types that are superior to ones previously available.***

Pre-historic man domesticated virtually all of our present-day crops. But, domestication of wild species continues even today and is likely to continue for a long time in the future. For example, in many countries there is an increasing emphasis on bringing forest trees producing timber, etc. under human management. A similar trend is seen for medicinal plants and plants producing specific compounds of interest, e.g., hydrocarbons, oils, etc. Kala jeera (*Bunium persicum*), a perennial spice, was domesticated during 1990s in Himachal Pradesh; it is being cultivated as an orchard crop. *One important aspect of domestication at present is the use of particular genes from wild relatives in the improvement of cultivated species. When a plant breeder transfers one or a few desirable genes from a wild relative to a cultivated type, he is, in a sense, domesticating the wild species in part.*

8.2 Purpose of Domestication

Man has brought wild plants under his control for various purposes. Plants have been domesticated primarily for five main purposes, *viz*., (i) food, (ii) clothes, (iii) shelter, (iv) medicines and (v) aesthetic value. They are briefly discussed below:

i) Food: Those plants that are domesticated for large-scale food production are generally called food crops including food grains, vegetables and fruits.

ii) Clothes: Those plants that are domesticated for large-scale cloth production are referred to as fibre crops including cotton, jute, sunnhemp, kenaf, etc.

iii) Shelter: Those plants that are domesticated for large-scale wood production are known as timber trees. Such plants are grown for in house use for windows, doors and furniture.

iv) Medicines: Those plants that are domesticated for large-scale medicine production are called medicinal plants. They are used for treatment of various human and animal diseases.

v) Aesthetics: Those plants that are domesticated for decoration purposes are called as ornamental plants. Such plants are grown in and around home, and usually called house plants or ornamentals.

8.3 Selection Under Domestication

Selection operates over the genetic variability. For instance, in a mixed population some types may get favoured over the other; but in a pureline, no selection, natural or artificial (human), can operate.

8.3.1 Natural Selection

Forces of nature consisting of abiotic (temperature, water, etc.) and biotic (disease, insect-pest, etc.) stresses operate. The genotype which is fit to survive is favoured and remains viable in the next generation. The one that is not fit cannot compete and disappears in due course of time.

8.3.2 Artificial Selection

Artificial selection depends on the choice of man, the selector. It may be (i) directional, (ii) stabilizing or (iii) disruptive. They are discussed below:

i) Directional Selection: It leads to a population one extreme direction. It can change the phenotype or mode of reproduction of the population in the direction of the selection. For example, tomato was cross-pollinated in its place of origin (Peru-Bolivia). But, under European glass-house domestication it became self-pollinated.

ii) Stabilizing Selection: It does not lead to extreme type but to intermediate phenotype. It favours those genotypes whose phenotypic expression clusters around the population mean. Extreme types are not favoured.

iii) Disruptive Selection: It operates on those populations that are subjected to distinct ecological niches. In each ecology, a different phenotype gets selected so that the population ultimately consists of more than one recognizable form. Thus, disruptive selection maintains polymorphism in a population.

8.4 Evolution versus Domestication

Plant breeding and evolution are similar processes. The mechanisms that govern them are essentially the same, except that the creation of variation and the selection among the variability are conducted by humans in the case of the former. Compared to evolution, plant breeding occurs in the twinkle of an eye. Evolution and domestication are about change. The mechanisms of change (e.g., mutation, selection, hybridization, polyploidization) are similar for both processes. The major difference is regarding objective. Evolution brings about change for better reproduction in the natural environment (wild), whereas domestication changes wild species to make them adapted to better reproduction in cultural environments (agricultural production). Evolution changes plants to become dependent on nature while domestication changes plants to become dependent on humans.

8.5 Centres of Plant Domestication

Plant breeders are interested in centres of plant domestication as regions of genetic diversity. De Candolle (1886) was the first to suggest that a crop plant originates from the area where its wild progenitor occurs. He considered archaeological evidence to be direct proof of the ancient existence of a crop species in a geographic area.

Several scientists, notably, N.I. Vavilov and J.R. Harlan have provided the two most enduring views of plant domestication. Vavilov noticed that extensive genetic variability within a crop species occurred in clusters within small geographic regions separated by geographic features such as mountains, rivers and deserts. Vavilov proposed the concept of centres of diversity to summarize his observations. **He defined the centre of origin of a crop plant as the geographic area(s) where it exhibits maximum diversity of botanical varieties. Additional evidence is that the presumed ancestral species occurs in that region in the wild.** Vavilov identified eight major centres of diversity. In the 1960s many more scientists came with modifications and proposed other and additional areas, some of which were subdivided (sub-centres). Centres of domestication always seem to be the presumed regions of domestication of several crop species. Centres of domestication are almost always also centres of diversity, but the reverse is not true. Some centres of diversity do not coincide with centre of domestication. For example, China is a (secondary) region of diversity for waxy corn, which was domesticated in Central America. Important centres of diversity are shown in Table 8.1. The table gives mostly the areas where crops have been domesticated. These are called **"primary centres of diversity"** and usually show the highest diversity. However, the region where crops are domesticated is not always unequivocal. Again, some crops may have been domesticated simultaneously in several regions.

Further, Vavilov found that the maximum amount of variability and the concentration of dominant genes for crops occurred at the centre and decreased towards the periphery of the cluster of diversity. He also discovered that there were parallelisms (common features) in variability among related species and genera. He called this **the law of homologous series in heritable variation** (or **parallel variation**). In other words, species and genera that are genetically closely related are usually characterized by a similar series of heritable variations. So, it is possible to predict what parallel forms would occur in one species or genus, from observations of the series of forms in another related species or genus. The breeding implication is that if a desirable gene is found in one species, it likely would occur in another related species.

Table 8.1. Important centres of diversity (Zeven and de Wet,1982)

Region of diversity	Crops having their primary centre of diversity in the region
1. Chinese-Japanese	Rhubarb, Chinese cabbage, tea.
2. Indochinese-Indonesian	Ginger, *Piper*-pepper, taro, banana, wax gourd, cardamom, various *Citrus* crops.
3. Australian	Macadamia nut, *Eucalyptus.*
4. Hindustani	Mango, okra, black (*Piper*) pepper, eggplant, cucumber.
5. Central Asia	Onion, garlic, pistachio, spinach, walnut, carrot, almond.
6. Near Eastern	Pea, cabbage, leek.
7. Mediterranean	Celery, beet kale, radish, olive, poppy, grape.
8. African	Coffee, watermelon, melon, yam, bean, cowpea, oil palm, date palm.
9. European-Siberian	Lettuce, *Ribes* berries, hemp, hops, beet, chicory, turnip, peppermint, apple, pear, strawberry.
10. South-American	Cashew, rubber, passion fruit, cacao, tomato, potato.
11. Central American and Mexican	Potato, sweet potato, pumpkin, zucchini, *Opuntia*, papaya, avocado, *Phaseolus* bean, runner bean, vanilla, *Capsicum*-pepper.
12. North American	Sunflower, pecan, strawberry.

8.6 Effects of Plant Domestication

Various plant characteristics have been modified under domestication in relation to their usefulness for humans. The major effects of plant domestication include:

1. Higher germination rates.
2. Greater germination predictability.
3. More uniform timing of germination.
4. Increased size of reproductive organ.
5. Reduced complexity of reproductive organs.
6. Reduction of toxicity.
7. Change in biomass allocation.
8. Change in life-cycle.
9. Change in quality of produce in terms of colour, taste, texture and odour.
10. Improvement in the productivity of the economic plant part.

8.7 Limitations of Plant Domestication

Despite a lot of progress in farming, few crops have been domesticated. It is estimated that 230 plant species belonging to 180 genera and 64 families have been domesticated. Some families, namely, Poaceae, Fabaceae, Brassicaceae and Solanaceae have yielded more domesticates than others. The main limitations of domestication are:

1. Domestication is a very slow process and takes hundreds of years to bring a wild plant under domestication (cultivation).
2. Wild plants have several undesirable traits and eliminating them requires continuous breeding efforts.
3. The domestication leads to reduction in the resistance of plants to biotic and abiotic stresses.
4. The domestication makes the plants human-dependent for their reproduction.

8.8 Achievements of Plant Domestication

A number of horticultural crops have been domesticated in addition to agricultural crops. However, the place and date of domestication is available only for very few plant species (Table 8.2).

1. Fruit Crops: This group includes apple, pear, plum, peach, cherries, strawberry, mango, guava, litchi, loquat, banana, grapes, pomegranate, fig, jackfruit, papaya, lime, lemon, oranges, grapefruit, almond, cashew, chestnut, walnut, etc.

2. Vegetable Crops: This group includes potato, onion, garlic, cabbage, cauliflower, knol-khol, radish, turnip, carrot, chilli, tomato, brinjal, bottle gourd, sponge gourd, bitter gourd, spinach, okra, cucumber, etc.
3. Flowering Plants: This group includes several annual, biennial and perennial flowering plants.
4. Medicinal Plants: This group includes belladona, cacao, cannabis, coffee, cola, opium poppy, quinine, tea, etc.

Table 8.2. Place and year of domestication of some plants

S.No.	Crop	Place of domestication	Year of domestication
1.	Bottle gourd	Central America	8000 BC
2.	Chilli	Central America	4000 BC
3.	Potato	South America	5000 BC
4.	Avocado	Central America	5000 BC
5.	Sweet Potato	Peru	2500 BC
6.	Watermelon	Near East	4000 BC
7.	Chenopod	North America	750 BC
8.	Peas	Middle East	9000 BC
9.	Cassava (Tapioca)	South America	5500 BC
10.	Figtree	Near East	9000 BC

8.9 Brief References

Acquaah 2012, Allard 1960, Chaudhary 2014, Mohanan 2010, Singh BD 2011, 2018; Singh Phundan 2011, 2013a, Zeven and de Wet, 1982.

Review Questions

Part A

Answer the following questions true or false.

1. Harlan put forth the law of homologous series in heritable variation. (True/False)
2. Around 230 plant species have been domesticated. (True/False)
3. Centres of domestication are almost always also those of diversity. (True/False)
4. Domestication causes increase in toxicity in plants. (True/False)
5. Vavilov proposed the concept of centres of diversity. (True/False)

Part B

Answer the following questions.

1. Who defined centre of origin of crop plant?
2. What is natural selection?
3. Domestication is a very slow / quick process.
4. De Candolle (1886) suggested ____________
5. Evolution brings about change for better reproduction in the ____________ environment.

Part C

Write a brief note on each of the following topics.

1. Limitations of plant domestication.
2. Effects of plant domestication.
3. Evolution versus domestication.
4. Centres of plant domestication
5. Artificial selection.

9

Plant Genetic Resources Collection, Conservation and Utilization

9.1 Introduction

The sum total of genes in a plant species is called as **plant genetic resources** (also called **germplasm**, **gene pool**, or **genetic stock**). In other words, plant genetic resources refer to a whole library of different alleles of a plant (crop) species. Germplasm is the lifeblood of plant breeding without which breeding is impossible to conduct. It is the genetic material that can be used to perpetuate a species or population. It not only has reproductive value, but through genetic manipulation (plant breeding), germplasm can be improved for better performance of the crop. Germplasm provides the materials (parents) used to initiate breeding programme. Sometimes, what a plant breeder does is to evaluate plant genetic resources and make a selection out of existing biological variation. Promising genotypes that are adapted to the production region are then released for cultivation by the farmers. Other times, a plant breeder generates new variability by using different methods such as hybridization, mutagenesis and more recently gene transfer. This base population is then subjected to appropriate selection methods, leading to the identification and further evaluation of promising genotypes for release as cultivars. When a plant breeder needs to improve crop plants, he has to find a source of germplasm that would supply the genes needed to undertake the plant breeding programme. To facilitate the use of genetic resources (germplasm), germplasm (gene) banks are given the responsibility of collecting, evaluating, cataloguing, storing and managing large numbers of germplasm. This strategy allows a plant breeder ready and quick access to genetic resources when he needs it. ***The fact that Sir Otto Frankel coined the term 'genetic resources' only in 1968 shows that the plant breeders, though aware of the gradual loss of germplasm, failed to recognize the urgency of protecting the genetic resources of crop plants prior to a point of no return.***

9.2 Kinds of Plant Genetic Resources

The plant genetic resources consist of various plant materials of a crop such as (1) land races, (2) obsolete cultivars, (3) modern cultivars, (4) advanced (homozygous) breeding lines, (5) wild forms (races) of cultivated species, (6) wild relatives and (7) mutants. They are briefly discussed below:

9.2.1 Land Races

Land races are primitive cultivars that were selected and cultivated by the farmers for many generations. They are not bred like modern cultivars by plant breeders. They evolved under subsistence agriculture. They have high level of genetic diversity that provides them high degree of resistance to biotic and abiotic stresses. They have broad genetic base that provides them wider adaptability and protection from epidemic of diseases and insect-pests. They possess valuable alleles. So, their conservation is essential. The main drawbacks of land races are that they are less uniform and low yielders.

9.2.2 Obsolete Cultivars

Improved cultivars of recent past are known as **obsolete cultivars**.

9.2.3 Modern Cultivars

The presently cultivated high yielding cultivars are known as **modern cultivars** (also called **improved cultivars, advanced cultivars**). They have higher yield potential and uniformity than obsolete cultivars and land races. However, they have narrow genetic base and low adaptability as compared to land races.

9.2.4 Advanced Breeding Lines

Pre-released plants that have been developed by plant breeders are known as **advanced lines** (also called **cultures, stocks**).

9.2.5 Wild Forms of Cultivated Species

Wild forms of cultivated species are available in many crops. Such forms have usually high degree of resistance to biotic and abiotic stresses. They can easily cross with cultivated species. However, wild forms of many crops are extinct.

9.2.6 Wild Relatives

The naturally occurring plant species that have common ancestry with crops and can cross with crop species are known as **wild relatives** (also called **wild species**). They are important sources of resistance to biotic and abiotic stresses. But they are used as the last resort in plant breeding programmes due to hybrid sterility, hybrid inviability and transfer of several undesirable genes to the cultivated species along with desirable alleles.

9.2.7 Mutants

Mutation breeding is used when a desired character is not available in the genetic stock of cultivated species and their wild relatives. Mutations occur in nature and can be artificially induced. The extra variability which is created through mutagenesis constitutes important components of gene pool.

9.3 Gene Pool System of Classification

A gene pool consists of all the genes and their alleles present in all such individuals including all cultivars (obsolete and current), wild species and wild relatives, which hybridize or can hybridize with each other. J.R. Harlan and J.M.J. de Wet (1971) proposed a categorization of genes pools of cultivated crops according to the feasibility of gene transfer or gene flow from those species to the crop species. Three categories were defined: primary, secondary and tertiary gene pools.

9.3.1 Primary Gene Pool (GP1)

The gene pool in which intermating (crossing) is easy leading to production of fertile hybrids is known as **primary gene pool** (also called **gene pool one, GP1**). So, there is no restriction to gene exchange between members of the group; genes can be exchanged between lines by making normal crosses. This group includes plants of the same species or closely related species which produce completely fertile offspring on intermating. *Members of GP1 are the most commonly used in breeding programmes.*

9.3.2 Secondary Gene Pool (GP2)

The genetic material that leads to partial fertility on crossing with GP1 is known as **secondary gene pool** (also called **gene pool two, GP2**). It includes plants that belong to related species, both cultivated and wild. Members of GP2 can be crossed with those of GP1, but usually the hybrids are sterile and some of the progeny to some extent are fertile. Gene transfers from GP2 to GP1 are possible, but usually difficult. *Members of GP2 are often used in breeding programmes.*

9.3.3 Tertiary Gene Pool (GP3)

The genetic material that leads to production of sterile hybrids on crossing with GP1 is known as **tertiary gene pool** (also called **gene pool three, GP3**). It includes material which can be crossed with GP1, but the hybrids are sterile. Transfer of genes from such material to GP1 is possible with the help of special techniques such as embryo rescue and bridge crossing. Gene transfers from GP3 to GP2 are comparatively easier. *Members of GP3 are only occasionally* *used in breeding programmes, and that too by a group of workers having the competence and the patience for handling the associated problems.*

9.3.4 Quaternary Gene Pool (GP4)

The recent developments in technology of genetic engineering have brought the entire world of living organisms within the limits of germplasm of any crop plant. The genes can now be transferred not only across species or genera but also from different types of organisms through special techniques of recombinant DNA technology. The transfer of 'Bt' gene from a bacterium, namely, ***Bacillus*** *thuringiensis*, to cotton, maize, rice, **tomato** and other plant species extends the limits of gene pool to cross human imagination. A new type of gene pool 'GP4' which may be referred to as '**gene-ocean**' has thus been made available by the modern science. Such gene pool is limited to neither centres of origin nor gene banks but to the imagination and scientific acumen of its user.

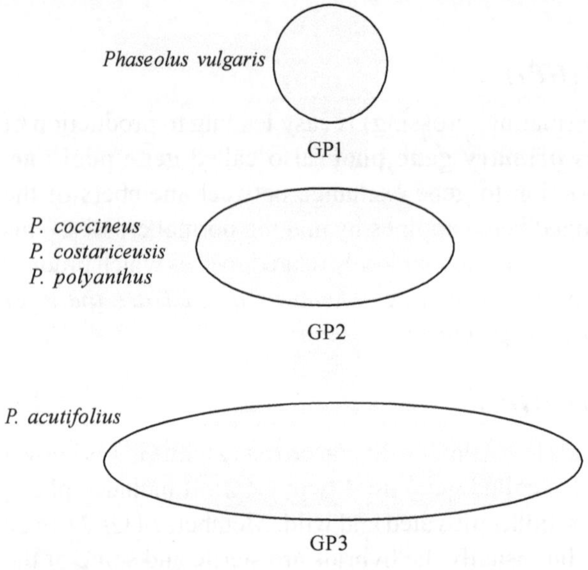

Figure 9.1. Gene pools of French bean (*Phaseolus vulgaris*)

A classification of French bean (also called common bean, dry bean), *Phaseolus vulgaris*, is presented in Figure 9.1 for illustration of gene pool concept. In collecting germplasm for a plant breeding programme, the general rule is to start by searching the domesticated germplasm collection first, before considering other sources, for reasons previously stated. However, there are occasions when the gene of interest occurs in undomesticated germplasm, or even outside the species. Gene transfer techniques enable breeders to transfer genes beyond the tertiary gene pool. Whereas all crop plants have a GP1 that includes the cultivated forms, all crops do not have wild forms in GP1 (e.g., broad bean, cassava and onions whose wild types are yet to be identified). Again, occasionally, the GP1 may contain taxa of other crop plants (e.g., almond belongs to the GP1 of peach). Most crop plants have a GP2, which consists primarily of species of the same genus. Some crop plants have no GP2 (e.g., barley, soybean, onion, broad bean).

9.4 Types of Seed Collections

Seed collections are of three types: base collections, active collections and working collections that are discussed below:

9.4.1 Base Collections

Base collections include maximum number of accessions available in a crop. They are meant for long term conservation (up to 50 years or more) and are stored at –18 or –20 °C in sealed containers. The seeds are dried to 5 ± 1 % moisture and have more than 85 % initial seed viability. They are disturbed only for the purpose of regeneration. They are used only when germplasm from other sources is not available for use in plant breeding programmes.

9.4.2 Active Collections

This category of germplasm is actively utilized in plant breeding programmes. They are conserved for medium term (8-10 years or more) and are stored at 0 °C with moisture content around 8 %. Germination test is carried out after every 5-10 years to assess the reduction in seed viability.

9.4.3 Woking Collections

These collections are frequently utilized by plant breeders in their plant breeding programmes. They are stored for short term (3-5 years). The seeds are stored at 5-10 °C with moisture content of 8-10 %.

9.5 Germplasm Activities

There are six important germplasm activities, *viz.*, exploration and collection, conservation, evaluation, documentation, distribution and utilization.

9.5.1 Exploration and Collection

Exploration means collection trips, and collection refers to tapping of genetic diversity from various sources and assembling the same at one place. The exploration and collection are highly scientific process. The germplasm is collected from different sources such as centres of diversity, gene banks, gene sanctuaries, seed companies, and farmers' fields. Moreover, collections can be made by local exploration trips to the regions of crop diversity.

9.5.2 Conservation

Conservation refers to protection of genetic diversity of crop plants from genetic erosion. There are two important methods of germplasm conservation, namely, *in-situ* conservation and *ex-situ* conservation.

9.5.2.1 In-situ Conservation

Conservation of germplasm in its natural habitat is known as in-situ conservation. It requires establishment of natural or biosphere reserves, national parks or protection of endangered areas or species. In this method, the wild species and the complete natural or semi-natural ecosystems are considered together. However, it is a costly method of germplasm conservation.

9.5.2.2 Ex-situ Conservation

Conservation of germplasm away from its natural habitat is known as ex-situ conservation. It can be done through seed banks, plant (or field) banks, shoot-tip banks, cell and organ banks, and DNA banks. This is the most practical method of germplasm conservation.

9.5.2.2.1 Seed Banks

In seed banks, germplasm is stored as seeds. The conservation of seed is easy, safe and requiring minimum space. Seeds of many species can be stored up to 50-100 years. Containers of glass, tin, plastic or a combination of them are used for seed storage. Roberts (1973) has classified seeds into two major groups for storage purpose, namely, orthodox seeds and recalcitrant seeds.

1. Orthodox Seeds: *Seeds that can be dried to low moisture content (5% or lower) and stored at low temperature without losing viability are named as orthodox seeds.* This category includes most of the crop species like corn (maize), wheat, rice, **carrot, beets, papaya, pepper,** chickpea (Bengal gram), lentil, soybean, cotton, sunflower, **various beans, egg plant** (brinjal), and **all the brassicas**. They can be easily conserved for long period.
2. Recalcitrant Seeds: *Seeds that show very drastic loss in viability with a decrease in moisture content below 12-13 % are known as recalcitrant seeds.* Examples are seeds of many forest and fruit trees like ***Citrus*, mango, jackfruit, cocoa, coffee, tea, rubber, coconut, oil palm** etc. Such seeds cannot be stored in seed banks. They are conserved in field gene banks.

9.5.2.2.2 Plant or Field Banks

Field Banks (also called Plant Banks) are orchards of fruit trees and vegetatively propagated crops where their germplasms are grown. They have, however, various limitations such as requiring large areas, expensive to establish and maintain, prone to damage from biotic stresses and man-made or natural disasters, and human errors in handling.

9.5.2.2.3 Shoot Tip Banks

In such gene banks, germplasms are conserved as slow-growth cultures of shoot tips and nodal segments. Their regeneration consists of sub-culturing the cultures that may be done every 6 months to 3 years. They are suitable for the crops that do not produce seeds or produce recalcitrant seeds.

9.5.2.2.4 Cell and Organ Banks

A germplasm collection based on cryopreserved (at –196 ^{0}C in liquid nitrogen) embryogenic cell cultures, shoot-tips and/or somatic/zygotic embryos may be termed as a cell and organ bank.

9.5.2.2.5 DNA Banks

In these banks, DNA segments of the accessions of germplasm are maintained as cosmid clones or phage lysates. These DNA segments may be evaluated and the desired ones may be used to produce transgenic plants. This approach is applicable for the conservation of genetic material of already extinct species since DNA extracted from the preserved herbarium specimens can often be cloned.

9.5.3 Evaluation

Evaluation means assessment of germplasm for different traits that are useful in plant breeding programmes. Germplasm accessions are evaluated for morphological, physiological, biochemical, plant pathological, entomological and other features. The traits evaluated must be as per need of the plant breeders so that they may utilize in their breeding programmes.

9.5.4 Documentation

Documentation (also called information system) consists of compilation, analysis, classification, storage and dissemination of information. In plant genetic resources, documentation means dissemination of information about various activities such as collection, evaluation, conservation, storage and retrieval of data. It helps in making genetic resources accessible to plant breeders. For this purpose, catalogues of the germplasm collections of different crops are regularly published.

9.5.5 Distribution

The distribution of germplasm is one of the important activities of the plant genetic resources centres. The accessions of germplasm asked for by the plant breeders are supplied by the centres for utilization in the plant breeding programmes.

9.5.6 Utilization

Utilization means use of germplasm in plant breeding programmes. The cultivated germplasm may be used in three main ways: as a cultivar, as a parent in crossing and as a variant in the gene pool. On the other hand, the wild germplasm is used to transfer resistance, adaptability and sometimes quality.

9.6 Organizations Associated with Plant Genetic Resources

International and national organizations are associated with germplasm. International Plant Genetic Resources Institute (IPGRI), Rome, Italy coordinates at global level. Various International Crop Research Institutes deal with germplasm of concerned crops. In India, National Bureau of Plant Genetic Resources (NBPGR), New Delhi deals with various aspects of germplasm of agricultural and horticultural crops. Forest Research Institute, Dehradun deals with germplasm of forest species and Botanical Survey of India, Kolkata deals with germplasm of remaining plant species.

9.7 Quarantine

To prevent the entry of new diseases, insects and weeds from other countries an Act was formulated in India in 1914 which is called as Destructive Insects and Pest Act. *Quarantine refers to prophylactic means to prevent the entry of new diseases, insects and weeds from other countries.* The quarantine offices are located at all seaports and international airports. Their duty is to examine imported plant materials for diseases, insects and weeds. They also check the phytosanitary certificate of the plant materials issued by the sender which states that the plant material does not contain any disease, insect or weed. In event of non-availability of the phytosanitary certificate, quarantine department may reject and destroy the plant material.

9.8 Brief References

Acquaah 2012, Chahal and Gosal 2002, Dabholkar 2006, Kumari and Gowthami 2015, Ram 2014, Roy 2012a, Sharma 1994, Singh BD 2011,2018; Singh Phundan 2013a, Singh and Pawar 2006.

Review Questions

Part A

Answer the following questions true or false.

1. Orthodox seeds lose viability when moisture is decreased below 12-13 %. True/False
2. Plant bank is also known as field bank. True/False
3. Gene banks are virtually seed banks. True/False
4. Conservation of germplasm in its natural habitat is *ex-situ* conservation. True/False
5. Occasionally, GP1 may contain taxa of other crop plants. True/False.

Part B

Answer the following questions.

1. IPGRI stands for ____________.
2. What is gene pool?
3. Give an example of a recalcitrant seed.
4. Define plant genetic resources.
5. NBPGR deals with ____________ and ____________ crops.

Part C

Write a brief note on each of the following topics.

1. Primary gene pool.
2. Types of seed collections.
3. Quarantine.
4. GP4.
5. Kinds of plant genetic resources.

Part D

1. Descibe the *ex-situ* conservation of plant genetic resources.
2. Discuss the gene pool system of classification.

10

Variation Types and Estimation

10.1 Introduction

Plant breeders come across two categories of characters (traits) for improvement, namely, qualitative characters and quantitative characters. Red vs white flower, round vs wrinkled seed, determinate plant growth vs indeterminate plant growth etc. are examples of **qualitative trait differences**. There is no overlapping of characters and no confusion during classification. Such trait differences are usually governed by one or two gene pairs, each with large effect that can be measured even in the presence of segregation at other loci and even in the presence of effects of environmental factors. These characters are very less influenced by the environmental factors.

On the other hand, most of the traits of economic importance like yield of fruits of vegetable and fruit crops, yield of vegetative parts in fodder crops, yield of lint in cotton, yield of grains in cereals, pulses, oilseeds etc., plant height, days to maturity etc. are examples of quantitative characters. Variation observed for such traits is continuous. Again, there is **overlapping of trait differences**. So, one cannot assign individuals to a specific class unambiguously. Such characters are influenced by environmental factors, such that phenotypes are not same when grown over locations and/or seasons.

When different plants/lines of a species exhibit trait differences, it is called variation. Variation is essential for any improvement in a crop. It may be present in the population to be improved or else it may be created by different means such as hybridization, mutation, polyploidy, domestication, plant introduction, somaclonal variation, genetic engineering etc. Now, we shall discuss the types and estimation of variation.

10.2 Measures of Variation

The extent of variation present for a trait among the plants of a population or a species is known as variability. *The amount of variation is estimated by the measures of dispersion; the most common ones are range, variance and coefficient of variation.* They are applied to quantitative characters. They are estimated usually from the data of a replicated trial. Each measurement is called an **observation** and all the observations for that trait from a given trial together form a **sample**.

10.2.1 Range

Range is the difference between the largest and the smallest observations of a sample. It is shown by the values of the largest and the smallest observations, e.g., the range of plant height is 30-52 cm.

Range = largest - smallest

= 52cm - 30cm = 22cm

As range has number of limitations, it has limited application.

10.2.2 Variance (s^2 or v)

Due to its several unique properties, it is used universally as the estimate of variability. The variance of a sample is worked out by dividing the sum of squared deviations of all the observations in the sample from the sample mean by the number of observations minus one. Mathematically it is expressed as:

$$v = \frac{\Sigma(X - \overline{X})^2}{n-1}$$

Where, v = variance, Σ = summation, X = an observation in the sample, $\overline{X}$ = sample mean, and n= number of observations in the sample. Estimation of variance is relatively easy. *Variance has the same unit as mean or observation, its magnitude is dependent on that of the mean.*

10.2.3 Coefficient of Variation (CV)

Coefficient of variation (CV) is estimated by dividing the standard deviation (SD, $\sqrt{v}$) of a sample by its mean. It is mathematically expressed as:

$$CV = \frac{\sqrt{v}}{\overline{X}}$$

where, $\sqrt{v}$ = standard deviation (SD is square root of variance),= sample mean.

CV may be expressed as percentage by multiplying the estimate with 100. *It is used to compare the variability for a character in different samples from the same population or from different populations or for different characters in the same sample or in different samples.* Such comparison is not possible with variance or standard deviation.

10.3 Types of Variation

The expression of a quantitative trait is affected by genotype (G), environment (E) and genotype × environment (G×E) interaction. Hence, the value (X) of a quantitative trait may be expressed as:

$$X = \mu + G + E + (G \times E)$$

where, μ = population mean

The effects of G, E and G×E cause the value of X to deviate from μ. *The variation of quantitative characters is divided into three types: genetic, environmental and genotype × environment interaction variations.* The variability is usually measured as variance. So, the three types of variations are called as **genotypic (genetic), environmental and G×E interaction variances,** respectively. The sum of genotypic (v_g), environmental (v_e) and G×E interaction ($v_{g \times e}$) variances for a character in a sample renders the value of its phenotypic variance (v_p). So, mathematically:

$$v_p = v_g + v_e + v_{g \times e}$$

10.3.1 Genetic Variation

The portion of phenotypic variation (also called **total variation) that is produced by genetic factors is called genetic variation.** It may arise due to nuclear genes, plasmagenes, and/or chromosomal aberrations (structural and numerical changes in chromosome). It is heritable and useful in plant breeding.

10.3.2 Environmental Variation

The portion of phenotypic variation that is produced by environmental factors (biotic and abiotic) is called environmental variation. It is nonheritable and so, of no use in plant breeding.

10.3.3 G × E Interaction Variation

When the relative performances of different cultivars of a crop change due to environments, it is known as genotype × environment (G×E) interaction, and the variation produced is called G×E interaction variation. This variation is also nonheritable and so, of no use in plant breeding.

10.4 Estimation of Phenotypic Variance (v_p) and Its Components

Phenotypic variance (v_p) is estimated from a replicated or non-replicated trial based on several genotypes (lines/cultivars) of the crop concerned.

10.4.1 Non-Replicated Trial

When variance for a trait is estimated from the data of a non-replicated trial based on several genotypes (lines/cultivars), an estimate of is obtained. *But it cannot be partitioned further into* V_g, V_e and V_{gxe} *components.*

10.4.2 Replicated Trial

The V_g *and* V_e *components of* V_p *can be estimated from the data of a simple replicated trial consisting of several genotypes (lines/cultivars) of a crop plant.* The field trial is conducted at one location, in a single year under a single environment. The data from such a trial are analysed statistically as per the design of experiment. From the analysis of variance (ANOVA), the variance components can be estimated. The elaboration is as follows:

ANOVA (According to RBD)

Source of Variation	Degrees of Freedom(df)	Sum of squares(SS)	Mean Squares (MS) = SS/df	Expectation of MS
Replication	(r – 1)	SS_r	MS_r	v_e+nv_r
Genotype	(n – 1)	SS_g	MS_g	v_e+rv_g
Error	(r – 1) (n – 1)	SS_e	MS_e	v_e

NB: r = number of replications, n= number of genotypes, v_e= environmental variance, v_r= replication variance, v_g= genotypic variance.

10.4.2.1 Environmental Variance (v_e)

Environmental variance (v_e) is estimated from MSe (error mean squares) of the ANOVA table. The expectation of MSe is:

$MS_e = v_e$ = environmental variance

10.4.2.2 Genotypic Variance (v_g)

Genotypic variance (v_g) is estimated from MSg (genotype mean squares) of the ANOVA table. The expectation of MSg is:

$$MS_g = v_e + r v_g$$

$$or, v_g = \frac{MS_g - v_e}{r}$$

$$or, v_g = \frac{MS_g - MS_e}{r}$$

Estimates of v_g from such trials contain v_{gxe} also. So, such v_g estimates are higher than their actual values.

10.4.2.3 Phenotypic Variance (v_p)

Phenotypic variance (v_p)from replicated trial data is estimated by adding the v_g and v_e estimates and mathematically:

$$v_p = v_g + v_e$$

10.4.3 Replicated Trials in several Environments

For estimation of G × E interaction, the replicated trial must be conducted under two or, preferably, more different environments. *Different locations and/or years usually represent the environments.* The data are analysed statistically as per design. The various MS estimates from the ANOVA table are used to estimate different variance components. *The estimates of* V_g *obtained from such trials are more precise than those estimated from trials of single environment.* The elaboration is as follows:

Pooled ANOVA

Source of Variation	**Degrees of Freedom(df)**	**Sum of squares(SS)**	**Mean Squares (MS) = SS/df**	**Expectation of MS**
Replication (R)	(R–1)	SS_r	MS_r	
Genotype(N)	(N–1)	SS_g	MS_g	$v_e + Rv_{gxe} + RLv_g$
Environment(L)	(L–1)	SS_L	MS_L	$v_e + Rv_{gxe} + RNv_L$
G×E Interaction	(N–1) (L–1)	$SS_{g\times e}$	$MS_{g\times e}$	$v_e + Rv_{gxe}$
Pooled Error	L (R–1) (N–1)	SS_e	MS_e	v_e

NB: v_e= pooled error variance, v_g= genotypic variance, v_L= environmental (location or season) variance, v_{gxe} = G × E interaction variance.

$$MS_g - MS_{g\times e} = RL\, v_g$$

$$or, v_g = \frac{MS_g - MS_{g\times e}}{RL}$$

$$MS_L - MS_{g\times e} = RN\, v_L$$

$$or, v_L = \frac{MS_L - MS_{g\times e}}{RN}$$

$$MS_{g\times e} - MS_e = R\, v_{g\times e}$$

$$or, v_{g\times e} = \frac{MS_{g\times e} - MS_e}{R}$$

10.5 Brief References

Dabholkar 2006, Kute and Aher 2013, Kute and Shinde 2016, Nadarajan and Gunasekaran 2005, Singh BD 2011.

Review Questions

Part A

Answer the following questions true or false.

1. Phenotypic variation is also called total variation. True/False
2. Vgxe can be estimated from a non-replicated trial. True/False
3. One cannot assign individuals to a specific class unambiguously in case of qualitative characters. True/False
4. Variation observed for qualitative traits is continuous. True/False
5. Genotypic variance and genetic variance are the same. True/False

Part B

Answer the following questions.

1. What are the three types of variation in quantitative traits?
2. Define G × E interaction variation.
3. Explain Vgxe.
4. The expression of a quantitative trait is affected by ______, _____ and ______.
5. Mathematically represent phenotypic variance (v_p).

Part C

Write a brief note on each of the following.

1. Measures of variation.
2. Derivation of Vgxe from analysis of variance table.
3. Types of variation.

11

Genetic Components of Polygenic Variation and Breeding Strategies

11.1 Introduction

In chapter 10, we have discussed the different types of variation, namely, genetic, environmental and genotype × environment interaction variations. We have also learnt how to estimate the phenotypic variance and its components including genotypic (also called genetic), environmental and g × e interaction variances from trial data. Now in this chapter, we shall discuss on the genetic components of polygenic (quantitative) variation and its implications on breeding strategies.

11.2 Genetic Components of Polygenic Variation

In this chapter whenever we call genetic variation, we mean polygenic (quantitative) variation, and we know that variation is usually measured in term of variance. However, variation, variability and variance are almost used synonymously. The genetic variance has been classified into different components by various workers. Fisher (1918) divided genetic variance into three components: *additive, dominance* and *epistatic* (interaction) *variances*. Wright (1935) classified genetic variance into two components: *additive variance* and *non-additive variance* (dominance and epistatic variances). Mather (1949) classified genetic variance into two components: *heritable fixable variance* (additive variance and additive × additive epistatic variance) and *heritable non-fixable variance* (dominance variance, and additive × dominance and dominance × dominance epistatic variances). Later, Hayman and Mather further divided the epistatic variance into three categories: additive × additive, additive × dominance and dominance × dominance.

11.2.1 Additive Variance (V_A)

The additive genetic variance, or in short, *additive variance (V_A) arises from the differences between phenotypes of two homozygotes for the genes affecting a polygenic character.* For instance, it represents the difference between phenotypes of the homozygotes AA and aa (for single gene). The genetic variance estimated from a replicated trial of homozygous genotypes is mainly V_A. There may be little additive × additive epistatic variance. So, V_g of variable homozygous populations is mainly V_A. Additive variance is fixable. So, effectiveness of selections depends on the magnitude of V_A.

11.2.2 Dominance Variance (v_D)

The dominance genetic variance, or in short, *dominance variance* (v_D) *arises from deviations in the phenotypic values of heterozygotes from the respective mid-parent values for each of the genes affecting a polygenic character.* The deviation of phenotypic value of Aa from the mid-parent value, i.e. mean of the phenotypic values of AA and aa, is dominance variance. It arises due to such deviations for all the genes affecting a given character. V_D is expressed only when the genes concerned are in heterozygous state. So, *it is found in natural populations of cross-pollinated crops and segregating generations of self-pollinated crops.* So, it does not contribute to effectiveness of selection. On the other hand, it reduces. It is utilized in hybrids.

11.2.3 Epistatic (Interaction) Variance

Epistatic (interaction) genetic variance, or in short, *epistatic (interaction) variance arises from the phenotypic effects produced by interaction between two or more genes affecting the same character.* The magnitude of epistatic variance is smaller than v_A and v_D. It is of three types: additive × additive (v_{AxA}), additive × dominance (v_{AxD}) and dominance × dominance (v_{DxD}).

11.2.3.1 Additive × Additive Epistatic Variance (v_{AxA})

It arises due to interaction between homozygous genotypes of two genes, e.g., interaction between AA and BB, etc. It is fixable and so utilized by selection.

11.2.3.2 Additive × Dominance Epistatic Variance (v_{AxD})

It arises due to interaction between a homozygous genotype for one gene and the heterozygous one for the other, e.g., interaction between AA and Bb, Aa and BB, etc. It is non-fixable and so reduces the effectiveness of selection. It is used in hybrids.

11.2.3.3 Dominance × Dominance Epistatic Variance (v_{DxD})

It arises due to interaction between heterozygous genotypes of the interacting genes, e.g., interaction between Aa and Bb. It is non-fixable and so reduces the effectiveness of selection. It is utilized in hybrids.

11.3 Heritability

Genetic variance (v_g) is the only component of phenotypic variance (v_p) that is heritable and useful in plant breeding. **The ratio of genetic variance (v_g) to the phenotypic variance (v_p) is called heritability** (H_{bs}). So mathematically:

$$H_{bs}(\%) = \frac{v_g}{v_p} \times 100 = \frac{v_g}{v_g + v_e} \times 100$$

($v_{g\times e}$, if known, is also included in the denominator)

where v_g, v_p, v_e, and $v_{g\times e}$ are genetic, phenotypic, environmental, and G×E interaction variances, respectively. This estimate is called broad sense heritability. It is valid for homozygous populations and lines. H_{bs} is estimated as follows:

1. It can be computed from the estimates of V_g and V_e obtained from a replicated trial based on a large number of homozygous lines/cultivars.
2. The estimates of Vg and Ve may be obtained from the variances of F_2, F_1, P_1 and P_2 generations as follows:

   $$v_e = \frac{v_{F1} + v_{P1} + v_{P2}}{3}$$

   $$v_g = v_{F2} - v_e$$

   These estimates are then used for computing heritability.
3. Heritability estimate may also be obtained from the estimate of regression of progeny means on the parental values.

In cross-pollinated crop plants,

$H_{bs} = 2b$, when the parents are non-inbreds.

In self-pollinated crops, the estimate depends on the degree of inbreeding of the population and the formula (Smith and Kinman, 1965; Kuhn *et al.*, 1984) is:

$$H_{bs} = \left(\frac{\overline{X}}{\overline{Y}}\right)\left(\frac{b}{2\gamma_{xy}}\right)$$

where, H_{bs} = broad sense heritability, $\overline{X}$ = mean of parental population, $\overline{Y}$ = unweighted mean of the progeny population, b = regression of progeny means on the parental values, γ_{xy} = co-efficient of inbreeding of the progeny (The value will be $\frac{1}{2}, \frac{3}{4}, \frac{7}{8}, \frac{15}{16}, \frac{31}{32}$ etc. for F_2, F_3, F_4, F_5, F_6, etc. generations).

The segregating generations of self-pollinated crops show additive, dominance and, to some extent epistatic components of genetic variance. From practical point of view, only the additive variance (v_A) is fixable and can be utilized by selection in plant breeding programme. Therefore, for segregating populations, heritability is estimated as the ratio of additive variance (v_A) to the phenotypic variance (v_P). Such an estimate is called as narrow sense heritability (H_{ns}) and mathematically:

$$H_{ns}(\%) = \frac{v_A}{v_P} \times 100$$

The additive variance (v_A) is estimated by a suitable technique. Now, the fraction of V_A appropriate for the generation, i.e., the generation of which the phenotypic variance (v_P) is used, is used for estimating H_{ns}. Variances of parents (P_1, P_2), F_1, F_2, and back-crosses (B_1, B_2) are used to obtain additive (v_A) and dominance (v_D) components of genetic variance. They are derived as follows:

$$v_{F2} = \frac{1}{2}v_A + \frac{1}{4}v_D + v_e = v_P$$

$$v_{B1} + v_{B2} = \frac{1}{2}v_A + \frac{1}{2}v_D + 2v_e$$

$$v_e = \frac{v_{P1} + v_{P2} + v_{F1}}{3}$$

$$v_A = 4v_{F2} - 2(v_{B1} + v_{B2})$$

$$v_D = 4(v_{B1} + v_{B2} - v_{F2}) - 4v_e$$

where v_{P1}, v_{P2}, v_{F1}, v_{F2}, v_{B1}, and are variances of P_1, P_2, F_1, F_2, B_1, and B_2 generations respectively; v_e= environmental variance, v_P= phenotypic variance.

The variance of F_2 generation (v_{F2}) of a cross has $1/2V_A$. Therefore, the estimate of narrow sense heritability (H_{ns}) for F_2 generation will be:

$$H_{ns}(\%) = \frac{\frac{1}{2}v_A}{v_{F2}} \times 100 \, (v_{F2} = \textit{phenotypic variance of } F_2)$$

Johnson *et al.* (1955) classified broad sense heritability (H_{bs}) as low (< 30 %), moderate (30-60 %) and high (> 60 %), while Robinson *et al.* (1966) categorized narrow sense heritability (H_{ns}) as low (5-10 %), medium (10-30 %) and high (> 30 %). Heritability provides an indication of the expected effectiveness of selection for the concerned trait. If a character has high heritability, selection for this trait is fairly easy and quite effective. However, selection for a trait with low heritability would be considerably difficult and almost impractical.

11.4 Genetic Advance

When a genetically variable population is subjected to selection, a proportion of the population with extreme phenotype is selected. The mean performance of the progeny of the selected plants is expected to be higher than that of the original or base population. This increase in performance is called genetic advance. **Improvement in the mean genotypic value of the selected families over that of the base population is known as genetic advance (GA).** It is, mathematically, expressed as:

$$GA = k \times \sqrt{v_P} \times H_{bs}$$

where, k = selection differential, $\sqrt{v_P}$ = phenotypic standard deviation (SD) of the base population for the concerned trait, H_{bs} = broad sense heritability of the trait (which is not in per cent).

The value of selection differential varies with intensity of selection (i.e. the percentage of plants selected) only. Usually, the best 1, 2 or 5 % plants are selected from the original population. In that case the value of k would be 2.64, 2.42 and 2.06, respectively. Phenotypic standard deviation is estimated on the original population, that simply measures the amount variability present in that population. Obviously, if heritability (H_{bs}) is high, phenotypic standard deviation (SD) is high (meaning variability is more) and the top 1% plants are selected, it will give maximum genetic advance. Such GA estimates are applicable to homozygous base populations.

In case of segregating generation, GA is estimated using narrow sense heritability (H_{ns}) and the expression of GA is as follows:

$$GA = k \times \sqrt{v_P} \times H_{ns}$$

In general, the realized values of GA are lower than the expected ones, particularly in case of segregating generations. Hence, ordinarily GA estimates indicate upper limit attainable through selection.

GA has the same unit as of the concerned trait. It may also be expressed as percentage of the base population mean. It is mathematically:

$$\text{GA (as \% of base population mean)} = \frac{\text{GA}}{\overline{\text{X}}} \times 100$$

where $\overline{X}$ = mean of the concerned trait in the base population.

The range of genetic advance is classified by Johnson *et al.* (1955) as low (< 10 %), moderate (10-20 %) and high (> 20 %).

Heritability and genetic advance are important selection parameters. Heritability along with genetic advance are normally more helpful in predicting the gain under selection than heritability alone. However, it is not necessary that a trait showing high heritability will also exhibit high genetic advance (Johnson *et al.* 1955).

11.5 Interpretation of Heritability along with Genetic Advance

The heritability along with genetic advance is interpreted as follows (Table 11.1):

1. High heritability accompanied with high genetic advance indicates that heritability is due to additive (or fixable) gene action and selection may be effective.
2. High heritability accompanied with low genetic advance indicates that heritability is due to non-additive gene action and selection may be ineffective.
3. Low heritability accompanied with high genetic advance indicates that heritability is due to additive gene action and low heritability is due to high environmental influence; and selection may be effective.
4. Low heritability accompanied with low genetic advance indicates that heritability is due to non-additive gene action and selection may be ineffective.

Table 11.1. Interpretation of heritability along with genetic advance

Heritability	Genetic Advance	Gene Action	Selection
High	High	Additive	Effective
High	Low	Non-additive	Ineffective
Low	High	Additive	Effective
Low	Low	Non-additive	Ineffective

11.6 Gene Action

Gene action may be defined as the manner in which a gene or set of genes affects the phenotypic expression of a polygenic character. It is affected by a number of factors such as nature of the material, mode of pollination, linkage and the design of study. A study of homozygous genotypes will reveal only additive and additive × additive gene actions. Dominance, additive × dominance and dominance × dominance gene actions are possible when some of the concerned genes are in heterozygous state. Self-pollinated populations are naturally homozygous, and exhibit only additive and additive × additive gene actions. On the other hand, cross-pollinated populations are highly heterozygous, and exhibit all types of gene actions. In the coupling phase of linkage, the estimates of the additive and dominance components are greater than their actual values. In the repulsion phase of linkage, the estimate of dominance component is greater and that of additive one is smaller than the actual value. The mating designs/

schemes of estimation differ to some extent in the types and the precision of estimates obtained.

11.7 Breeding Strategies

The choice of appropriate crop improvement methods depends on the gene action involved in the expression of a polygenic character. The breeding strategies adopted in different gene actions are:

1. Additive gene action is pre-requisite for genetic gain under selection, because this is the only genetic variance that responds to selection. *If there is preponderance of additive gene action, breeding strategies should be pureline selection and mass selection in **self-pollinated crops** and development of synthetic and composite cultivars in **cross-pollinated crops**.*
2. Non-additive gene action is pre-requisite for heterosis breeding. *If there is preponderance of non-additive gene action, the breeding strategy should be development of hybrids.*
3. *If both additive and non-additive gene actions are of equal magnitude, the breeding strategy should be population improvement for development of superior lines with several desirable genes.*

11.8 Brief References

Dabholkar 1999, Falconer 1989, Jain 1992, Kute and Aher 2013, Kute and Shinde 2016, Mather and Jinks 1982, Nadarajan and Gunasekaran 2005, Narain 1990, Singh BD 2011, Singh Phundan 2011, 2013a, Singh and Narayanan 2013, Singh and Chaudhary 1985, Singh and Pawar 2005, Thirugnanakumar *et al.* 2012.

Review Questions

Part A

Answer the following questions true or false.

1. Dominance variance is utilized in hybrids. True/False
2. Additive variance is non-fixable. True/False
3. Fisher (1918) divided genetic variance into two components. True/False
4. Heritable fixable variance includes only additive variance. True/False
5. Linkage does not influence gene action. True/False

Part B

Answer the following questions.

1. Define gene action.
2. Mather (1949) classified genetic variance into ________ and ________ components.
3. What is fixable variance?
4. ________ gene action is pre-requisite for genetic gain under selection.
5. Non-additive gene action is pre-requisite for recombination/heterosis breeding.

Part C

Write a brief note on each of the following.

1. Breeding strategies.
2. Gene action.
3. Heritability.
4. Dominance variance.
5. Additive variance.

12

Selection Basis of Crop Breeding

'All the crops we are familiar with today are the result of human selection and intervention.'

—VIB

12.1 Introduction

Allowing certain plants from the population to reproduce, while barring others, comprises selection. This permission for reproduction may be granted by nature or by the plant breeder. In the other words, both nature and breeder exert selection pressure. Natural selection operates through differences of fertility among the adult individuals of parent generation as well as through differences of viability among the individuals of offspring generation. Artificial selection, i.e., the selection from action of the plant breeder in the choice of parents, produces its change of gene frequency by separating the adult individuals of parent generation into two groups, *viz*., the selected and the discarded, that differ in gene frequency. So, **selection simply implies the retaining of desirable types or culling of unwanted or undesirable types.**

In connection with selection it is important to recognize that:

i) Selection can act effectively on heritable differences.

ii) Selection cannot create new variability but can act on variability that already exists in the population.

Selection tends to change the gene frequency. Effect of selection is to change the frequencies with which various genes or combinations of genes occur in the population. According to Allard (1960) three phenotypically observable responses to selection are:

i) Change in the proportion of previously existing genotypes (genotype frequency), accompanied by shift in the population mean.

ii) Appearance of new genotypes in the selected population, as compared to the base population.

iii) Reduction in the variability for selected character in the selected population

12.2 Prerequisites of (Phenotypic) Selection

Selection (phenotypic) can operate subject to fulfilment of certain conditions. They are called prerequisites of selection and are discussed below.

12.2.1 Presence of Character

The required character must be present in the population. For example, if we are breeding for wilt resistance in tomato, the character must be present in some accession(s) of the germplasm. Otherwise, selection is not possible.

12.2.2 Genetic Diversity of the Character

Selection is effective when the character shows sufficient variability. If the magnitude of variability is more, the rate of success is higher.

12.2.3 Heritability of the Character

Selection works on heritable variation of a character. Heritability indicates the heritable portion of value of the character. If the heritability is higher, selection will be more effective because of low environmental effect and more genetic advance per generation.

12.2.4 Population Size

To obtain full representation of whole of the variation in a character a large population is required. In a cross, if 10 genes are segregating, to get homozygous individuals for all the genes a minimum size of F_2 population is 1024 plants (Allard 1961). Yield and yield contributing characters are polygenic ones. So, a quite large size of population must be maintained. Moreover, in hybridization programme, selection of such characters is delayed until F_3 or F_4 generation when they attain almost homozygosity.

12.3 Selection Methods

The choice of selection methods depends on the technical and financial resources at hand. *The primary objective of any selection method is to increase the frequency of desirable genes and genotypes in the breeding population*. Technical steps vary depending on the mode of reproduction of the crop. **Single plant selection is commonly practised in self-pollinated crops, population improvement in cross-pollinated crops and clonal selection in vegetatively propagated crops.** The selection approach may vary depending on the crop and breeding method being followed.

12.4 Character-Based Selection Approach

Each character is product of single or many genes. The characters that are governed by one or few genes are called **qualitative characters**. On the other hand, the characters that are governed by many genes are known as **quantitative (polygenic) characters**.

12.4.1 Qualitative Characters

They produce discrete phenotypic classes and are little influenced by genetic background and environment. Hence, simple phenotypic selection is very effective in improvement of such characters, e.g., resistance to certain disease, pest, and abiotic stresses, and quality traits.

12.4.2 Quantitative (Polygenic) Characters

Most of the yield and yield contributing characters are quantitative in nature. They are governed by polygenes that are highly influenced by environment. So, they do not show distinct classes. Unlike qualitative character, here the effect of individual gene is very small and for expression of a character, genes act cumulatively. These genes may be scattered over a population in different individuals. They can be assembled by hybridization. *While handling segregating populations, the selection of polygenic traits is postponed till* F_3 *or* F_4 *generation when many of them become homozygous.*

12.4.3 Selection of Linked Characters

When two or more characters are linked, it may be due to pleiotropism or close association on the chromosome. In pleiotropism, selection for dissociating the genes is futile. In true linkage, the linkage may be broken by hybridization with a suitable parent and growing larger F_3 and F_4 generations with a view to recovering higher number of recombinants. In few recombinants the linkage might be broken and desirable genotypes are obtained and selected. *Sometimes hybridization is supplemented with mutagenesis to break the linkage.*

12.4.4 Selection Based on Correlation of Characters

Often correlations of main character with contributing characters are estimated. The positively correlated contributing characters are selected to increase the main character. Selection is further assisted by partitioning of correlation coefficients into direct and indirect effects of the characters by path-coefficient analysis. High direct effect indicates simple selection of the character. In case the direct effect is low, other characters influencing the character through indirect effects must be simultaneously considered in selection to bring improvement in the target character.

12.5 Marker-Assisted Selection (MAS)

The selection step in a hybridization programme is the most crucial. Certain characters such as disease resistance cannot be evaluated based on the plant's appearance. For this, the disease must first appear in the field. But this does not happen always. So, breeders often resort to creation of artificial epiphytotics to be able to select the least sensitive plants from the population. Alongside the presence or absence of a particular qualitative character—which is relatively easy to select during a breeding programme—there are also quantitative characters such yield or growth speed. These quantitative characters are often governed by polygenes, which makes breeding a lot more difficult and above all, time-consuming.

Due to increasing knowledge about how plant genes work and their role in plant growth and development, it is known for a large number of characters which genes are encoding. By detecting a **specific DNA fragment** (also called a '**marker**') which is linked to a character (e.g. disease resistance) in a segregating population, it can be determined at a very early stage whether or not the plant will be disease resistant, without needing to infect the plant. Therefore, hybridization programmes can be conducted much more efficiently by **selection based on DNA** (also called '**marker-assisted selection**') than **selection based on appearance** (also called '**phenotypic selection**'). This is possible when the genes that have a direct influence on the character are known or, the **neighbouring genes** (also called '**flanking genes**') are known. For each progeny of a certain hybrid, it can be thus determined, with the help of molecular DNA marker techniques, which combination of genes is present in the DNA. The progeny with the most beneficial combination of genes can thus be quickly identified and used for further breeding. So, *marker-assisted selection (MAS) makes use of biotechnological know-how to diagnose the presence or absence of specific genes* (Figure 12.1).

MAS can also select recessive characters hidden in heterozygous genotypes. MABC (marker-assisted back-crossing) is a simple application of MAS that is widely used in breeding programmes. One or a few targeted quantitative trait loci (QTLs) can be easily introgressed into elite lines through MABC. In conventional breeding, during backcrossing, the progeny is selected based on the target phenotype, whereas in MABC, the selection is based on the DNA marker genotypes. In general, both foreground and background selections are conducted at the same time. Foreground selection confirms the presence of the desired allele from the donor parents until the final backcross, whereas background selection eliminates unwanted genomic introgression from the donor parent. Thus, in background selection, markers distributed across the whole genome are required. A well-known example is IRRI's efforts to develop rice cultivars through backcross breeding using MABC.

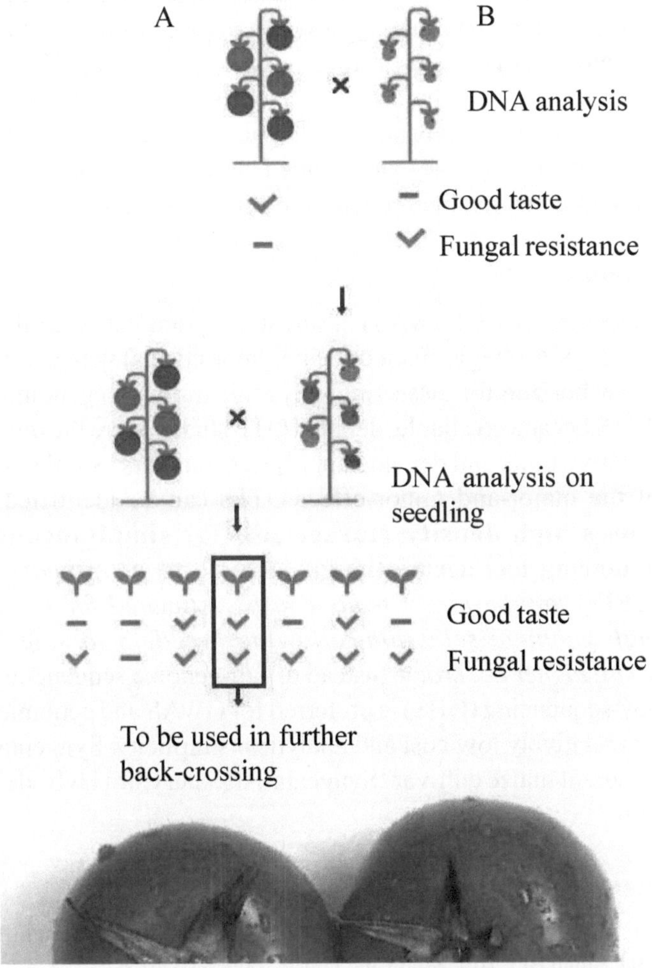

Figure 12.1. Schematic representation of marker-assisted selection in tomato; A = fungus-susceptible but good tasting tomato plant, B = fungus-resistant but poor tasting tomato plant. A is crossed with B. Using DNA analysis techniques tomato seedlings can be selected that are both fungus-resistant and good tasting. Source: VIB 2016.

Many DNA markers (gene-based or simply linked to the trait) have been developed, mainly for qualitative traits. Genome sequences for most major crops and great progress in NGS (next-generation sequencing) technologies have accelerated the systematic development of useful markers on a large scale. However, DNA markers have limited applications for selecting polygenic traits of low heritability.

Now-a-days, MAS has almost become standard in crop improvement programmes. In most cases, the DNA to be analysed is isolated from a piece leaf tissue from young plant. It can even be done at earlier stages. Breeders can now use 'seed chipping': a technology that allows a small piece of the seed to be shaved off to extract the DNA without harming the embryo or its capacity to germinate. With seed chipping, it can therefore be determined at an extremely early stage, which seeds contain which beneficial characters. The less suitable seeds can be discarded before even they are planted. It speeds up breeding programmes and makes for considerable reduction in selection costs.

12.6 Genomic Selection

The genome-wide association study (GWAS), a statistical examination of the association of an array of SNPs (single-nucleotide polymorphisms) with a trait of interest, opened a new horizon for dissecting polygenic traits into genomic information. Until GWAS became available, detailed QTL studies were the only way to interpret quantitative traits and develop appropriate markers. Similar to GWAS, where all of the major-and minor-effect QTLs can be identified, **genomic selection uses high-density marker sets for simultaneous selection of trait-enhancing loci across the genome**. *Genomic-estimated breeding values (GEBVs) for the target traits can be estimated for each breeding line through genomic selection, allowing breeders to select promising genotypes with higher accuracy.* Instead of full-genome sequencing by NGS, genotyping-by-sequencing (GBS) is preferred for GWAS and genomic selection because of its relatively low cost and analytical simplicity. Syngenta developed a drought-tolerant maize cultivar 'Syngenta-Artesian Corn Hybrids' using genomic selection.

12.7 Brief References

Dabholkar 2006, Koh *et al.* 2015, Nadarajan and Gunasekaran 2005, Ram 2014, Sharma JP 2009, Sharma JR 1994, Singh BD 2011, Singh and Pawar 2006, VIB 2016.

Review Questions

Part A

Answer the following questions true or false.

1. Single plant selection is commonly practised in cross-pollinated crops. True/False
2. Marker-assisted selection makes use of biotechnological know-how. True/False
3. Genomic selection uses low-density marker sets. True/False

4. Selection can act effectively on heritable differences. True/False
5. Syngenta-Artesian Corn Hybrids are developed using MAS. True/False

Part B

Answer the following questions.

1. What is selection?
2. GBS stands for______________.
3. What do you understand by marker-assisted selection?
4. Why do you select polygenic traits in F_3 or F_4?
5. Each trait is product of __________ or _____________ genes.

Part C

Write a brief note on each of the following.

1. Marker-assisted selection.
2. Genomic selection.
3. Character-based selection approach.
4. Prerequisites of selection.
5. Responses to selection.

13

Heterosis and Inbreeding Depression

13.1 Introduction

F_1 plants produced by crossing two parents with contrasting characters are called hybrids. Hybrids generally exhibit increased vigour and fitness over the parents. This phenomenon is called **hybrid vigour** or **heterosis**. Thus, *heterosis is the phenomenon in which the hybrid of two genetically dissimilar parents shows increased vigour at least over the mid-parental value*. Heterosis is just the reverse of inbreeding depression, which results after selfing the plants of the cross-pollinated crops. Kolreuter (1761-66) was the first to report hybrid vigour in the hybrids of tobacco, *Datura*, etc. Most of the early plant hybridizers including Mendel (1865) had noticed this phenomenon. Shull (1914) coined the term heterosis from two Greek words, *hetero* (different) and *osis* (condition), to replace his own earlier word 'stimulus of heterozygosis'. *In current usage, heterosis and hybrid vigour are used as synonyms and interchangeable.*

13.2 Types of Heterosis

Depending on the nature, origin, adaptability, reproducibility and non-reproducibility, heterosis can be categorized into two: (1) euheterosis (true heterosis) and (2) pseudo-heterosis (false heterosis, i.e., luxuriance).

13.2.1 Euheterosis

This is the true heterosis that is inherited. It is further divided into (i) mutational euheterosis and (ii) balanced euheterosis.

13.2.1.1 Mutational Euheterosis

This is the simplest type of euheterosis. It results from the sheltering or shadowing of the deleterious recessive mutant genes by their adaptively superior dominant alleles in a population of sexually reproducing cross-pollinated crop. Most mutations are lethal, recessive and adaptively inferior. But mutational euheterosis protects them from quick elimination from population.

13.2.1.2 Balanced Euheterosis

This is the true type of heterosis. It arises out of balanced gene combinations with better adaptive value and usefulness. It is utilized in crop breeding programme for evolving hybrid cultivars.

13.2.2 Pseudo-Heterosis

This is basically a phenomenon by which the crossing of the two parental forms results in an accidental, excessive and unadaptable expression of temporary vigour and vegetative growth (luxuriance). Luxuriance cannot be utilized for breeding hybrid cultivars.

13.3 Estimation of Heterosis

Heterosis is estimated in three different ways, namely, (1) over mid parent, (2) over better parent, and (3) over standard commercial cultivar/hybrid. Thus, on the basis of estimation, heterosis is of three types as given below:

13.3.1 Mid Parent / Relative / Average Heterosis

When the heterosis is estimated over mid parent (mean value of the two parents), it is known as mid parent heterosis, relative heterosis or average heterosis.

13.3.2 Better Parent Heterosis / Heterobeltiosis

When the heterosis is estimated over the superior or better parent, it is known as better parent heterosis or heterobeltiosis. The term heterobeltiosis was used by Bitzer *et al.* (1968) to describe the improvement of heterozygote over the better parent of the cross.

13.3.3 Standard / Useful / Economic Heterosis

When the heterosis is estimated over standard commercial cultivar/hybrid, it is known as standard heterosis, useful heterosis or economic heterosis. The term useful heterosis was used by Meredith and Bridge (1972).

Heterosis is usually expressed as percentage and computed as follow:

$$\text{Mid parent heterosis (\%)} = \frac{\overline{F_1} - \overline{MP}}{\overline{MP}} \times 100$$

$$\text{Better parent heterosis (\%)} = \frac{\overline{F_1} - \overline{BP}}{\overline{BP}} \times 100$$

$$\text{Standard heterosis (\%)} = \frac{\overline{F_1} - \overline{CC}}{\overline{CC}} \times 100 \text{ (over standard commercial cultivar)}$$

$$\text{Standard heterosis (\%)} = \frac{\overline{F_1}\text{-}\overline{SH}}{\overline{SH}} \times 100 \text{ (over standard commercial hybrid)}$$

Where, $\overline{F_1}$ = *Mean value of F_1 hybrid*

$\overline{MP}$ = *Mean value of average of two parents involved in the cross*

$\overline{BP}$ = *Mean value of better parents involved in the cross*

$\overline{CC}$ = *Mean value of standard commercial cultivar*

$\overline{SH}$ = *Mean value of standard commercial hybrid*

Testing of significance of heterosis by 't' test:

$$\text{Standard Error (Relative Heterosis)} = \sqrt{\frac{3EMS}{2r}}$$

$$\text{Standard Error(Heterobeltiosis and Standard Heterosis)} = \sqrt{\frac{2EMS}{r}}$$

where, EMS = Error mean square

r = Number of replications

$$\text{t value (Relative Heterosis)} = \frac{\overline{F_1} - \overline{MP}}{SE}$$

$$\text{t value (Heterobeltiosis)} = \frac{\overline{F_1} - \overline{BP}}{SE}$$

$$\text{t value (Standard Heterosis)} = \frac{\overline{F_1} - \overline{CC}}{SE} \; or \; \frac{\overline{F_1} - \overline{SH}}{SE}$$

Compare the calculated 't' value with table 't' value at 5 % or 1 % at error degree of freedom.

or

The significance of heterosis is tested with the help of critical difference (CD) as below:

$$CD = SE_{(Diff)} \times t_{\text{at 5\% or 1\% at error df}}$$

The difference between F_1, and parent (mid parent, better parent) or standard cultivar/hybrid used for estimation of heterosis is taken into account cross-wise. When the difference between mean value of F_1, and parent or standard cultivar/hybrid is greater than CD, *or* calculated 't' value is greater than table 't' value, it is considered significant.

13.4 Manifestations of Heterosis

Heterosis is manifested in different ways. The main characters of economic importance affected by heterosis are briefly summarized below:

1. Increased yield is the main effect of heterosis.
2. Increased reproduction ability is always the result of heterosis.
3. Hybrids are usually more vigorous, faster growing and larger in size than their parents.
4. Many hybrids exhibit improved quality.
5. Many hybrids exhibit earlier flowering and maturity.
6. Some hybrids show increased resistance to biotic stresses.
7. Hybrids are usually more adapted to environmental variations.

13.5 Genetic Factors Affecting Heterosis

There are four main genetic factors that affect the amount of heterosis in crop plants.

i) *Mode of Pollination*: The level of heterosis is usually higher in cross-pollinated species than self-pollinated species.

ii) *Genetic Diversity (Distance) of Parents:* Maximum heterosis occurs at an optimal or intermediate level of parental diversity.

iii) *Genetic Base of Parents*: The higher heterosis occurs when parents have broad genetic base.

iv) *Adaptability of Parents*: In many crops, heterosis is associated with wide adaptability because there is close association between adaptability and genetic base.

13.6 Fixation of Heterosis

Heterosis is a property exhibited by F_1 and it is difficult to fix it. However, four methods are usually used to fix heterosis.

a) *Vegetative Propagation*: Vegetative propagation is an effective method to conserve heterosis. This method can be used in potato, sweet potato, banana, etc.

b) *Apomixis*: Apomictic progeny developed from maternal diploid cells usually conserve heterosis, since the progeny is identical to the mother plant. Fixation of heterosis by apomixis is possible in citrus, rose, etc.

c) *Balanced Lethal System*: Certain types of balanced lethal systems, in which homozygous organisms do not survive, result in the survival of populations consisting of heterozygous organisms only. The heterozygous organisms are usually as vigorous as F_1. This type of fixation of heterosis has been reported from *Oenothera*.

d) *Polyploidy*: Heterosis can be fixed by chromosome doubling or polyploidization, especially in interspecific and intergeneric hybrids.

13.7 Genetic Basis of Heterosis

There are three possible genetic causes of heterosis, *viz*. dominance, overdominance and epistasis.

13.7.1 Dominance Hypothesis

First proposed by Davenport (1908) and later expanded by Bruce (1910) and Keeble and Pellew (1910), this hypothesis is based on the hiding of deleterious effects of recessive alleles by the dominant alleles in the hybrid. Thus, the dominant alleles, that are favourable, mask the effect of recessive alleles and bring in heterotic effect in the hybrid.

For example,

Parent 1 × Parent 2
(AABBccdd) (aabbCCDD)
2 dominant genes 2 dominant genes
↓
F_1
(AaBbCcDd)
4 dominant genes

Each parent has only two dominant genes, but the F_1 has four dominant genes and they mask the effect of recessive alleles and show heterosis.

Objections: Two major objections have been raised.

a) **Why not true breeding heterotic lines?** If above hypothesis is correct, it should be possible to recover a plant with genotype AABBCCDD (as per example) that is highly vigorous as F_1 but true breeding.

But this objection was removed by Jones (1917) in his theory '**Dominance of Linked Genes Hypothesis**'. He pointed out that a large number of dominant genes distributed over different chromosomes are involved in producing heterosis. The favourable and unfavourable genes may be linked. So, true breeding plants with all favourable (dominant) loci cannot be obtained.

b) **Why symmetrical distribution?** If heterosis is due to dominance of independent genes, the F_2 distribution curve for heterotic character should be skewed towards dominant genes rather than symmetrical.

The objection was removed by Collins (1921) who pointed out that heterosis is governed by large number of genes, so that even in the absence of linkage skewness will not appear.

13.7.2 Overdominance Hypothesis

This hypothesis was first proposed by Shull (1908) and East (1908) independently and supported by East (1936) and Hull (1945). The basis of this hypothesis is that the heterozygote (Aa) is superior to both homozygotes (AA or aa) and the amount of heterosis depends on the amount of heterozygosity in the hybrid. Hull (1945) observed in maize that if two inbreds A and B yielded 10 and 12 quintals, their F_1 yielded more than 22, say 25 to 27 quintals. This could not possibly be explained with dominance hypothesis. But this could perhaps be better explained on the basis of overdominance hypothesis.

Objections: Most support to this hypothesis comes from single locus heterosis. Doubts have been raised that what is true for qualitative genes may not hold good for quantitative genes, i.e., heterosis in quantitative traits like yield may not be single locus heterosis.

13.7.3 Epistasis

It is well established that the occurrence and magnitude of heterosis has positive association with presence and magnitude of non-allelic interaction, i.e., epistasis. Particularly, dominance × dominance non-allelic interaction may contribute to heterosis.

Out of above three explanations of heterosis, the dominance hypothesis is most widely accepted. Overdominance and epistasis also operate in the manifestation of heterosis.

13.8 Inbreeding Depression

The mating between closely related individuals is called ***inbreeding***. Self-pollination is the closest form of inbreeding. Other forms are sibmating and half-sibmating. **Inbreeding depression refers to decrease in fitness and vigour due to inbreeding.** The main features of inbreeding depression are: inbreeding results (1) from mating between closely related individuals and (2) due to fixation of unfavourable recessive genes in F_2.

13.8.1 Estimation of Inbreeding Depression

Inbreeding depression is estimated when both F_1 and F_2 populations of the same cross are available. It is measured as follows:

$$\text{Inbreeding Depression (\%)} = \frac{\overline{F_1} - \overline{F_2}}{\overline{F_1}} \times 100$$

Where, $\bar{F}_1$ = Mean value of F_1 hybrid

$\bar{F}_2$ = Mean value of F_2

The significance of inbreeding depression is tested with the help of 't' test or critical difference (CD) as follows:

$$\text{Standard Error (SE)} = \sqrt{\frac{2\text{EMS}}{r}}$$

$$\text{t value for Inbreeding Depression} = \frac{\overline{F_1} - \overline{F_2}}{SE}$$

Or

$$\text{CD} = \text{SE}_{(\text{diff})} \times t_{\text{at 5 \% or 1 \% at error df}}$$

The difference between $\overline{F_1}$ and $\overline{F_2}$ is taken cross-wise. When the difference is greater than CD or 't' calculated is greater than 't' table value, it is considered as significant.

13.9 Inferences

The following inferences about type of gene action involved in the expression of the quantitative characters can be drawn from the estimates of heterosis and inbreeding depression:

1. If high heterosis is followed by high inbreeding depression, it indicates presence of non-additive gene action.
2. If high heterosis is followed by low inbreeding depression, it indicates presence of additive gene action.
3. If high heterosis is negative in F_1 and there is increase in F_2, it also indicates presence of additive gene action.
4. The genes with lack of dominance will not exhibit heterosis in F_1 but may show increase in performance in F_2 (low inbreeding depression) due to fixation of genes, i.e., additive gene action.

13.10 Brief References

Allard 1960, Chaudhary 2014, Dabholkar 2006, Kute and Aher 2013, Kute and Shinde 2016, Mukherjee 1995, Mohanan 2010, Nadarajan and Gunasekaran 2005, Ram 2014, Sharma JP 2009, Sharma JR 1994, Singh BD 2011, Singh Phundan 2011, 2013a, Singh and Narayanan 2013, Singh and Pawar 2006.

Review Questions

Part A

Answer the following questions true or false.

1. Inbreeding depression is estimated when F_1 or F_2 populations are available. True/False
2. Dominance hypothesis was first proposed by Bruce (1910). True/False
3. Mid parent heterosis is also known as relative heterosis. True/False
4. Vegetative propagation is not an effective method to conserve heterosis. True/False
5. The dominance hypothesis is most widely accepted. True/False

Part B

Answer the following questions

1. What is luxuriance?
2. Heterosis is estimated in three/five different ways.
3. Define heterobeltiosis.
4. What is inbreeding?
5. What do you understand by useful heterosis?

Part C

Write a brief note on each of the following.

1. Types of heterosis.
2. Estimation of heterosis.
3. Fixation of heterosis.
4. Dominance hypothesis.
5. Inbreeding depression.

14

Hybridization Techniques and Consequences

14.1 Introduction

The mating or crossing of genetically dissimilar individuals is called ***hybridization***. Hybridization results in recombination that in turn widens the spectrum of variation among individuals. So, the reason for genetic diversity in progenies of crosses is the formation of new combinations of genes. This newly created variability is then channelized into development of better cultivars.

The importance of pure line selection as a method of breeding started decreasing when plant breeders felt that the genetic variability existing in the land races of self-pollinated crops had been almost completely exploited and there was little chance for further improvement in these crops by using pure line breeding. Therefore, no option was left for plant breeders except to cross diverse individuals and combine together desirable characters in a single genotype from two or more parents. The planned hybridization which could avoid self-pollination and chance cross-pollination, soon became the main feature of almost all plant breeding programmes.

14.2 Goals of Hybridization

The goals of hybridization are briefly discussed below:

14.2.1 Creation of Genetic Variation

This is the most common goal of hybridization. The F_1 plants obtained from a cross are heterozygous for all the genes for which its parents were dissimilar. Segregation of and recombination among the heterozygous genes of F_1 will generate genetic variability in F_2 and subsequent generations. This variability is utilized to achieve the following goals:

14.2.1.1 Combination Breeding

When the main goal of hybridization is to transfer (or combine) into a single cultivar one or more traits from another cultivar/other cultivars, it is called ***combination breeding***. This increased yield of the cultivar improved in this approach results from the correction of weakness or defects, e.g., susceptibility to a disease, of the parent cultivar. **Backcross method is the most suited to achieve this goal.**

14.2.1.2 Transgressive Breeding

The main goal of this approach is to obtain transgressive segregants for yield and/or other important traits. This is called ***transgressive breeding***. Hence, the parents used in such a programme must be genetically diverse and combine well with each other. **The pedigree method of breeding is the most commonly used for this purpose.**

14.2.2 Hybrid Cultivars

Hybrid cultivars are the F_1 generation of a cross between genetically dissimilar genotypes or lines with good combining ability. Thus, process of hybridization is an integral component of hybrid cultivars. Heterosis is best utilized by hybrid cultivars. In case of hybrid cultivars, the F_1 generation from a suitable cross is used for raising the commercial crop. Therefore, seeds of hybrid cultivars have to be produced every year by crossing their respective parents.

14.3 Types of Hybridization

Hybridization is classified into two groups on the basis of taxonomic relationship between the parents used for hybridization: (1) intraspecific and (2) distant hybridization.

14.3.1 Intraspecific Hybridization

Hybridization between lines/strains/ cultivars of the same species is called ***intraspecific*** *or* ***intervarietal hybridization***. It is the most commonly used hybridization approach in plant breeding. On the basis of the number of parents used, it is of two types: (i) simple and (ii) complex crosses.

14.3.1.1 Simple Cross

When two parents are crossed to produce the F_1 {e.g. (A × B)}, it is called a ***simple cross***. In case of a hybrid cultivar, such a cross is commonly referred to as **single cross**.

14.3.1.2 Complex Cross

When more than two parents are crossed to produce the hybrid, it is known as ***complex, convergent, composite*** *or* ***multiple cross***. In **three-way** or **triple**

cross three parents are involved to produce a hybrid {e.g. (A × B) × C}. In **double cross**, four parents are involved to give rise to a hybrid {e.g. (A × B) × (C × D)}. Multiple crosses are done to bring the desirable genes from several parents into a single hybrid.

14.3.2 Distant Hybridization

Hybridization between the members of different species of a genus or hybridization beyond species level is called ***distant*** *or* ***wide hybridization***. Thus, it may be **interspecific (intrageneric)** or **intergeneric**. When conventional methods of hybridization fail, para-sexual methods are used in such cases.

14.4 Techniques of Hybridization

The action plan for hybridization consists of two phases: (a) planning phase and (b) execution phase.

14.4.1 Planning Phase

It includes socio-economical importantce of the problem, resources at disposal, number of crosses to be attempted, raising and handling of segregating material.

a) Problem or objectives

Hybridization is a part of long-term crop improvement programme. Its results will be available after spending a lot of time and resources. So, thorough analysis of problems to be rectified or improvements to be incorporated should be made before proceeding for this large investment-based project.

b) Resources at disposal

Material should be generated to the magnitude for which land, water, management and scientific supervision can be adequately provided.

c) Number of crosses to be undertaken

Too many crosses may generate large amount of segregating material in subsequent generations creating problem of handling. Too fewer crosses may not generate desired recombinants leading to failure. Optimum number will come from experience of the plant breeder. The number is also related to objectives (one or many), nature of crop (self-or cross-pollinated) and genetics of the character (monogenic, oligogenic or polygenic) in addition to resources available.

d) Handling of Segregating Material

This should be pre-decided that when the selection for desired genotypes has to be carried out. Selection in F_2 generation should need limited space for further generations. In case of polygenic characters or characters having low heritability, selection is postponed until F_3 or F_4 generations where comparatively larger space would be required to raise the segregating material. Early generation (F_2) selection has been advised as imperative step but must be exercised carefully giving the segregants competitive and appropriate biotic and abiotic environments to make the selection process more effective and reliable.

14.4.2 Execution Phase

Selection of parents, evaluation of parents and making of crosses are important steps of execution phase.

14.4.2.1 Selection of Parents

It is the most important step in hybridization. Mostly success or failure of a crossing programme may be attached with this step. The parents selected for hybridization must together possess all the characters, and in sufficient intensity, which the breeder wishes to improve in the new cultivar. At least one parent must be well adapted in the region for which the cultivar is to be developed. The other parent(s) should possess the traits which one wishes to improve in this well adapted cultivar. The parents should be genetically diverse if the aim is to obtain transgressive segregants. The general combining ability of the parents should also be considered, if the estimates for the same are available.

14.4.2.2 Evaluation of Parents

It is desirable to evaluate the performance of the selected parents in the area for which breeding is to be done. Evaluation provides information about the parents regarding: (1) performance in terms of yield and yield contributing traits, (2) reaction to the prevalent diseases and insect-pests, (3) presence of mechanical mixture and (4) existence of heterozygosity. Evaluation is particularly necessary in the case of introduced material.

14.4.2.3 Making Cross Combination

Raising of parental populations needs certain management cares so that seed setting is enhanced. Site for crossing block should be ploughed well and organic manures adequately incorporated. If required, micronutrients may be applied. Plants in hybridization blocks must be spaced wider than normal recommendation. There should be facility of irrigation with good quality water. It is desirable that parental populations are raised under most congenial management environment. If necessary, plant growth regulators and/or fortified micronutrient formulations may be used for enhancing seed set in crosses.

14.4.2.3.1 Emasculation

The removal stamens/anthers or inactivation of pollen grains, before pollen grains become mature without any harm to the gynoecium, is called ***emasculation.*** The purpose of emasculation is to prevent self-pollination in the flowers of the female parent. The different techniques of emasculation are: (i) hand emasculation, (ii) suction method, (iii) hot water method, (iv) cold water treatment, (v) selective alcohol treatment and (vi) genetic emasculation. They are briefly discussed below:

i) Hand Emasculation

In species with relatively large flowers, stamens or anthers are removed with a pair of forceps. Emasculation is usually done one day before the anthesis of flowers, ordinarily between 4-6 PM. Generally, younger buds and older flowers, even fruits close to the floral bud selected for emasculation are removed to avoid confusion. The selected bud is then opened with a pair of forceps and all anthers/stamens are carefully removed.

ii) Suction Method

This method is useful in species with small flowers, where hand emasculation is tedious. Emasculation is done in the morning on the day of anthesis just before or immediately after the flowers open. If needed, petals may be removed with a pair of forceps. The emasculation is done with a suction pump. A considerable amount (up to 15 %) of self-pollination may occur with this method.

iii) Hot Water Method

Treatment with hot water of a suitable temperature inactivates pollen grains without reducing female fertility. Hot water is carried in a thermos flask. It is generally highly effective but regulating water temperature is a problem.

iv) Alcohol Treatment

The entire inflorescence is immersed in alcohol for a brief period, then washed with water. The duration of the treatment should be strictly controlled. Otherwise, even slightly longer duration would damage gynoecium and reduce seed set.

v) Cold Water Treatment

Generally, the pollen grains are more sensitive to lower temperature and can be killed by giving selective low temperature treatment.

vi) Genetic Emasculation

Prevention of self-pollination by genetic means is called genetic emas-culation. This is achieved by using genetic or cytoplasmic-genetic male sterility, self-incompatibility or pistillate condition.

14.4.2.3.2 Bagging

The enclosing of the emasculated flowers or the whole inflorescence with suitable bags is called ***bagging***. Bagging is done to prevent chance cross-pollination. Bags may be made of paper, butter paper, glassine or fine cloth. Butter paper or vegetable parchment bags are the most commonly used. The bags are tied to the base of inflorescence or to the stalk of flower with thread, wire or pin. The bags are usually removed 2-3 days after pollination to prevent fungal growth.

14.4.2.3.3 Tagging

The emasculated flowers are tagged just after bagging. In most crops, circular tags of about 3 cm diameter or rectangular tags of 3 × 2 cm are used. The tags are attached to the base of the inflorescence or the stalk of flowers with the help of thread. Date of emasculation, date of pollination and names of ♀ and ♂ parents are mentioned on the tags with carbon pencil. The ♀ parent is written first.

14.4.2.3.4 Pollination

Transfer of mature and active pollen from the flowers of ♂ parent to the stigmas of emasculated ♀ flowers is called ***pollination***. Pollination is usually done between 7.30-10 AM the next day following emasculation.

14.4.2.3.5 Harvesting and Storing of F_1 Seed

The crossed heads/spikes/pods are harvested, threshed, dried and stored properly. Care should be taken to protect them from storage pests and moisture. Every precaution should be taken to prevent contamination from other seed. Seeds from each cross should be handled separately, and the original tags should be kept with them.

14.4.2.4 Raising of F_1 Generation

When the parents of a cross differ for a seedling character, the parent that has dominant form should be used as ♂. This allows identification of selfed seeds in F_1 itself. All the seedlings with recessive form of the seedling trait will be selfs. The number of F_1 plants should be a large as possible. But some authors feel that as few as 12 F_1 plants in a cross may be adequate. The F_1s of self-pollinated crops are usually allowed to self-pollinate. But in a backcross programme, the F_1s are backcrossed to one of their parents.

14.4.2.5 Further Handling of the Plants

Further handling the hybrids depends on the objective of the cross. In the case of hybrid seed production, the F_1 seeds are directly released to farmers. In case of combination breeding and transgressive breeding, F_2 is raised and the most appropriate selection programme is used.

14.5 Genetic Consequences of Hybridization

The F_1 hybrids are heterozygous for all the genes for which their two parents differ. Hence, heterotic F_1s are used as hybrid cultivars, if feasible. The F_1 and subsequent generations may be subjected to either selfing or backcrossing. Segregation of and recombination among heterozygous genes produce a large number of new gene combinations. The number of genotypes produced in F_2 increases very rapidly with an increase in the number of segregating genes; it is 3^n for n genes segregating in the F_1. The size of perfect F_2 population is 4^n. In a perfect F_2 population, each genotype is expected to be represented by at least one plant. Hence, it is virtually impossible to grow a prefect population since it may be expected to segregate for more than 10 genes. Consequently, recovery of a transgressive sergeant in F_2 or a later generation is only a matter of chance. There is a rapid decline in heterozygosity. Each generation of selfing reduces heterozygosity to ½ of that in the previous generation. So, there is a corresponding increase in homozygosity. The non-fixable components (dominance, additive × dominance and dominance × dominance) of genetic variance decline progressively with selfing. By the 7th or 8th generation of selfing, only additive and additive × additive components remain. Linkage between the segregating genes would not affect the degree of homozygosity. But it will increase the frequency of completely homozygous plants in a given generation.

14.6 Brief References

Mohanan 2010, Sharma JP 2009, Singh BD 2011, Singh and Pawar 2006.

Review Questions

Part A

Answer the following questions true or false.

1. Seeds of hybrid cultivars have to be produced alternate year. True/False
2. The pollen grains are less sensitive to lower temperature. True/False
3. A considerable amount of self-pollination may occur with suction method. True/False
4. Hybridization widens the spectrum of variation among individuals. True/False
5. The emasculated flowers are tagged just before bagging. True/False

Part B

Answer the following questions.

1. What is combination breeding?
2. Hybridization is classified into two/four groups on the basis of taxonomic relationship.
3. Define intraspecific hybridization.
4. What is a complex cross?
5. Bagging: explain.

Part C

Write a brief note on each of the following.

1. Emasculation.
2. Genetic consequences of hybridization.
3. Selection of parents.
4. Planning phase of hybridization.
5. Raising of F_1 generation.

15

Timelines in Conventional Crop Improvement Pre-Breeding and Breeding Procedures*

15.1 Introduction

Plant breeding aims at developing genetically improved crop cultivars with economic benefits for small-scale and commercial farmers. Population growth, dwindling agricultural land and global climate change present increasing risk to crop production. Consequently, plant breeding aims to constantly develop crop cultivars with improved yield and quality and tolerant to drought, diseases and pests. Use of genetically improved crop cultivars and better management practices are among the best strategies to increase food production and meet a projected doubling of food demand in the next 40 years. Both the public and private plant breeding sectors face several constraints, including a shortage of professional plant breeders and breeding technicians, limited budget, inadequate facilities and lack of access to information. Successful plant breeding requires adequate infrastructural investment, in addition to the actual breeding activities. The benefits of plant breeding research and development can only be realized on a long-term basis because of the inherent nature of the crops and the eminent breeding procedures involved. **Plant breeding involves two main activities, i.e., pre-breeding/ germplasm enhancement and cultivar development *per se*.** These interdependent activities are the controlling factors that determine the pace at which cultivars are released timeously and consistently to farmers. The two phases, in turn, depend upon various factors, e.g., breeding goals, genetics and agronomy of the crop, breeder's long-term objectives, availability of testing facilities and national cultivar-registration requirements. Despite these factors, there are certain clear steps and breeding procedures found in any conventional breeding programme. These include choice of parents, making crosses among

---

*Adapted with permission from Shimelis H and Laing M 2012. Timelines in conventional crop improvement: pre-breeding and breeding procedures. *Australian Journal of Crop Science* 6 (11):1542-1549.

chosen parents, and selection from recombined parents followed by extensive field testing of the candidate cultivars. Maintenance, multiplication and distribution of the seed are the ultimate stages of a breeding programme. Systematically outlined conventional breeding procedures, timeframes and estimated breeding generations are essential to objectively consider the ultimate timescales required to release new and improved cultivars of all crops. The information may assist plant breeding- students, educators, researchers, project advisors, administrators and funding agencies. Dear students, in this chapter we outline the timeframes of conventional breeding procedures of **clonally** (also called **vegetatively**) **propagated**, **self-fertilizing** (also called **self-pollinated**) and **cross-fertilizing** (also called **cross-pollinated**) crops.

15.2 Pre-Breeding

The Global Partnership Initiative for Plant Breeding Capacity Building (GIPB)/ FAO and Biodiversity International *use the term* ***'pre-breeding'*** *to describe the various activities of plant breeding research that have to precede the stages involved in cultivar development, testing and release*. Further, the Global Crop Diversity Trust *defined* ***pre-breeding*** as *'the art of identifying desired traits, and incorporation of these into modern breeding materials'*.

Pre-breeding aims to reduce genetic uniformity in crops through the use of a wider pool of genetic material to increase yield, resistance to pests and diseases, and other quality traits. Pre-breeding is routinely applied in commercial breeding programmes where desired traits are constantly sought and identified from source genotypes for use in cultivar development. Overall, pre-breeding includes all activities directed at identification of desirable crop traits and/or genes, and their subsequent transfer into a suitable set of parents for further selection. The procedure identifies useful character(s) or genes that can be exploited in cultivar development. Pre-breeding is involved in the following activities, among others:

15.2.1 Characterization of Landrace Populations

Landraces are often referred to as **'farmers varieties'** or **'cultivated native varieties'** that are adapted to a specific agro-ecological and farming system without a scientific form of selection. Landraces are invariably heterogeneous and an excellent source of genetic variation for crop breeding programmes. *Landraces harbour useful genes such as genes for early maturity, yield potential, disease and pest resistance and other desired traits.* Landraces are most prevalent in the centres of diversity. They can be characterized using various markers (morphological, biochemical, physiological and molecular markers) for breeding and to determine the level of genetic variation. In developing countries landrace varieties are predominantly grown for their farmers-preferred traits and lack of access to well-balanced modern cultivars *via* an effective production and delivery mechanism of commercial seeds.

15.2.2 Creation of New Parent Populations to be used as Breeding Material, with the Long-Term Goal of using the Best Parents for Cultivar Development following Progeny Testing

The success of a crop breeding programme relies on choice of the best parents possessing complementary and desired traits. Thus, *breeders continuously select potential parent populations from diverse sources including landraces, modern cultivars, obsolete or primitive cultivars, wild or semi-wild species.* Parents with high specific or general combining abilities are selected *via* progeny testing through well-designed recombination. The progenies are evaluated to determine the genetic potential of parents for subsequent breeding and to discern the type of cultivar to be developed, i.e., pure line, hybrid, or open-pollinated. Progeny testing is performed in a set of target and representative environments with half-sibs, full-sibs, testcrosses or recombinant inbreds.

15.2.3 Introgression of New Traits from Other Useful Sources, usually Exotic Germplasm or a Landrace or Related Species

The plant breeder transfers one or more desirable traits from unrelated, exotic or semi-exotic, landrace or related germplasm into an intermediate variety with good agronomic potential but lacking a specific trait. Thus, the new variety will be developed with the introduced novel gene(s) in the existing genetic background. Often the unrelated or exotic germplasm does not have immediate usefulness and as such it has to be selected for adaptation to the target production environment. Exotic germplasm may constitute races, populations, clones, inbred lines, or other forms of genetic structure. *When introgressing genes from unrelated, exotic, primitive or wild germplasm, both the desired gene(s) and a considerable amount of undesirable genetic material is introduced into the progeny that has to be removed through a series of backcrosses to the recurrent parent.*

15.2.4 Creation of Novel Traits, for instance, through Mutation Breeding followed by Backcrosses to Good Parents

Mutations lead to spontaneous changes of the genetics of individuals that are often heritable. Naturally, mutational events occur at low frequencies, i.e., 10^{-5} to 10^{-8} per locus. Induced mutagenesis through the use of artificial mutagenic agents is an important tool in plant breeding and functional genomics to increase the frequency of mutations and consequently to broaden genetic variation. Induced genetic variations have been used successfully in several crops to create useful mutants. The technique can be regarded as an efficient option for germplasm enhancement towards important agronomic traits. *The novel mutational events can either be directly developed as essentially derived varieties or novel genes introgressed into candidate parents through a backcross programme.*

15.2.5 Creation of Polyploids

The breeder may create new variability through changing the number of chromosomes in a species, either by altering the basic chromosome set or addition or deletion of specific chromosome(s). Individuals with altered chromosome set (euploids) are developed by doubling the number of genomes of a species or by crossing unrelated species followed by chromosome doubling of the inter-specific hybrid. *Polyploids can be artificially induced by various means such as exposing plant materials to environmental shock (e.g. low or high temperature treatment, x-ray irradiation) or with chemicals (e.g. colchicine) that disrupt normal chromosome division.*

15.2.6 Acquisition of New Information on Crop Genetics

The breeder constantly looks for new genes from diverse sources for enhanced nutritional qualities, early maturity, high yield potential, and biotic and abiotic stress tolerance. Understanding the candidate genes and the pattern of inheritance of the genes in controlling these characters is profoundly significant for effective transfer and to improve the efficiency of selection in cultivar development.

15.2.7 Development of New Plant Breeding Techniques

New and modern breeding techniques can assist in improving selection response. These include development of more efficient conventional selection procedures, biotechnology, molecular marker technologies and identification of markers linked to traits of interest, effective gametocides and cytoplasmic sterility systems with a desired genetic background.

15.3 Cultivar Development

Cultivar development embraces well-defined breeding procedures directed at the production of improved cultivars with respect to the mating system of the crop. In self-fertilized and vegetatively reproduced (propagated) crops, the various breeding procedures or phases can be differentiated to include selection of parents, crosses, progeny selections, finishing off the selections to create the new cultivar, followed by its maintenance, multiplication and distribution. However, these stages are less lucid with cross-fertilizing crops because crosses and selection are closely connected. In cross-fertilizing crops the controlled crosses are performed by the breeder or happens randomly, i.e., in each selection cycle many recombination events occur. Therefore, in cross-fertilizing crops it is not possible to discern a basic procedure, but the breeder identifies the best selection method(s) depending on the crop and the breeding objectives. Cultivar development requires well-developed and elite breeding material, generated from an established and relational pre-breeding programme. If present, locally screened and adapted germplasm are an ideal starting material in cultivar development. A plant breeder can spend many years on pre-breeding to ensure that locally adapted and the best available parental material is used to meet the demands of a defined set of agro-ecological zone and day-length requirements of a given latitude. Alternatively, the breeder may use elite material (e.g., inbred lines) that other breeders have developed, such as the germplasm available from the CGIAR centres that is not locally adapted. This will allow a plant breeder to release new candidate cultivars, even if they are not adapted to the local agro-ecological and cropping system into which they are to be released. The released cultivar may have some superior traits, but the local farmers often reject the new cultivars because they are not bred to include farmers-preferred traits. Farmers' trait preferences are diverse and complex such that the total crop value is more important than absolute crop yield. Other farmers-preferred traits include: cooking quality, taste, market acceptability, storability and the quantity of utilizable parts of the crop left after harvesting, processing and storage. McGuire (2008) indicated that despite 25 years of sorghum breeding in Ethiopia most of the released cultivars had been poorly adopted by the small-scale farmers. Thus, a balance between farmers-preferred traits and solutions to production constraints should be the breeders' goal in order to enhance cultivar uptake by farmers. Crop breeding units with adequate resources should run two parallel programmes concurrently. Firstly, a pre-breeding programme to continually develop new and improved parent materials with superior traits of commercial importance. This is vital to the ongoing improvement of cultivars. Without the pre-breeding programme, the development of new cultivars will stall (hamper) because there will be no new genetic material. Secondly, the breeding programme should have a cultivar-development programme that will take the selected superior parents and develops this into cultivars, followed by multi-location testing in the target growing environments. If there is no cultivar-development programme, none of the best parent materials of the pre-breeding programme will reach the farmers. Based on their natural modes of reproduction, crops are classified as clonally propagated, self-fertilizing or cross-fertilizing. To these, hybrids are included as man-made entities. The predominantly common mating systems and propagation of the selected horticultural crops are summarized in Table 15.1. The timeframes of the traditional breeding procedures of the common horticultural crops are schematically presented in Figures 15.1, 15.2 and 15.3. It takes at least 7 (Figure 15.1) or 9 (Figure 15.2) or 17 (Figure 15.3) minimum breeding generations to create an improved cultivar of vegetatively reproducing or self-fertilizing or cross-fertilizing crop, respectively. *Typically, the estimated minimum timeframes of breeding different crops would be fairly similar unless different selection approaches are adopted to shorten the timeframe for releasing new crop cultivars, e.g., use of double haploid technology and marker-assisted breeding, which are briefly described below:*

Year/season	**Phase/activity**
1	**Identify parents** • Germplasm evaluation and characterization ↓
2	**Cross selected parents** • Random polycross, controlled crosses ↓
3	**Select superior clones** • Seedling trials ↓
4-6	**Yield test of superior clones at representative sites** • Clonal selection: first year, second year, third year ↓
7	**Release** • Multiplication of recommended clones • Maintenance

Figure 15.1. Scheme showing the minimum breeding generations and phases/activity in clonally propagated crops.

Year/season	**Phase/activity**
1	**Identify parents** • germplasm evaluation and characterization ↓
2	**Cross selected parents** ↓
3	**Self F_1 individuals** ↓
4	**Select superior F_2 plants** ↓
5-8	**Select within and between F_3, F_4, F_5, F_6 and F_7 lines across representative sites** • Test under prevailing field condition ↓
9	**Release** • Maintenance and multiplication: breeder's seed → foundation seed → certified seed → farmers

Figure 15.2. Scheme showing the minimum breeding generations and phases/ activity in self-fertilizing crops.

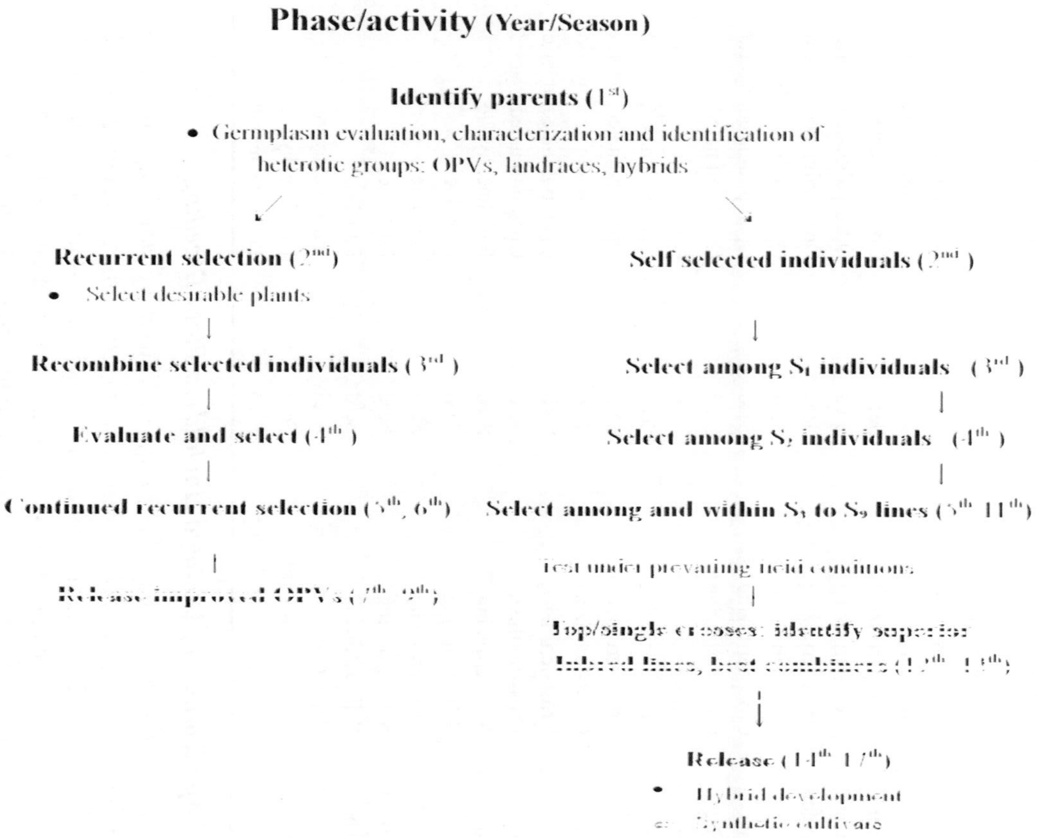

Figure15.3. Scheme showing the minimum breeding generations and phases/ activity in cross-fertilizing crops

Table 15. 1. Mating systems and propagation in selected horticultural crops.

Crop	Mating system	Commercial Propagation	Cultivar
Cassava	Cross-fertilizing	Vegetative	Clone
Sweet potato	Cross-fertilizing	Vegetative	Clone
Potato	Cross-fertilizing	Vegetative	Clone
Banana	Cross-fertilizing	Vegetative	Clone
Pineapple	Cross-fertilizing	Vegetative	Clone
Roses	Cross-fertilizing	Vegetative	Clone
Papaya	Dioecious Cross- fertilizing	Seed	Open pollinated, hybrid
Onion	Cross-fertilizing	Seed	Open pollinated, hybrid
Coriander	Cross-fertilizing	Seed	Open pollinated
Cucumber	Cross-fertilizing	Seed	Open pollinated, hybrid
Watermelon	Cross-fertilizing	Seed	Open pollinated, hybrid
Pumpkin	Cross-fertilizing	Seed	Open pollinated, hybrid
Marigold	Cross-fertilizing	Seed	Open pollinated, hybrid
Hollyhock	Cross-fertilizing	Seed	Open pollinated
Petunia	Cross-fertilizing	Seed	Open pollinated, hybrid
Begonia	Cross-fertilizing	Seed	Open pollinated, hybrid
Cauliflower	Cross-fertilizing	Seed	Open pollinated, hybrid
Cabbage	Cross-fertilizing	Seed	Open pollinated, hybrid
Carrot	Cross-fertilizing	Seed	Open pollinated, hybrid
Radish	Cross-fertilizing	Seed	Open pollinated, hybrid
Tomato	Self- fertilizing	Seed	Pure line, hybrid
Pea	Self- fertilizing	Seed	Pure line
Cowpea	Self- fertilizing	Seed	Pure line
Indian bean	Self- fertilizing	Seed	Pure line
French bean	Self- fertilizing	Seed	Pure line
Okra	Often Cross- fertilizing	Seed	Pure line, hybrid
Brinjal	Often Cross- fertilizing	Seed	Pure line, hybrid
Chilli	Often Cross- fertilizing	Seed	Pure line, hybrid

15.3.1 Applications of Doubled Haploids in Plant Breeding

In-vitro production of haploid plants followed by doubling of somatic chromosomes is the quickest means to produce pure breeding doubled haploids (DHs). Haploids are produced through various means, such as anther culture, or genome elimination following distant hybridisation. During wide (distant) crosses the entire genome of one of the parents are lost from a hybrid embryo and endosperm in the early cell divisions. In wheat breeding, genome elimination, in particular the use of a wheat × maize cross, is usually more reliable than anther culture. *The DH lines are fully fertile and manifest only additive genetic variance which provides a basis for efficient selection with a minimum sequence of a breeding cycle.* The doubled haploid method has several advantages in crop breeding programmes. Firstly, production of doubled haploids leads to homozygosity in a single generation after recombination of selected parents. This is unlike the conventional selection method that requires six to seven selfing generations to achieve a practical level of homozygosity (Figure 15. 2). Secondly, selection is more efficient for oligogenic or polygenic traits in DHs because the genes are fixed in a homozygous background, limiting dominance genetic variation and segregation. Thirdly, the DH method prevents losses of valuable genetic variations better than the conventional selection method. Traditionally, early generation segregating populations are selected in a single environment where certain genotypes perform poorly are discarded. These genotypes might have carried useful genes that would be expressed in other target environments. Owing to the gains in speeded up cultivar development and the creation of desirable genetic backgrounds, doubled haploids are widely utilized in breeding as well as in genetic studies of various crops and traits. *Overall, the DH technology allows for the creation of stable haploids after recombination of parents with broad genetic variation. Thus, DH derivatives can be selected for improved traits such as yield, earliness, plant height, nutritional quality and pest and disease resistance, in a fully homozygous state. Selected genotypes can be used as homogenous varieties or as breeding parents in the ensuing crosses and selection cycles.*

15.3.2 DNA Based Molecular Markers and their Applications in Plant Breeding

Molecular markers reveal genetic differences in the primary structure of DNA between individuals. *Compared to protein markers, DNA based polymorphisms are more stable, and can reveal subtle changes in the genomic DNA.* Different DNA based marker techniques have been successfully used such as restriction fragment length polymorphism (RFLP), random amplified polymorphic DNA (RAPD), amplified fragment length polymorphism (AFLP), simple sequence repeats (SSR) and single nucleotide polymorphisms (SNP). **Molecular markers are 'landmarks' on chromosomes that serve as reference points to the location of other genes when a genetic map becomes available.** If genetic maps are constructed, the plant breeder establishes association between markers and desirable phenotypic traits. The trait of interest is then selected by indirectly selecting for the marker which is readily assayed or observed. In plant breeding, markers are used to locate the chromosomal positions of candidate genes, to determine genomic organization among different gene pools and to conduct marker-assisted breeding. Identification of DNA markers associated with traits of interest may be facilitated by comparative mapping, i.e., by cross-referencing to the maps of model crop species, owing to gene synteny. These markers may facilitate inter-generic gene transfers and help minimize linkage drag.

15.3.3 Applications of Molecular Markers in Plant Breeding

When molecular markers are available, conveniently co-segregating with candidate genes, **marker-assisted selection (MAS)** or **marker-aided selection** may improve the efficiency of selections of simple traits in conventional plant breeding programmes. Broadly, molecular markers are applied in plant breeding in the following areas:

1. To screen for useful single gene traits, e.g., disease resistance. This may facilitate the introgression of new genes from a non-adapted parent and in pyramiding desired alleles into enhanced lines of candidate cultivars.
2. To accelerate backcross breeding programmes through identification of the gene of interest and to eliminate the undesirable genome of the donor parent. Unlike conventional backcrossing, this method reduces linkage drag and requires few numbers of repeated backcrosses to recover the genotype of the recurrent parent.
3. To characterize diverse germplasm and establish heterotic patterns. Markers are useful to determine the magnitude of genetic diversity for crop improvement and to assign exotic (or non-adapted) germplasm into an appropriate breeding pool. In inbred lines, markers assist in establishing heterotic patterns in order to guide the selection of parents for use in a hybrid breeding programme. Marker information may be used in combination with phenotypic and pedigree analyses to ascertain genetic differences between lines of different heterotic groups to enable the breeder to predict the performance of hybrids to be developed from different intergroup crosses.
4. To identify and protect commercial cultivars through fingerprinting.

15.4 Summary

Let us summarize the timeframes involved in the two interdependent breeding activities in crop-improvement programmes, namely, pre-breeding/germplasm enhancement, and the actual breeding aimed at cultivar development. It would take at least 7, 9 or 17 generations to release an improved cultivar of a crop that is vegetatively reproducing (e.g. as tubers or suckers), self-fertilizing (pure line, open-pollinated or hybrids) or cross-fertilizing (open-pollinated or hybrids), respectively. As such, both private and public breeding programmes require a pool of highly trained and motivated plant breeders to run both activities, which is contingent upon continued and sustained funding support over these time scales. Additionally, the following points should be taken into consideration during planning plant breeding programmes:

- The indicated timeframes should be seen as a generic (general, common) framework for pre-breeding and cultivar development of some horticultural crops, leading to the release of superior cultivars for small-scale farmers;
- The new cultivars need to be locally adapted for the specific agro-ecological systems and latitudes (day-lengths) of the region in which they are to be grown. Very few modern cultivars of relatively few crops are so adaptable that they can be grown across wide agro-ecological zones. Breeders need to develop novel cultivars of each crop to be specifically adapted to each and every agro-ecological systems and latitude, particularly for day length;
- In self-pollinating and vegetatively propagated crops, it is easy to discern and follow the described breeding procedures. However, the procedures are more varied and difficult to predict with cross-pollinating crops. Thus, a breeder may apply varied selection procedures, such as recurrent selection, half-sib family selection, full-sib family selection or mass selection for the development of improved open-pollinated varieties or to create synthetic or hybrid cultivars. If male gametocides or male-sterility systems are available for a self-pollinating crop, these procedures can also be applied to produce hybrid cultivars.
- Marker-assisted selection (MAS) should be integrated with traditional breeding methods to enhance the efficiency of cultivar development. The application of MAS is currently limited to Mendelian traits, whereas it is less efficient for complex quantitative traits such as yield and drought tolerance.
- In our discussion, specifics on the detailed breeding methods and mating and experimental designs used for different crops are omitted to maintain a focus on the overall breeding timelines.

In conclusion, the speed at which novel crop cultivars can be released by breeders depends upon many variables, in particular:

- The reproductive biology and growth habit of crop plants;
- The resources available to the breeder;
- The scale of pre-breeding that is needed, and the timescales of the pre-breeding operations that have to be followed, before the actual breeding or cultivar development can take place;
- The timescales of the cultivar-development to be followed (e.g., an open-pollinated versus a hybrid cultivar);

- Institutional requirements unique to each crop in each country (e.g., how many years [seasons] and locations [sites] of field testing that are required before registration of a cultivar).

15.5 Brief References

Shimelis and Laing 2012.

Review Questions

Part A

Answer the following questions true or false.

1. Molecular markers reveal differences in the primary structure of DNA between individuals. Yes/No
2. The minimum breeding generations required to develop an improved cultivar of vegetatively propagated crop is nine. Yes/No
3. MAS cannot accelerate backcross breeding programmes. Yes/No
4. The DH lines manifest only non-additive genetic variance. Yes/No
5. The DH method prevents losses of valuable genetic variations better than the conventional selection method. Yes/No

Part B

Answer the following questions.

1. What is pre-breeding?
2. Marker-assisted selection and marker-aided selection are the same/different.
3. How can polyploids be artificially induced?
4. RAPD stands for ___________.
5. What are landraces?

Part C

Write a short note on each of the following topics.

1. Doubled haploids in plant breeding.
2. Applications of molecular markers in plant breeding.
3. Pre-breeding.

Section III
Breeding Practices

Part A: Conventional Breeding Practices

Part B: Non-Conventional Breeding Practices

Part C: Mission-Oriented Breeding Practices

Part A
Conventional Breeding Practices

16

Plant Introduction and Acclimatization Sexually and Asexually Propagated Crops

16.1 Introduction

Plant introduction is the process of introducing plants from their growing locality to a new locality. Introduction of plants from a foreign country is called **foreign introduction or exotic collection**. Introduction from one state to another within a country is called **indigenous collection or indigenous introduction**. With introduction, the process of acclimatization is associated. *Acclimatization is the adaptation or adjustment of an individual plant or population of plants under the changed climate for a number of generations.* **Plant introduction is an oldest method of plant breeding.** It is used in self-pollinated, cross-pollinated and asexually propagated crops. The material of seed-propagated crops is introduced in the form of seed and that of vegetatively propagated crops in the form of cuttings or propagules. The introduced material may be a new crop, a wild relative or a new cultivar of a crop already grown in the area. There are five important sources of plant collection, namely, (1) centres of diversity, (2) gene banks, (3) gene sanctuaries, (4) seed companies, and (5) farmers' fields.

16.2 Types of Plant Introduction

Plant introduction is classified into two types, namely

1. Primary introduction and
2. Secondary introduction.

16.2.1 Primary Introduction

Introduction that can be used for commercial cultivation (after proper evaluation) as a cultivar without any change in the original genotype is called as **primary introduction**. Apple cultivars Red Chief, Oregon Spur, Silver Spur, Top Red and Vance Delicious are some examples. Primary introductions are less common than secondary ones.

16.2.2 Secondary Introduction

Introduction that can be used as a cultivar after selection from the original genotype or used for transfer of some desirable gene to the cultivar is called **secondary introduction**. Secondary introductions are more common than primary ones. Wheat cultivars Kalyan Sona and Sonalika are examples of secondary introduction.

16.3 Purpose of Plant Introduction

Plant materials are introduced for five main purposes. They are briefly discussed below:

16.3.1 Economic Use

New field crops, horticultural crops, forest species and medicinal plants are introduced from other countries as new sources of food, vegetable, fruits, oil, fibre and wood.

16.3.2 Study of Origin and Evolution

Plant introduction is done for study of origin and evolution of crop plants. Based on world collection of crop plants, Vavilov put forth the idea of centres of origin of cultivated species and other workers identified progenitors of different crop plants.

16.3.3 Conservation of Diversity

The genetic diversity of crop plants is gradually being eroded. Introduction and collection of crop plants is useful in conservation of genetic diversity of crop plants.

16.3.4 Use in Hybridization

New plants are also introduced for improvement of crop plants. Introduced material is used in hybridization programmes for transferring desirable genes.

16.3.5 Aesthetic Interest

Various flowering plants and ornamentals are used for beautification of gardens, parks, offices, houses, bungalows and roads. New ornamentals are introduced for this purpose. Most of the flowers and garden plants are introductions from far and near.

16.4 Agencies for Plant Introduction

International Plant Genetic Resources Institute (IPGRI), Rome and various international crop research institutes undertake plant introduction on global level. In India, National Bureau of Plant Genetic Resources (for agricultural and horticultural crops), Forest Research Institute (for forest trees) and Botanical Survey of India (for medicinal and botanical plants) deal with plant introduction.

16.5 Procedure of Plant Introduction

The procedure of plant introduction includes procurement, quarantine, cataloguing, evaluation, multiplication and distribution.

16.5.1 Procurement

The material that is required to be introduced from other countries is procured through National Bureau of Plant Genetic Resources (NBPGR). In most of the cases, free and mutual exchange of genetic resources is possible between countries. Plant propagules like seeds, clonal propagules or cultured tissues can be exchanged. Proper packaging of the material being exchanged is very important. Now-a-days, *in vitro* exchange of germplasm is becoming popular.

16.5.2 Quarantine

Quarantine is the screening of the living organisms introduced from foreign sources for the presence of pathogens, pests, weeds etc. Plant materials introduced from such sources are screened and suspected materials are treated appropriately. All the plant materials being introduced should carry an authentic phytosanitary certificate.

Quarantine control is usually exercised by the agencies authorized for introduction, at prescribed ports of entry. Materials contaminated by pathogens or pests are destroyed or returned. Materials conforming to quarantine laws are fumigated against pathogens and pests. Sometimes they are grown in isolation or quarantine plots for evaluation. Restrictions are imposed on the introduction of plant materials by different countries due to several reasons.

16.5.3 Cataloguing

When a material is introduced, it is given an entry number and information regarding the name of cultivar, species, place of origin and major characteristics are recorded.

16.5.4 Evaluation

The newly introduced materials are usually assessed and evaluated by the respective Crop Research Institutes. Resistance to pests and diseases is evaluated under appropriate conditions.

The acclimatization of an introduced cultivar or species to the environmental conditions of the new geographical area is a serious problem. However, since the introduced lot may contain genetically different plant types, some of them may show higher levels of adaptability to the new environment. Such genotypes are multiplied and distributed widely.

16.5.5 Multiplication and Distribution

Superior genotypes are selected, multiplied and made available on commercial scale. They are subjected to field trials and released for cultivation. Some of them are used as parents in hybridization programmes so as to produce improved hybrids.

16.6 Merits of Plant Introduction

Plant introduction is the second oldest method of plant breeding, the first one being plant domestication. It continues to be an important activity even today. It has the following merits in crop improvement:

1. It provides entirely new crop species. For example, maize, potato, tomato, groundnut, papaya, pineapple, etc. were introduced into India from other countries.
2. Sometimes it provides superior cultivars very rapidly and cheaply.
3. It enriches the genetic diversity of a crop plant.
4. It is very quick and economical method of plant breeding, especially when introduced material is directly released as a cultivar.
5. It could protect a crop from a devastating disease/pest, e.g., rubber and coffee.
6. It is applicable in all self-pollinated, cross-pollinated and vegetatively propagated crops.

16.7 Demerits of Plant Introduction

The main demerit is that there are chances of entry of new diseases, insects and weeds in the country along with introduced material. For example, *Argimone mexicana*, *Phalaris miner*, *Parthenium*, etc. weeds; late blight of potato, bunchy top of banana, coffee rust, etc. diseases; wooly aphis of apple, fluted scales of citrus, potato tuber moth, etc. insects were introduced into India along with introduced materials. However, now-a-days chances of entry of new insects, diseases, and weeds are remote due to quarantine regulations.

16.8 Achievements

Introduced materials have provided new crops, new cultivars by direct release, parents for hybridization programmes and new cultivars following selections. Many of our crops were obtained by introduction, e.g., potato, tomato, pea, chilli, coffee, guava, grape, pineapple, papaya etc. Many introductions have been released directly as cultivars, e.g., Bonneville, Arkel, Early Badger in peas, Sioux in tomato. Several popular cultivars have been developed through selection from plant introductions, e.g., Kalyan Sona, Sonalika in wheat; Pusa Lal, Pusa Sunehari in sweet potato. A large number of cultivars have been developed through hybridization incorporating introduced materials as parents, e.g., semi-dwarf cultivars in wheat and rice.

16.9 Brief References

Chaudhary 2014, Kute and Aher 2013, Mohanan 2010, Ram M 2014, Sharma JR 1994, Singh BD 2011, Singh Phundan 2013a.

Review Questions

Part A

Answer the following questions true or false.

1. Introduction of plants from a foreign country is called indigenous introduction. True/False
2. *Argimone mexicana* was introduced into India. True/False
3. Introduction may provide entirely a new crop species. True/False
4. Tomato cultivar Sioux was introduced into India. True/False
5. Plant introduction erodes the genetic diversity of a crop plant. True/False

Part B

Answer the following questions.

1. Define secondary introduction.
2. Give one example of primary introduction.
3. The screening of the living organisms introduced from foreign sources for the presence of pathogens, pests, weeds etc. is known as _______.

4. NBPGR stands for _______.
5. Botanical Survey of India deals with introduction of _______ and _______ plants.

Part C

Write a short note on each of the following topics.

1. Purpose of plant introduction.
2. Procedure of plant introduction.
3. Primary introduction.
4. Merits of plant introduction.
5. Achievements of introduction.

17

Methods of Selection Heterogeneous, Homozygous Self-Pollinated Populations

17.1 Introduction

Self-pollinated vegetable, spice and ornamental crops may be heterogeneous. It is exemplified by landrace cultivars. They may have originated from single plant selections but have been grown over a long period of time. Landrace cultivars are a mixture of genotypes, perhaps because of breeders' lack of appreciation of selection techniques to produce a pure line, spontaneous mutations, chance hybridization, or a combination of factors. Natural selection in landraces has resulted in a cultivar that is well adapted to the local environment. Depending on amount of human selection, differences among genotypes in a landrace cultivar may be small and inconspicuous. On the other hand, a lack of human intervention may mean that differences among plants in a landrace cultivar will be conspicuous. After years of self-pollination, individual plants are homozygous. Selection of such individuals would result in a true-to-type reproduction of their phenotype within the limitations imposed by the environment. Selection of different plants in the heterogeneous homozygous population, however, could result in contrastingly different plants. Selection consists of the following three operations:

1. Isolation of plants having desirable traits from a population. Seeds of selected plants are harvested and saved.
2. The rest of the plants are culled out.
3. Seeds of the selected plants are sown to raise the next crop. The lines/population so obtained constitute the selected lines/population.

The selected lines may be kept separately or mixed together. On the basis of this activity, the methods of selection applicable to heterogeneous, homozygous, self-pollinated populations are divided into two types: mass selection and pureline selection.

17.2 Mass Selection

Mass selection is one of the oldest methods of crop breeding. This is a very simple procedure. A large number of plants with similar and superior phenotype are selected from heterogeneous homozygous population. Their seeds are bulked to form a new population that may be released as a cultivar. Selection is usually done for easily observable traits, e.g., plant height, fruit size, flower size, biotic stress resistance, etc. Sometimes yield may also be the basis selection.

17.2.1 Applications of Mass Selection

Mass selection has two main applications, namely, (i) improvement of local land races and (ii) purification of pureline cultivars.

17.2.1.1 Improvement of Local Land Races

Local land races are usually mixtures of a number of purelines. Many of them are inferior. Culling of the inferior ones through mass selection will improve the performance of the cultivar, make it more uniform and maintain the adaptability and stability of the parent cultivar. This application was important in the past but is not much relevant now-a-days since variations in land races have been exploited in most of the crops.

17.2.1.2 Purification of Pureline Cultivars

Purelines become genetically variable in course of time. Hence, it is necessary to maintain the purity of such cultivars by mass selection. Mass selection is used for the maintenance of purity of nucleus seed of pureline cultivars.

17.2.2 **Procedure of Mass Selection**

Plants selected in mass selection need not be subjected to progeny test. However, if progeny test is included, it increases the efficiency. A simplified description of mass selection without and with progeny test is given below:

17.2.2.1 Mass Selection without Progeny Test

The different steps involved are (Figure 17.1):

First Year

Large number of phenotypically similar and superior plants are selected. The number may be 200-2000. The seeds of the selected plants are composited to raise the next generation.

Second Year

Composited seed population is sown and evaluated in a preliminary yield trial (PYT) along with standard checks. The cultivar from which the selection was made should also be taken as one of the checks. Phenotypic characteristics of the selected line are critically examined and evaluated. If the new line is superior to the checks, it is included in multilocation trial.

Third to Fifth Year

The new line is evaluated in All India Coordinated Multilocation Trial at several locations with suitable checks. It is evaluated in an initial evaluation trial for one year. If found superior, it is promoted to advance trials for another 2 years. If the new line is superior, it is identified for release as a cultivar.

Sixth Year

Once the new line is identified for release, its seed multiplication is initiated. After its release and notification as a cultivar, certified seed can be produced.

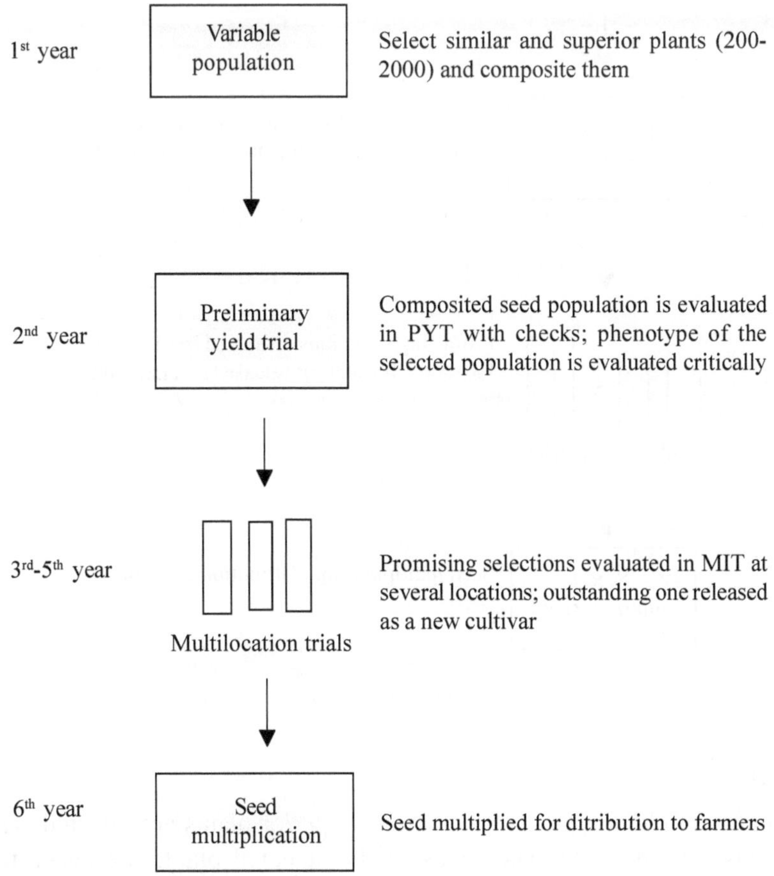

Figure 17.1. A generalized outline of mass selection without progeny test

17.2.2.2 Mass Selection with Progeny Test

The different steps involved are (Figure 17.2):

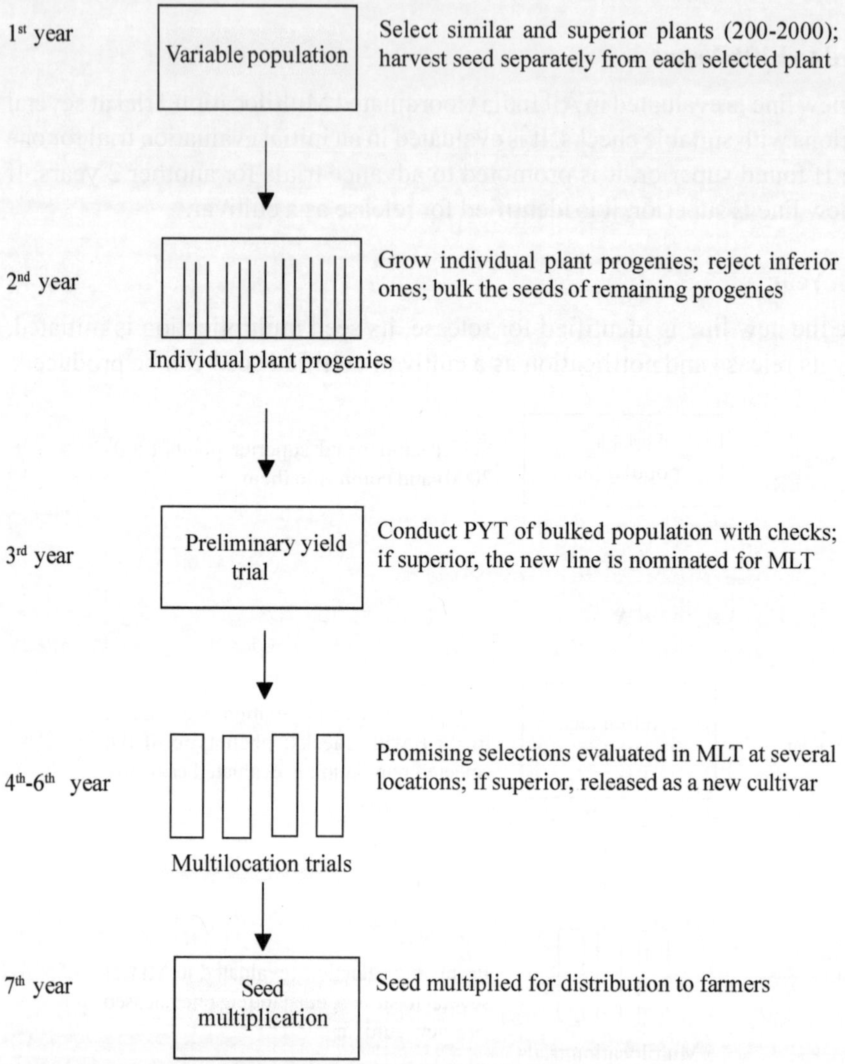

Figure 17.2. A generalized outline of mass selection with progeny test

First Year

Large number of phenotypically similar and superior plants are selected. The number may be 200-2000. The seeds of the selected plants are harvested separately.

Second Year

Individual plant progenies are grown for progeny test. Inferior and segregating progenies are discarded. The rest progenies are harvested and seeds are bulked.

Third Year

Bulked seed population is sown and evaluated in a preliminary yield trial (PYT) along with standard checks. The cultivar from which the selection was made should also be taken as one of the checks. Phenotypic characteristics of the selected line are critically examined and evaluated. If the new line is superior to the checks, it is included in multilocation trial.

Fourth to Sixth Year

The new line is evaluated in All India Coordinated Multilocation Trial at several locations with suitable checks. It is evaluated in an initial evaluation trial for one year. If found superior, it is promoted to advance trials for another 2 years. If the new line is superior, it is identified for release as a cultivar.

Seventh Year

Once the new line is identified for release, its seed multiplication is initiated. After its release and notification as a cultivar, certified seed can be produced.

17.2.3 Merits of Mass Selection

1. The new cultivar has the same adaptation and stability as the parent cultivar.
2. It retains considerable genetic variability and hence further improvement is possible in future by selection.
3. Useful for purification of pureline cultivars.
4. Improvement of traits governed by few genes with high heritability is possible.
5. Less time-consuming and less expensive.

17.2.4 Demerits of Mass Selection

1. Cultivars are not uniform.
2. The improvement obtained is less than that produced by pureline selection.
3. Seed certification of cultivars is difficult due to presence genetic variability in them.
4. It can be applied only to heterogeneous homozygous populations.
5. In the absence of progeny test, heterozygous and inferior plants may be selected.

6. Improvement of traits governed by large number of genes with low heritability is not possible.
7. It cannot create any new genotype but utilizes existing genetic variability.

In view of the above, mass selection is no more used for the development of new cultivars. However, it is routinely used for maintaining the genetic purity of pureline cultivars.

17.2.5 Achievements

Mass selection must have been used by pre-historic man to develop crops from their wild parents. It was also used extensively before pureline selection came into existence. Below are few examples of cultivars developed through mass selection:

Cotton	:	American Cotton
Groundnut	:	TMV 1, TMV 2, etc.
Rice	:	SLO 13, MTU 15, etc.
Potato	:	K 122, etc.

17.3 Pureline Selection

The term pureline was first proposed by the Danish botanist W.L. Johannsen (1903) while working with Princess cultivar of French bean (also called common bean, dry bean, *rajma*), *Phaseolus vulgaris*. ***A pureline may be defined as the progeny of a single self-pollinated homozygous plant.*** In his work, Johannsen took a commercial sample of Princess cultivar. From this he selected large and small seeds and sowed them. He harvested the seeds from individual lines and found considerable variations in seed weight of these lines. Further, he selected 19 plants from the lightest to the heaviest seeds.

For six successive generations the smallest seeds of pureline, e.g. line 19, were found to grow into plants with seeds of the same mean weight of 351 mg (Table 17.1). Similarly, the largest seeds selected for six successive generations from pureline 1 were found to grow into plants with seeds of the same mean weight of 642 mg. Variations in seed size within individual lines were due to position of pod on plant or position of seed in pod. That is, 649 mg or 631 mg seeds of line 1 have the same genotype and the variation in their weight was due to non-hereditary or environmental reasons. On sowing, both the seeds gave progenies with the same mean seed weight, i.e., 642 mg. Thus, line 1 became a pureline and further selections in it for seed size were not effective. Johannsen concluded that differences in mean seed weight of different lines were hereditary but between individuals of a pureline were non-hereditary.

Table 17.1. Mean seed weight of some Johannsen's purelines in *Phaseolus vulgaris*

Pure line #	Mean seed weight (mg)	Range (mg)
1	642	649 - 631
10	465	469 - 421
19	351	358 - 348

Since French bean is a naturally self-pollinated crop, the sample of the Princess cultivar must have been a heterogeneous mixture of 19 homozygous lines. By one cycle of selection, these types were isolated into 19 lines. But once the homozygous lines were isolated, the variations within the lines were only environmental and thus selection was ineffective.

17.3.1 Applications of Pureline Selection

Pureline selection has the following applications:

1. Selection in landraces
2. Selection in introduced materials.
3. Improvement of old pureline cultivars
4. Selection of a new character in an existing pureline cultivar
5. Pureline selection method is also an integral part of other breeding methods such as pedigree selection and bulk population selection.

17.3.2 Procedure of Pureline Selection

The general procedure of pureline selection is outlined below (Figure 17.3). The breeder concerned may modify the scheme as per his need. However, there are three main steps of pureline selection that are:

1. Selection of individual plants from a heterogeneous (variable) homozygous population.
2. Raising and evaluation of individual plant progenies from the selected plants.
3. Multilocation yield trials of the superior plant progenies.

First Year

A large number of superior plants (200-3000) are selected from variable local land races (*Desi* cultivars) and harvested separately. The number of plants to be selected depends upon the breeder's discretion but should be as large as possible as per availability of time, land, capital, labour, etc. Plants are selected based on easily observable characters.

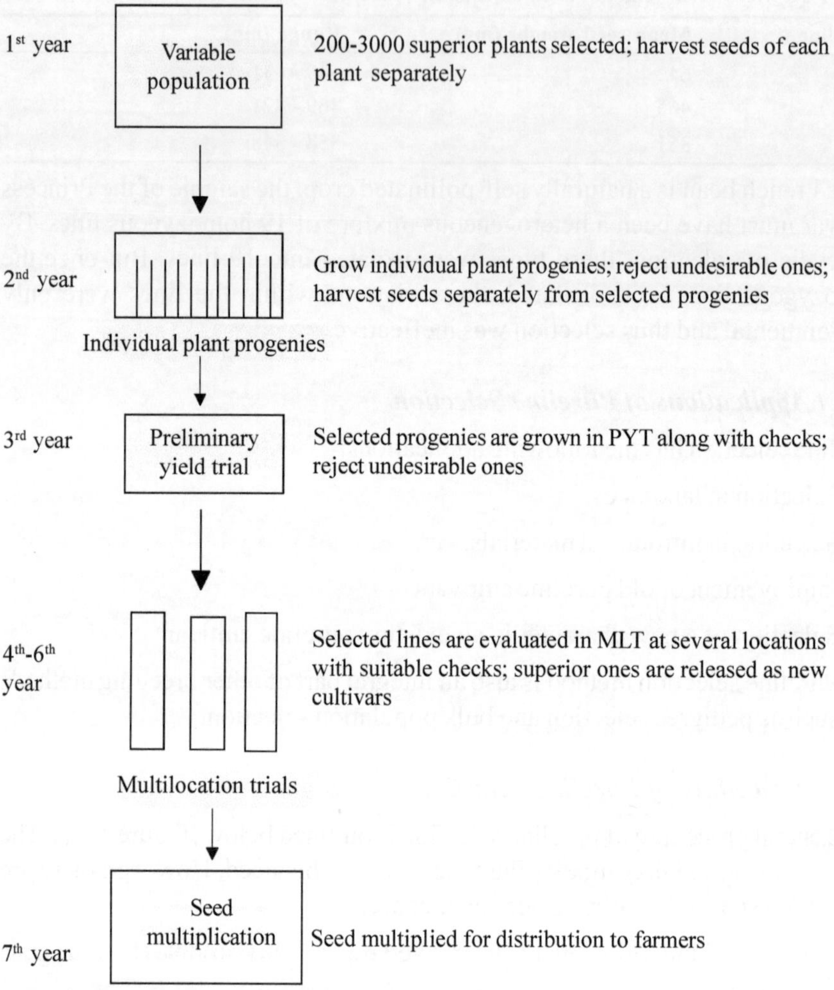

Figure 17.3. A generalized outline of pureline selection

Second Year

Individual plant progenies from selected plants are raised separately with proper spacing. They are evaluated visually by taking elaborate data on visual characters in addition to yield. The number of progenies should be reduced as far as possible. Disease epiphytotics may be created to test the progenies for disease resistance. Undesirable progenies such as poor, weak, diseased, insect-attacked, inferior, defective, and segregating ones are rejected. The superior progenies are harvested separately. If necessary, this process may be repeated for one or more years.

Third Year

Selected progenies are evaluated in a preliminary yield trial (PYT) along with local checks. The selected progenies, now called as cultures, are critically evaluated for yield and other observable traits. Superior cultures are selected drastically reducing the number.

Fourth to Sixth Year

The superior cultures are evaluated in All India Coordinated Multilocation Trial at several locations with suitable checks. Observations are recorded on many characters like days to flowering, days to maturity, plant height, characters of fruit, seed and flower, disease and insect resistance, yield, etc. Quality tests are also done. They are evaluated in an initial evaluation trial for one year. If found superior, they are promoted to advance trials for another 2 years. The superior culture is identified for release as a cultivar.

Seventh Year

Once the superior culture is identified for release, its seed multiplication is initiated. After its release and notification as a cultivar, certified seed can be produced.

17.3.3 Merits of Pureline Selection

1. The pureline cultivars are extremely uniform since all plants have the same genotype.
2. They are attractive and liked by the farmers and consumers.
3. Due to extreme uniformity of cultivar, seed certification is comparatively easy.

17.3.4 Demerits of Pureline Selection

1. New genotypes are not created by pureline selection.
2. The new cultivar may have less stability and adaptability than the parent cultivar.
3. It requires more time, space and effort than mass selection.
4. No more improvement is possible after isolation of the best genotype in the population.
5. The breeder has to devote more time than mass selection.

17.3.5 Achievements

Pureline selection was extensively used in early days of plant breeding in India. A large number of improved cultivars were developed through pureline selection both in local landraces and introduced materials. It is still used in crops where breeding work is in the initial stage. Some examples of cultivars developed through pureline selection are:

Tomato: CO 1, CO 2, Arka Alok, Arka Ashish, etc.

Brinjal: Pusa Purple Long, Pusa Purple Round, Pusa Purple Cluster, Arka Kusumkar, Arka Sheel, etc.

Chilli: Arka Mohini, Arka Gaurav, Arka Basant, G 1, G 2, G 3, Sabour Arun, etc.

Okra: Pusa Makhmali, Gujarat Bhendi 1, Kashi Mangali, Punjab No 13, etc.

Pea: Ooty 1, Harbhajan, Pant Uphar, etc.

French Bean: Arka Komal, YCD 1, Kashi Param, Pant Anupama, Pusa Himlata, Phule Surekha, etc.

17.4 Brief References

Agrawal 1998, Allard 1960, Chahal and Gosal 2002, Chaudhary 2014, Dabholkar 2006, Khan 2018, Kumar 2016, Kute and Aher 2013, Ram M 2014, Sharma JP 2009, Singh BD 2011, Singh Phundan 2011, Singh and Pawar 2006, Stoskopf, Tomes and Christie 2011.

Review Questions

Part A

Answer the following questions true or false.

1. Mass selection is a complex procedure. True/False
2. New genotypes are not created by pureline selection. True/False
3. The term pureline was first proposed by Mendel. True/False
4. A breeder has to devote less time in pureline selection than mass selection. True/False
5. Pureline cultivars are extremely uniform. True/False

Part B

Answer the following questions

1. Define pureline.
2. Why selection is not effective in a pureline?
3. What is the main application of mass selection now-a-days?
4. Why seed certification is easier in pureline cultivars?

5. The new cultivar developed through mass selection has the same adaptation and stability as the parent cultivar. Explain.

Part C

Write a short note on each of the following topics.

1. Merits and demerits of pureline selection.
2. Pureline theory of Johannsen.
3. You are going to work on improvement of a new crop. Land races are available. You are to develop an improved cultivar with little compromise on stability and adaptability but higher yield. Chalk out a scheme of work.

18

Methods of Selection Segregating, Self-Pollinated Populations

18.1 Introduction

The genetic variability present in populations of self-pollinated vegetable, spice and ornamental crops is rapidly exhausted by pureline selection. For further improvement, genetic variability must be created. This is generally achieved by hybridization. After the cross has been made, F_1 is grown and seeds set on it are harvested. This makes the seed source for raising F_2 population, and so on. The segregating populations from crosses are handled according to a suitable method. The three basic methods available for this purpose are: (1) pedigree method, (2) bulk method, and (3) backcross method. There are also certain modifications of these methods.

18.2 Pedigree Method

Pedigree method is developed from the pureline system and widely used method of plant breeding for self-pollinated crops. A main difference between pedigree selection and mass selection or pureline selection is that hybridization is used to create variability for the base population in case of the former, whereas in mass and pureline selections variability is naturally present in the base populations. The pedigree method was first described by H.H. Lowe (1927). This method predominated the crop improvement programme throughout the world until 1980s. Pedigree method is common in horticultural crops, especially where readily identifiable qualitative traits are involved. Vegetable crops in which pedigree system is used include tomato, cucumber, pepper, pea, lettuce, snap and field beans, all self-pollinated crops. In general, selection is done on F_2 plants. Yield is important in many horticultural plants, but quality is equally or even more important. Quality may be associated with identifiable traits, making the pedigree system ideal for self-pollinated vegetable crops.

In pedigree method

1. Individual plants are selected in F_2 and subsequent generations.
2. Individual plant progenies from the selected plants are grown for progeny test and selection.
3. A record of all parent-offspring relations (pedigree record) in all the generations is maintained.
4. Individual plant selection continues till the progeny shows no segregation. At this stage, selection is done among the progenies only.

18.2.1 Pedigree Record

Pedigree means ancestry. A record in which the pedigrees of individuals are maintained is called as pedigree record. With the help of this record, the origin of any plant of a given generation can be traced back sequentially through the individual plant progenies of the preceding generations to the F_2 plant from which it originated. It is important to take note of the fact that pedigree record does not establish a retraceable route enabling development of cultivar by repeating the same cross. The record provides basis for retaining selections or rejecting them on the basis of their origin from same family.

The breeder should develop an effective, easy to maintain system of record keeping. The most basic form is based on numbering of plants when they are selected, and developing an extension to indicate subsequent selections. Some breeders include letters to indicate parental sources or the kind of crop or some other useful information. The key is to keep it simple, manageable, and informative. For example, if five crosses of tomato are made at Chiplima in 2018 and 750 plants are selected in the F_2, a family could be designated CT 184-175 (meaning it was derived from plant # 175 selected from cross # 4 made during 2018 in tomato at Chiplima). If selection is subsequently made from this family, it can be named, e.g., CT 184-175-10. Here, CT indicates Chiplima Tomato; 18 indicates year of crossing, i.e., 2018; 4 is cross # in the year 2018; 175 is the plant # selected in F_2; 10 is plant # selected in the progeny of plant # 175.

18.2.2 Applications of Pedigree Method

Pedigree method is the most widely used method for handling the segregating populations of self-pollinated vegetable, spice and ornamental crops. It has the following two applications:

1. Correction of specific weaknesses of an established cultivar (**combination breeding**).
2. To improve yield and yield contributing characters by recovery of transgressive segregants (**transgressive breeding**).

18.2.3 Procedure of Pedigree Method

A generalized outline of the pedigree method is briefly narrated below (Figure 18.1). The breeder may modify the scheme according to his specific needs.

First Year

Crosses are made between selected parents possessing desirable characters.

Second Year

Sow 10-30 F_1 seeds with wide spacing so that F_1 plants produce more seeds. However, in case of vegetables like tomato, brinjal, chilli, okra, etc. even 3-5 F_1 plants are adequate to yield required quantity of seed for F_2 generation.

Third Year

Grow 2,000-10,000 plants of F_2 giving wide spacing for full expression of the characters, along with parents for comparison. Depending upon the facilities and objectives of the programme about 100-500 superior plants are selected. Generally, the size of F_2 population should be 10-100 times the number of F_2 plants the breeder plans to select from the population. Fifty percent of the genotypes in the F_2 are heterozygous. Hence, selection intensity should be moderate (about 10%) to select plants that would likely include those with the desired gene combinations. The actual number of plants selected depends on the heritability of the character and available resources. Characters with high heritability are more effectively selected and require lower numbers than for characters with low heritability. The selection in F_2 is done for simply inherited characters, and selection for characters governed by polygenes like yield will be reserved for later generations. The selected plants are harvested separately and assigned serial numbers, and their descriptions are entered in pedigree register.

Fourth Year

Individual plant progenies are space-planted along with parents and checks. Each progeny should have at least 30 plants. Inferior and defective progenies are culled out. Superior plants with desirable characters are selected from the rest progenies. The number of selected plants in F_3 should be smaller than F_2. They are harvested separately.

Fifth Year

Individual plant progenies are space-planted along with parents and checks. Inferior and defective progenies are culled out. Superior plants with desirable characters are selected from the rest progenies. They are harvested separately.

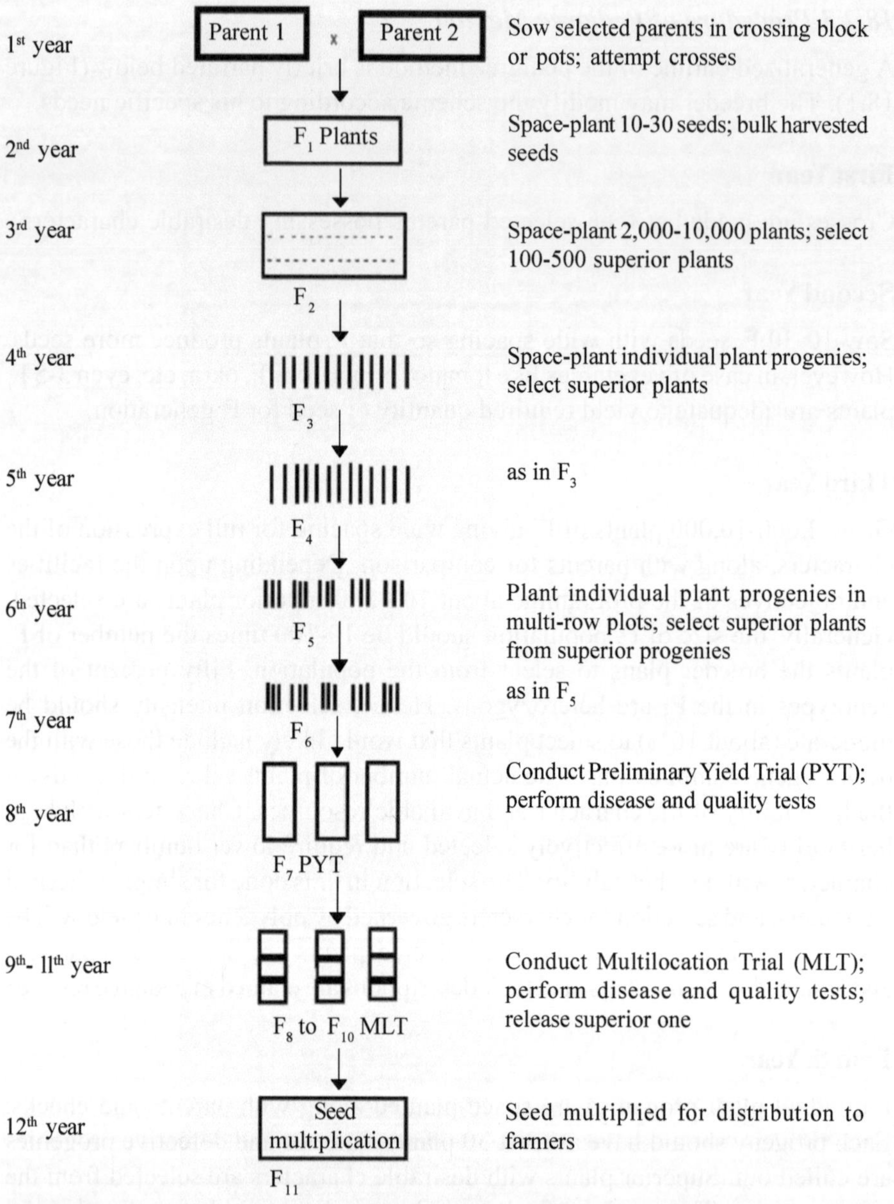

Figure 18.1: A generalized outline of pedigree method of selection

Sixth Year

Individual plant progenies are planted at commercial sowing rate in multi-row plots. Yield potential of the progenies is assessed visually. Inferior progenies are rejected. Superior homogeneous progenies are harvested in bulk. Individual plants may be selected from superior segregating progenies. The number of selected progenies (harvested in bulk) and plants should be 25-100 from a cross.

Seventh Year

Individual plant progenies are planted at commercial sowing rate in multi-row plots. Yield potential of the progenies is assessed visually. Inferior progenies are rejected. Superior homogeneous progenies are harvested in bulk.

Eighth Year

Conduct preliminary yield trial (PYT) of the selected progenies with 3 or more replications along with standard commercial cultivars as checks to identify superior lines. The progenies are evaluated for many characters including yield. Perform disease and quality tests. Progenies superior to checks are selected.

Ninth to Eleventh Year

The superior progenies are evaluated in All India Coordinated Multilocation Trial at several locations with suitable checks. Observations are recorded on many characters like days to flowering, days to maturity, plant height, characters of fruit, seed and flower, disease and insect resistance, yield, etc. Quality tests are also done. They are evaluated in an initial evaluation trial for one year. If found superior, they are promoted to advanced trials for another 2 years. The superior line is identified for release as a cultivar.

Twelfth Year

Once the superior line is identified for release, its seed multiplication is initiated. After its release and notification as a cultivar, certified seed can be produced.

18.2.4 Merits of Pedigree Method

1. It gives maximum opportunity to the breeder to use his skill and judgement in selection of plants.
2. It is well-suited for improvement of the characters that are easily identifiable and simply inherited.
3. Transgressive segregants may be recovered.
4. It takes less time than bulk method.
5. Inferior progenies are eliminated in early generations.
6. Information on inheritance of characters and pedigree of lines can be obtained.

18.2.5 Demerits of Pedigree Method

1. Valuable genotypes may be lost in early generations.
2. No opportunity for natural selection.
3. Difficult to handle many crosses.
4. Maintenance of pedigree records, selections, growing of progeny rows etc. are time-consuming and tedious.

18.2.6 Modifications

The pedigree method of selection is a continuous selection of individuals after hybridization. A discontinuous method has been proposed but is not considered practical enough for wide adoption. The breeder may modify the pedigree method to suit specific objectives and resources. Some specific ways are:

1. The number of plants selected at each step may be modified according to the crop, the breeding objective, and the genetics of the characters of interest, as well as the experience of the breeder with the crop, and resources available for the breeding programme.
2. The details of records kept are at the discretion of the breeder.
3. Off-season planting, use of the greenhouse, and multiple plantings a year (where possible) are ways of speeding up the breeding process.
4. Early generation selection for yield is ineffective. As a result, several modifications have been introduced by breeders to delay selection till later generations (e.g., F_5). Mass selection or bulk selection is practiced in the early generation.

18.2.7 Achievements

Pedigree method is the most widely used method of handling segregating populations of self-pollinated crops. Some examples of cultivars developed through pedigree method are:

- **Tomato**: Arka Meghali, Kashi Hemant, Kashi Sharad, Pusa Ageti, Pusa Ruby, Pusa Early Dwarf, Punjab Chhuhara, Selection 2, HS 101, HS 102, etc.
- **Brinjal**: Pusa Anupam, Pusa Kranti, H 4, HS, H 8, etc.
- **Chilli**: Bhaskar, Punjab Lal, Pant C-1, Pant C-2, etc.
- **Okra**: Varsha Uphar, Pusa Sawani, etc.
- **Pea**: Kashi Shakti, Swarna Mukti, VL Matar -3, Kasha Udai, Kashi Nandini, etc.

- **French Bean**: Arka Subidha, Arka Anoop, etc.
- **China Aster**: Kamini, Poornima, Shashank, Violet Cushion, etc.
- **Hollyhock**: Deepika, Dulhan, Gauri, Pusa Sweta, Pusa Krishna, Pusa Lalima, Pusa Gulabi, etc.

18.3 Bulk Method

Bulk method (also called '**bulk population method**' or '**bulk breeding method**') was developed by H. Nilsson-Ehle in 1908. H.V. Harlan and colleagues provided additional theoretical foundation for this method in 1940s. It is a strategy of plant breeding in which natural selection effect is solicited more directly in the early generations of the procedure by delaying stringent artificial selection until later generations. When seeds of several plants, especially the entire population, are harvested as a mixture, it is called **bulking** and the population so obtained is called as **bulk**. In bulk method, F_2 and later generations are maintained as bulks; at the end of **bulking period** individual plants are selected and their progenies are evaluated for 3-4 years. If a progeny is found superior to the check (s), it is released as a new cultivar. The duration of bulking may vary from 6-7 to 30 or more generations. Again, artificial selection may or may not be practised during the bulking period.

18.3.1 Applications of Bulk Method

It is most suitable for breeding crops that are normally closely spaced in production. It is used for field beans and soybeans. However, it is not suitable for improving fruit crops and many vegetables in which competitive ability is not desirable.

18.3.2 Procedure of Bulk Method

The exact procedure for the bulk method will vary depending upon the objective of the breeder. The following procedure is described for isolation of homozygous lines (Figure 18.2). The breeder may modify the scheme as per his needs.

First Year

Crosses are made between selected parents.

Second Year

F_1 is space-planted and harvested in bulk. The number of F_1 plants should be as large as possible; usually more than 20 plants should be raised.

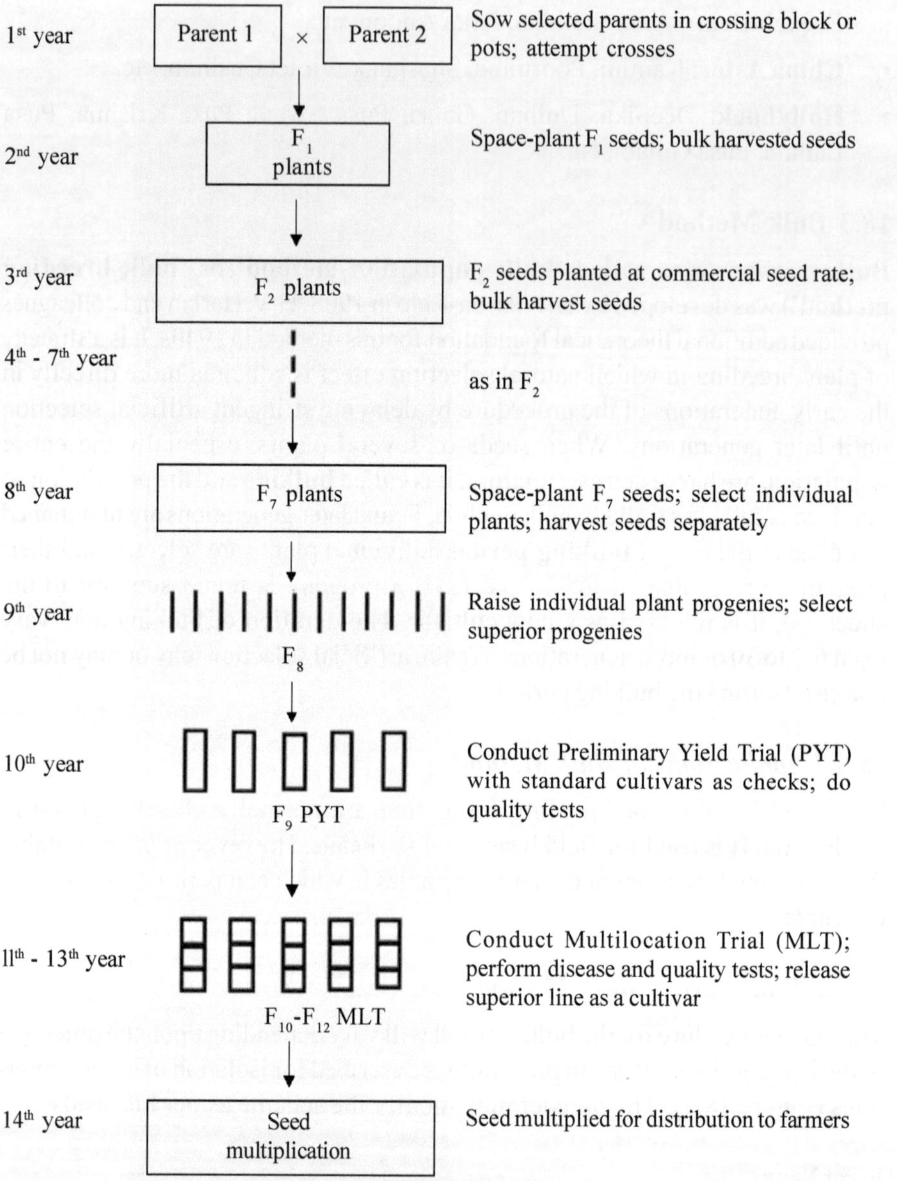

Figure 18.2. A generalized outline of bulk method of selection

Third to Seventh Year

F_2 to F_6 generations are planted at commercial seed rates and spacings. These generations are harvested in bulk. During this period, environmental factors, disease and pest outbreaks would change the frequencies of different genotypes in the population. Artificial selection is usually not done. The population size should be as large as possible, preferably 30,000-50,000 plants in each generation.

Eighth Year

About 30,000-50,000 plants are space-planted. Select 1000-5000 superior plants and harvest their seeds separately.

Ninth Year

Grow individual plant progenies in single or multi-row plots. Most of the progenies would be homogeneous; they are harvested in bulk. Weak and inferior ones are rejected. Select only 100-300 individual plant progenies with desirable traits. Segregating progenies are generally rejected unless they are of great promise. In promising progenies, select individual plants. In such cases, preliminary yield trial will be delayed for one year.

Tenth Year

Conduct preliminary yield trial (PYT) along with standard checks. Select the progenies that are superior to checks. Conduct quality test to further reject the undesirable ones.

Eleventh to Thirteenth Year

Conduct multilocation trials (MLTs) over several locations along with standard commercial cultivars as checks. The lines are evaluated for important characteristics in addition to yield, disease resistance and quality. If a line is superior to the standard checks, it would be identified for release as a cultivar.

Fourteenth Year

Once the superior line is identified for release, its seed multiplication is initiated. After its release and notification as a cultivar, certified seed can be produced.

18.3.3 Modified Bulk Method

In this case, selection can be practised in F_2 and F_3 generations. There will not be any pedigree record but superior plants are selected, bulked and carried forward. In F_4 superior plants are selected and harvested separately. In F_5 the single plants are studied in progeny rows and best progenies are selected and harvested. In F_6 PYT can be conducted to select best families. In subsequent generations regular MLT can be conducted.

This modification of bulk method provides an opportunity for the breeder to exercise his skill and judgement in selection. Again, there is no maintenance of pedigree record which is another advantage.

18.3.4 Mass Pedigree Method

This was proposed by Harrington in 1937. Mass pedigree method refers to raising segregating generations by **bulk (mass) method** when conditions are unfavourable for selection and use of **progeny testing (pedigree method)** when conditions are favourable for selection. The **bulking period** may vary from one to few generations depending upon the occurrence of favourable conditions. The bulking period is terminated as soon as favourable conditions for selection occur. It is a solution to one of the deficiencies in the pedigree method of breeding. For example, if the population is to be subjected to disease resistance screening like YMV (yellow mosaic virus) and if there is no method to create artificial epiphytotic conditions, it is wasteful to study the population in pedigree method. Instead we can carry the population as a mass and test them when there is occurrence of the disease. When conditions are favourable for the disease, we can terminate the bulking and resort to single plant selection.

18.3.5 Single Seed Descent Method

In principle the single seed descent (SSD) method is a modification of the bulk method of breeding. But the modification is such that it allows the equal survival of all segregants. This method was first suggested by Goulden (1941), but Grafius (1965) and Brim (1966) refined and advocated it. With this method one seed is randomly selected from each F_2 plant and then bulked to raise the F_3. Similar practice is continued till F_5 or F_6 or when single plant selections are made, and evaluation done. It is to preserve the total range of variation throughout the propagation period and to minimize the effects of natural selection in changing the genotypic array in the original population.

In one modification, i.e., the **multiple seed procedure** (also called **modified single seed descent**), 2-4 seeds are randomly selected from each plant. Some breeders simply harvest one multi-seeded pod per plant and so it is also referred to by some as **bulk-pod method**.

In another modification, *viz*., **single hill method**, progeny from individual plants are maintained as separate lines during the early generations by planting a few seeds in a hill. Seeds are harvested from the hill and planted in another hill the next generation. A plant is harvested from each line when homozygosity is attained.

18.3.6 Merits of Bulk Method

1. Thebulk method is simple, convenient and less expensive.
2. Since each F_2 plant is represented till F_6, no chance of elimination of good genotypes in early generations.
3. Artificial or natural disease epiphytotics, winter killing, high temperature, etc. eliminate undesirable types and increase the frequency of desirable types. Thus, isolation of desirable types becomes easier.
4. Selections from long-term bulk are superior to those from F_2 or short-term bulk.
5. Since little work and attention is needed in F_2 and subsequent generations, more crosses can be handled.
6. No pedigree record is maintained; it saves time.
7. Since large populations are grown, transgressive segregants are more likely to appear and increase due to natural selection. Hence, there is greater chance of isolation of good gegregants than pedigree method.

18.3.7 Demerits of Bulk Method

1. It takes longer time to develop a cultivar than pedigree method.
2. In short-term bulks, natural selection is unlikely to be important.
3. It provides little opportunity to the breeder to apply his skill and judgement in selection.
4. Sometimes natural selection may act against desirable types.
5. At the end of the bulking period, a large number of progenies are to be selected.
6. Information on genetics of traits cannot be obtained which is often possible in pedigree method.

18.3.8 Achievements

The bulk method has not been widely used. In USA, some barley cultivars have been developed by this method. In India, at least one cultivar has been developed by the bulk method, *viz*., Narendra Rai cultivar of *Brassica juncea*.

18.4 Backcross Method

Breeders of early 20th century engaged in the development of disease resistant cultivars observed that pureline selections with genes for resistance from intra- or inter-specific hybridization were inferior to the generally accepted superior parent in yield or quality characteristics. To overcome this problem, H.V. Harlan and M.N. Pope in 1922 proposed the backcross method by which an undesirable allele at a particular locus is replaced by the desirable allele in otherwise elite cultivar. In other words, backcross method conserves all good characteristics of a popular adapted cultivar and incorporates a desirable character from another cultivar or source. In principle backcross method does not improve the genotype of the parent, except the substituted gene(s).

18.4.1 Terminology

Backcross: A cross between a hybrid (F_1 or a segregating generation) and one of its parents is known as backcross.

Backcross Method: The method in which the hybrid and the progenies in the subsequent generations are repeatedly backcrossed to one of their parents is known as backcross method.

Recipient Parent (also called '**Recurrent Parent**'): Well-adapted high yielding cultivar lacking one or two characters and hence receives these genes from another cultivar.

Donor Parent (also called '**Non-Recurrent Parent**'): The cultivar which donates one or two useful gene(s).

18.4.2 Requirements of Backcross Method

1. Existence of a good recipient parent cultivar which requires improvement in some qualitatively inherited character or a quantitative character with high heritability.
2. A suitable donor parent must be available possessing the character or characters to be transferred in a highly intense form.
3. High expressivity of the character under transfer through several backcrosses in the genetic background of the recipient parent.
4. The character to be transferred must have high heritability and preferably governed by one or few genes.
5. Simple testing technique for detecting presence of the character under transfer.
6. Recovery of the recipient genotype in a reasonable number (usually, 5-7) of backcross generations.

18.4.3 Applications of Backcross Method

1. Inter-varietal transfer of simply (qualitatively) inherited characters (governed by one or two major genes).

2. Inter-varietal transfer of quantitative characters (with high heritability).
3. Inter-specific transfer of simply inherited characters.
4. Transfer of cytoplasm.
5. To obtain transgressive segregants.
6. Production of isogenic lines.
7. Germplasm conversion.

18.4.4 Procedure of Backcross Method

There are four schemes of back crossing depending on the situation. The number of back crosses varies from 3 to 10. But it should be continued only till the desired traits from the recipient parent are transferred.

1st year P	Recipient parent rr x Donor parent RR	Cross donor parent (DP) to recipient parent (RP)
2nd year F_1	Rr x rr	Backcross RP to F_1
3rd year BC_1F_1	rr, Rr x rr x	Discard homozygous recessive progeny; backcross RP to selected progeny; repeat the backcross of RP to selected progeny of BC_2F_1 to BC_5F_1
8th year BC_6F_1	rr, Rr x	Discard homozygous recessive progeny; allow selfing; harvest seed separately
9th year BC_6F_2		Grow individual plant progenies; discard homozygous recessive plants; select plants similar to RP; allow selfing
10th year BC_6F_3		Grow individual plant progenies; discard segregating line; select lines similar to RP with RR genotype

Replicated yield trials
Release new cultivar
Seed multiplication for certified seed production

Figure 18.3. A generalized outline of backcross method for transfer of character governed by dominant gene

18.4.4.1 Transfer of A Dominant Gene

A backcross programme consists of crossing recipient parent (RP) and donor parent (DP), and thereafter repeated crossing of selected plants from F_1 and latter back cross generations to RP (Figure 18.3). In each back cross generation, plants looking closer to RP but with trait (RR) of DP are crossed to RP. The same scheme can be used in the situation where the trait is governed by the recessive gene detectable before flowering due to some marker trait. Back crossing is continued up to BC_5 or BC_6. Finally, it is selfed to make selections for a fully homozygous plant even for transferred trait (RR). A number of such homozygous lines and similar to RP in look are selected in BC_6F_3. They may be evaluated in replicated trials and the superior one may be released as a new cultivar (improved version of RP).

18.4.4.2 Transfer of A Recessive Gene

If a recessive gene is to be transferred from DP, straight back cross cannot be done as a recessive gene would not express in a heterozygous condition. Suppose disease resistance governed by recessive gene from a DP is to be transferred to a superior cultivar. The F_1 would be susceptible but would be backcrossed to the RP. The BC_1F_1 generation would have RR and Rr plants in equal number, that cannot be distinguished, and thus would be selfed. Resistant plants (rr), but looking like RP, would again be backcrossed to RP followed by selfing (Figure 18.4). Thus, resistant lines that resemble RP may be selected after selfing in BC_5F_3. They may be evaluated in replicated trials and the superior one may be released as a new cultivar (improved version of RP).

1st year P	Recipient parent RR x Donor parent rr	Cross donor parent (DP) to recipient parent (RP)
2nd year F_1	Rr x RR	Backcross RP to F_1
3rd year BC_1F_1	RR, Rr	Allow selfing in all the plants
	↓ Selfing	
4th year BC_1F_2	RR, Rr, rr x RR x x	Select homozygous recesssive plants and backcross to RP
5th year BC_2F_1	Rr x RR	Select plants similar to RP and backcross to RP
6th year BC_3F_1	RR, Rr	Allow selfing in all the plants selfing
	↓ Selfing	
7th year BC_3F_2	RR, Rr, rr x RR x x	Select homozygous recessive plants and backcross to RP
8th year BC_4F_1	Rr x RR	Select plants similar to RP and backcross to RP
9th year BC_5F_1	RR, Rr	Allow selfing in all the plants
	↓ Selfing	
10th year BC_5F_2	RR, Rr, rr x x	Select plants similar to RP; allow selfing; harvest seeds separately
	↓ Selfing	
11th year BC_5F_3	| | | | | | | |	Grow individual plant progenies; select progenies similar to RP

↓

Replicated yield trials
Release as new cultlvar
Seed multiplication for certified seed production

Figure 18.4. A generalized outline of backcross method for transfer of character governed by recessive gene

18.4.4.3 Transfer of Quantitative Trait

When a quantitative trait governed by polygenes is to be transferred, each back cross generation should be selfed, and F_2 and F_3 generations should be grown. If the trait has low heritability, back cross F_2 and F_3 populations will have to be large (Briggs and Knowles 1967). Back crosses are made on plants selected in F_3, and BC_2 and BC_3 generations are grown. The system of back crossing and selfing alternates till the desired results are obtained.

18.4.4.4 Simultaneous Use of More Than One Donor Parent

Mac Key (1954) suggested a modified back cross programme where up to eight donor parents (DP) could be utilized first in multiple crossing and then in back crossing to recipient parent (RP) in the usual manner. Use of four DPs and one RP has been shown in Figure 18.5. Selections made from III cycle of crossing may be further backcrossed to RP to initiate a back cross programme with an RP.

DP1 x RP = DP1RP

x → DP1RP DP2RP

DP2 x RP = DP2RP

x → DP1RP DP2RP DP3RP DP4RP

DP3 x RP = DP3RP

x → DP3RP DP4RP

DP4 x RP = DP4RP

Figure 18.5. Backcross method to use four donor parents (DP) and one recipient parent (RP)

18.4.5 Merits of Backcross Method

1. The new cultivar is almost identical to the recipient parent (adapted cultivar to the region) except for the transferred character.
2. As the plants are evaluated for the character under transfer, multiple generations may be grown each year in off-season nurseries and green houses.
3. The size of population used in back crossing is quite small.
4. This is the only method to correct defects of an otherwise superior cultivar.
5. This is the only method to transfer a gene from wild species.
6. Extensive evaluation of the new cultivar is not required.
7. It can be modified to recover transgressive segregants.

18.4.6 Demerits of Backcross Method

1. Generally, the new cultivar is not superior to the recipient parent except for the transferred character.
2. There may be **linkage drag** when the gene is transferred from wild species.
3. Crossing is required in each backcross.
4. By the time the superior cultivar is improved, it may be replaced by another one.

18.4.7 Achievements

Kashi Vishesh, Kashi Amrit, Kashi Anupam in tomato, and Parbhani Kranti in okra are the cultivars developed through backcross method of breeding.

18.5 Brief References

Acquaah 2012, Agrawal 1998, Brown, Caligari and Campos 2014, Chaudhary 2014, Dabholkar 2006, De and Bhattacharjee 2011, Khan 2018, Kumar 2016, Priyadarshan 2019, Sharma JP 2009, Singh BD 2011, Singh and Shekhawat 2018, Singh Phundan 2011, Stoskopf, Tomes and Christie 2011.

Review Questions

Part A

Answer the following questions true or false.

1. The bulk method has been widely used. True/False
2. Donor parent is also known as recurrent. True/False
3. The size of population used in back crossing is quite small. True/False
4. In modified bulk method, selection may be done in F_2 and F_3. True/False
5. Nilsson-Ehle (1908) developed pedigree method. True/False

Part B

Answer the following questions

1. Who proposed back cross method of breeding?
2. Why selfing is essential while transferring a recessive gene in backcross method?
3. Bulk method may be called as __________ and ______________.
4. Why is recipient parent otherwise known as recurrent parent?
5. Who suggested first the single seed descent method?

Part C

Write a short note on each of the following topics.

1. Compare the pedigree and bulk methods of breeding.
2. Discuss the procedure of backcross breeding to transfer a desirable recessive character from a land race to an adapted and good cultivar of the region.
3. What are the modifications of bulk population method? Discuss them.
4. What are the advantages and limitations of pedigree method?
5. Enumerate the requirements that should be fulfilled before taking up a backcross breeding programme.

19

Methods of Selection Cross-Pollinated Populations

19.1 Introduction

In self-pollinated crops, the methods of selection aim at improving individual plants. On the other hand, in cross-pollinated crops they aim at improving a population of plants. *A population is a large group of interbreeding individuals.* Hence, **selection is called population improvement in cross-pollinated crops.** Selection in self-pollinated crops results in isolation of pure breeding lines which is the end product. But in cross-pollinated crops, it results in an improved population which is used as the commercial cultivar. Improved **inbred lines** (also called **inbreds**) for producing hybrids can be obtained from such populations. Again, like many rounds of selection being practised in different inbreeding generations in self-pollinated crops we are also practising many rounds of selection in cross-pollinated crops. But, unlike self-pollinated crops selection in cross-pollinated crops is followed by intermating to develop a new population before starting another round of selection. One cycle of selection consists of selection of individual/family and intermating of the selected units. Such cycle is repeated till we get the improved population. This method of practising the cycles of selection and recombination in a population is called **recurrent selection** or **population improvement**. Frey (1983) prefers the use of recurrent selection for closed populations where no new germplasm is added to the population during selection. Population improvement was used for a group of breeding techniques where all aspects are similar to recurrent selection but with the provision of adding new sources of germplasm whenever feasible. It is only a matter of coincidence that past history of application of recurrent selection has been limited to closed populations, but its basic aim has always been the accumulation of favourable genes, without any restriction on the migration of new genes during the course of selection. Apart from a slight nuance, population improvement and recurrent selection are two titles of the same principle and procedure of manipulating the genetic potential of populations. *Recurrent selection can also be applied in self-pollinated crops.*

Moll and Stuber (1974) divided the recurrent selection into two categories: intra-population improvement and inter-population improvement depending upon whether one is interested in improving one population or two populations simultaneously in order to isolate superior hybrids. Thus, the objective in intra-population improvement programme is the improvement of population mean, while in inter-population improvement, it is the improvement of the variety cross.

19.2 Types of Population Improvement

1. Intra-Population Improvement Methods
    - i) Individual Plant Selection Method
        - a) Mass Selection
    - ii) Family Selection Methods
        - a) Full-Sib Family Selection Method
        - b) Half-Sib Family Selection Method
        - c) Selfed (S_1 or S_2) Families Selection Method
    - iii) Combined Selection Method
    - iv) Top Cross Progeny Selection Method
        - a) Recurrent Selection for General Combining Ability (GCA)
        - b) Recurrent Selection for Specific Combining Ability (SCA)
2. Inter-Population Improvement Methods (Reciprocal Recurrent Selection)
    - i) Half-Sib Reciprocal Recurrent Selection
    - ii) Full- Sib Reciprocal Recurrent Selection
    - iii) Modified Reciprocal Recurrent Selection-1
    - iv) Modified Reciprocal Recurrent Selection-2

19.3 Intra-Population Improvement Methods

19.3.1 Individual Plant Selection: Mass Selection

A broad base population is grown in the field. Undesirable plants are culled out at the time of flowering. A number of plants with desirable characters are selected at the time of harvest and their seeds are bulked to raise the crop the next season. The recurrent mass selection continues till little or no improvement is observed over selection cycles (Figure 19.1). The starting or base population is called C_0 population and thus we can have C_1, C_2, C_3.....C_n generation population after n cycles of selection. Genetically mass selection takes advantage of $\frac{1}{2}v_A$

(additive genetic variance). The expected genetic advance (gain) due to mass selection, ΔG_m, takes the form:

$$\Delta G_m = k \times \frac{\frac{1}{2} v_A}{\sqrt{v_P}}$$

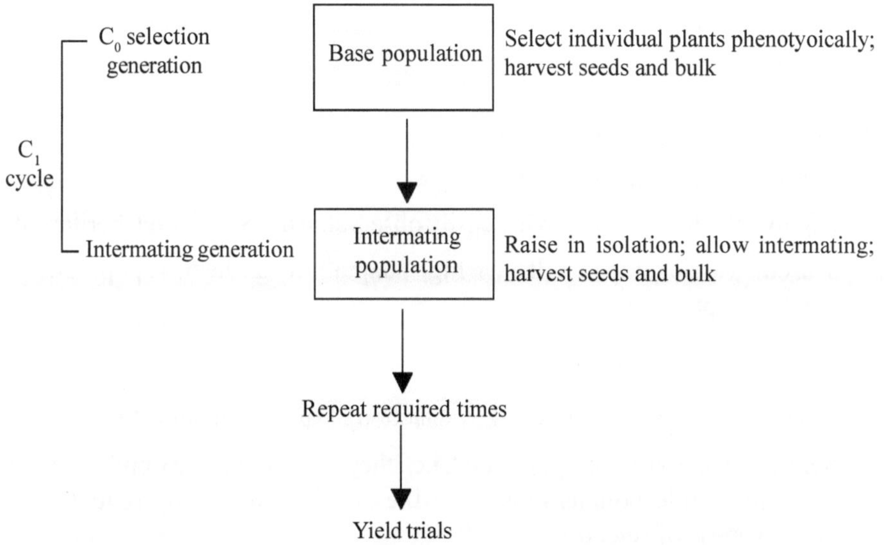

Figure 19.1. Recurrent mass selection as applied to cross-pollinated crops

where v_A is additive genetic variance, $\sqrt{v_P}$ is phenotypic standard deviation in the population and k is selection differential.

There may be another situation in mass selection. At flowering, individual plants are selected and they are mated randomly to produce the next generation population. Now, there is control over both ♀ and ♂ sexes whereas in earlier case, ♂ is not under control — the selected ♀ plants receive pollens from whole population. Genetic gain is higher when both the sexes are under control than when only one sex (♀ parents) is under control. The expected genetic advance (gain) due to mass selection ΔG_m, now takes the form:

$$\Delta G_m = k \times \frac{v_A}{\sqrt{v_P}}$$

Mass selection is effective when the characters with high heritability are under selection. For the characters with low heritability like yield, it will be effective only when the selection intensity is high.

19.3.1.1 Applications of Mass Selection

Mass selection has the following two applications in the improvement of cross-pollinated crops.

1. *Development of New Improved Open-Pollinated Cultivars*: This application is important in many crops where heterosis breeding is not feasible.
2. *Maintenance of Purity of Cultivars*: It is widely used to maintain the purity of open-pollinated, synthetic and composite cultivars of cross-pollinated crops.

19.3.1.2 Merits of Mass Selection

1. Simple and less time-consuming.
2. Highly effective for improving easily observable traits with high heritability.
3. The cultivar developed will have high adaptability as the base population is locally adapted one.

19.3.1.3 Demerits of Mass Selection

1. Using phenotypic selection makes selection of superior plants often difficult.
2. Selected plants are open-pollinated, i.e., they are pollinated by both desirable and undesirable pollens present in the pollination thereby reducing the effectiveness of selection.
3. If selection intensity is high (small population size advanced), the possibility of inbreeding depression as well as the probability of losing individuals with desirable combinations is increased.

19.3.1.4 Modifications of Mass Selection

They are proposed to overcome above defects.

1. **Detasseling**: It means removal of tassel before dehiscence of anthers. It is practised in maize. The inferior plants are detasseled. By this the inferior pollens are eliminated from the base population.
2. **Panmixis**: Pollens will be collected from selected plants and mixed together. This mix will be used to pollinate the selected plants. This ensures full control over pollen source.
3. **Stratified** or **Grid System**: The location effect may consist of a complex of sub-macro environmental effects at a given location. Some control over these variations can be realized by field stratification (Gardner 1961), i.e., by subdividing the area into a series of sub-blocks (small grids or subplots) with little environmental variance and practising selection within each unit.

An equal number of superior plants is selected from each sub-block for harvesting and bulking.

4. **Honeycomb Design**: Proposed by A. Fasoulas, the planting pattern is triangular rather than the conventional rectangular pattern. Each single plant is at the centre of a regular hexagon, with six equidistant plants, and is compared to the other six equidistant plants.
5. **Population Size and Selection Intensity**: Mass selection with a large effective population size (300-500 selected plants) and moderate selection intensity (20-40%) would minimize the loss of desirable alleles due to random drift.
6. **Selection Criterion and Mating System**: The genetic gain may be increased when heritability is low by adopting some criterion of selection other than individual's phenotype or by mating system other than that of random mating individuals selected on the basis of their phenotype.

19.3.1.5 Achievements

Some examples of mass selection in cross-pollinated horticultural crops are:

- **Cauliflower**: Pusa Katki, Early Kunwari, etc,
- **Cabbage**: Kalimpong English Ball, Kalimpong Eclipse Drumhead, Golden Acre, September, etc.
- **Radish**: Pusa Desi, Punjab Safed, Punjab Ageti, Arka Nishant, Punjab Pasand, Kashi Sweta, Kashi Hans, Pusa Jamuni, Pusa Gulabi, Pusa Mridula, Kalyanpur-1, etc.
- **Carrot**: Hisar Gairic, Pusa Rudhira, Pusa Asita, etc.
- **Onion**: Punjab Selection, Pusa Red, Arka Niketan, Udaipur 101, Udaipur 103, Arka Kalyan, Agrifound Dark Red, Nasik Red, Arka Pragati, Patna Red, N-53, Agrifound Light Red, Arka Bindu, Agrifound Rose, Punjab Naroya, Pusa White Round, Pusa White Flat, Baswant 780, Punjab 48, CO-2, Kalyanpur Red Round, Hisar-2, etc.
- **Beet Root**: Avon Early, etc.
- **Cassava**: Sree Prakash, etc.
- **Palak**: HS-23, etc.

19.3.2 Family Selection Method

In the (recurrent) family selection method, the unit of selection is not the individual plant but the families. Families can be either full-sib, half-sib or selfed progeny

(S_1 or S_2). In the family selection scheme, the plant breeder has to perform the following three operations:

1. Development of progeny.
2. Testing of the progenies and selection of superior ones.
3. Recombination of the superior families to synthesize a new population.

These three operations constitute one cycle of selection. One can go for few or more cycles of selection considering the observed realized improvement in population over cycles of selection.

19.3.2.1 Full-Sib Family Selection Method

A broad base population is grown in the field. About 10,000 plants are raised. Then undesirable plants are culled out. At flowering, a number of plants (say 1,000) are selected and 500 full-sibs are developed in the manner given below and shown in Figure 19.2. Reciprocal crosses are also developed.

Parents	Full-sibs
Plant # 1 × Plant # 2	Full-sib # 1
Plant # 3 × Plant # 4	Full-sib # 2
.	.
.	.
.	.
Plant # 999 × Plant # 1000	Full-sib # 500

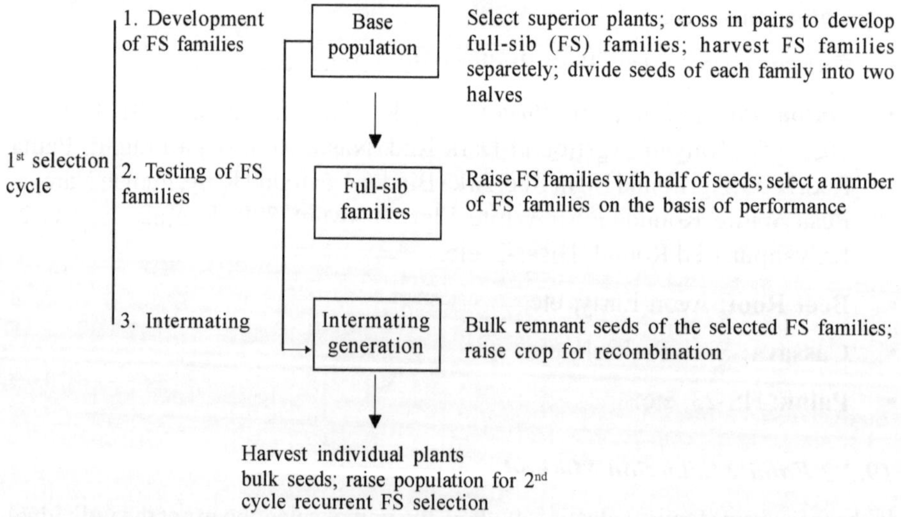

Figure 19.2. Recurrent full-sib family selection scheme in cross-pollinated crops

Depending upon the performance of full-sibs and reciprocal full-sibs, about 300-350 full-sibs are selected for testing. If one of the two crosses involving two parents yields poor, that full-sib is rejected.

After developing the progenies, multilocation replicated trial is conducted for testing. Observations on a number of traits are recorded. Half of the seeds is used for testing; the other half is stored for reconstituting population. Best 10-15 full-sib families across locations are selected and recombined to synthesize a population with wider adaptability for immediate commercial use. Similarly, best 10-15 full-sib families at a particular location are selected and a population is synthesized for use in that specific location. Thus, these are for short-term gain. Again, 25-30 full-sib families across environments are intermated to produce a population for starting the next cycle of selection. This is a minimum number based on the theoretical study. But, in practical breeding programme about 150 best full-sib (FS) families are selected for constructing a population for starting a new cycle of selection for long-term gain. The expected genetic advance (gain) due to full-sib family selection, ΔG_{FS}, takes the form:

$$\Delta G_{FS} = k \times \frac{\frac{1}{2} v_A}{\sqrt{v_{P\overline{FS}}}}$$

where v_A is additive genetic variance among FS families, $\sqrt{v_{P\overline{FS}}}$ is phenotypic standard deviation among FS families and k is selection differential.

19.3.2.2 Half-Sib Family Selection Method

Like full-sib family selection, in half-sib family selection scheme a broad base population of about 10,000 plants are raised and undesirable ones are culled out. At flowering, a number of plants (say 500) are selected and pollens from them are bulked. Each of the 500 selected plants is pollinated with these bulk pollens as given below:

Parents	**Half-sibs**
Plant # 1 × Bulk pollens	Half-sib # 1
Plant # 2 × Bulk pollens	Half-sib # 2
.	.
.	.
.	.
Plant # 500 × Bulk pollens	Half-sib # 500

At harvest, about 300-500 half-sibs are kept for testing. Again, poor performing half-sibs are rejected. The different steps involved are shown in Figure 19.3. The procedure of evaluation and recombination is the same as narrated in case of full-sib selection. The expected genetic advance (gain) due to half-sib family selection, ΔG_{HS}, takes the form:

$$\Delta G_{HS} = k \times \frac{\frac{1}{4} v_A}{\sqrt{v_{P\overline{HS}}}}$$

where v_A is additive genetic variance among HS families, $\sqrt{v_{P\overline{HS}}}$ is phenotypic standard deviation among HS families and k is selection differential.

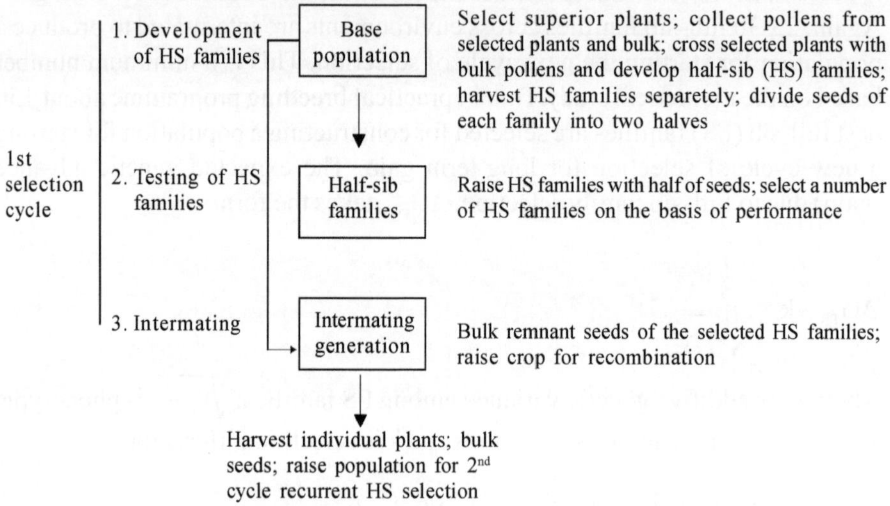

Figure 19.3. Recurrent half-sib family selection scheme in cross-pollinated crops

19.3.2.3 Selfed (S_1 or S_2) Families Selection Method

In case of S_1 family selection scheme, a broad base population of about 10,000 plants are grown. Undesirable ones are culled out. At flowering, a number of plants (say 500) are selected and selfed. At harvest, about 300-350 selfed progenies are kept for evaluation. However, in case of S_2 family selection scheme these S_1 progenies are further selfed to produce S_2 progenies that are then evaluated. Figure 19.4 shows various steps of S_1 family selection. The procedure of evaluation and recombination is the same as narrated in case of full-sib selection. The expected genetic advance (gain) due to S_1 family selection, ΔG_{S1}, takes the form:

$$\Delta G_{S1} = k \times \frac{v_{A1}}{\sqrt{v_{P\overline{S1}}}}$$

where v_{A1} is additive genetic variance among S_1 families, $\sqrt{v_{P\overline{S1}}}$ is phenotypic standard deviation among S_1 families and k is selection differential.

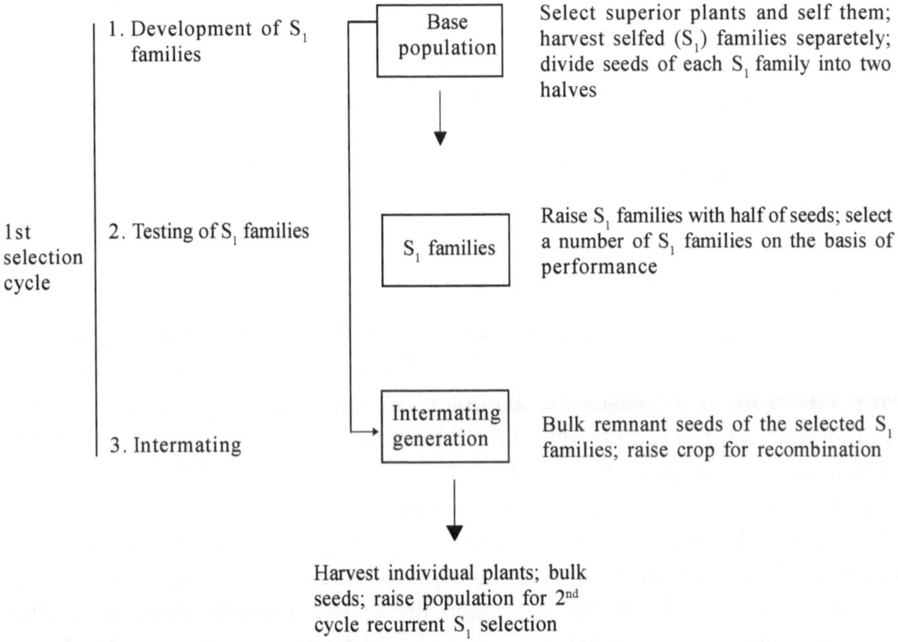

Figure 19.4. Recurrent S_1 family selection scheme in cross-pollinated crops

Further, the expected genetic advance (gain) due to S_2 family Selection ΔG_{S2}, takes the form:

$$\Delta G_{S2} = k \times \frac{\frac{3}{2} v_{A2}}{\sqrt{v_{P\overline{S2}}}}$$

where v_{A2} is additive genetic variance among S_2 families, $\sqrt{v_{P\overline{S2}}}$ is phenotypic standard deviation among S_2 families and k is selection differential.

When compared, full-sib (FS) selection will give more genetic gain than half-sib (HS) selection. But, for similar selection intensity one is sampling more population in HS selection scheme than in FS selection scheme, and also the FS correlation will be increased by the common environments. These two factors favour HS selection scheme. Again, it is easier to produce HS progenies. So, breeders prefer HS selection scheme to FS selection scheme. In comparison to mass selection, both FS and HS selection schemes utilize less additive genetic variance present in the population.

When one wants to eliminate the deleterious effects of undesirable genes, or when the ultimate aim is to isolate inbreds from the improved population, or when it is difficult to control environmental variations; one can resort to S_1 or S_2 family selection method for utilizing the additive genetic variance. This is the most efficient way of progeny testing. With inbreeding, variance among progeny means will increase whereas the variance within progeny will decrease and thus, discrimination of genotypes will be much better in S_2 than in S_1 generation, i.e., increase in genetic variability among families is more in S_2 ($\frac{3}{2}v_{A2}$) than in S_1 (v_{A1}).

19.3.2.4 Combined Selection Method

When two or more selection methods are used in the plant breeding programme, it is called as combined selection method. Hopkins (1899) proposed the **ear-to-row selection**. It increases the accuracy of intra-population selection. This method is essentially a between-and-within-family selection. It combines half-sib testing and mass selection within superior progenies. **Modified ear-to-row selection** of Webel and Lonnquist (1967) includes multilocation evaluation of HS families. High yielding ears are half-sibs. They are selected and planted in ear-to-row fashion in the field. The mean performance of three locations decides the selection of family. On the other hand, the selection of individuals within the selected families is dependent upon visual selection at one location. Modified ear-to-row method of selection uses $1/2_{vA}$ (additive genetic variance). The expected genetic advance (gain) due to modified ear-to-row selection, ΔG_{mHS}, takes the form:

$$\Delta G_{mHS} = \frac{k(1/8)v_A}{\sqrt{v_{PHS}}} + \frac{k(3/8)v_A}{\sqrt{v_{PFS}}}$$

where v_A is additive genetic variance, $\sqrt{v_{PHS}}$ is phenotypic standard deviation among HS families, $\sqrt{v_{PFS}}$ is phenotypic standard deviation among FS families, and k is selection differential.

In the combined selection methods, the expected progress will be the sum of progress expected from each method. The combined selection that uses information from HS and S_1 progeny increases the frequency of both alleles showing additive effects and alleles showing dominance or heterotic effects. Within-family selection can be applied in the inbreeding generation like S_1 selection or HS selection where the selfed progenies are recombined on the basis of HS performance. Within-family selection can also be applied in FS and S_1 or S_2 progeny selection scheme as well.

19.3.2.4.1 Achievements

Ear-to-row method is also known by other names such as plant-to-row selection, line breeding, family selection, mass-pedigree, etc. A little variation is associated with each name. Single plant selection with progeny testing has been commonly employed for improvement of cross-pollinated vegetables. Long list of vegetables developed through ear-to-row method or its modifications is available. Kalloo (1988) has discussed the prospect of this method in vegetable improvement. Qualitative as well as quantitative traits with high heritability can be improved by this method. Heterozygosity and heterogeneity can be maintained at high levels by selecting larger number of lines. Inferior genes can be eliminated by repeating cycles of selection. Some examples of cultivars developed through ear-to-row method or its modifications are cited below:

- **Amaranthus**: Badi Chaulai, Chhoti Chaulai, Kannara Local, Pusa Kiran, Pusa Kirti, CO-1, CO-2, CO-3, Arka Suguna, etc.
- **Bitter Gourd**: Coimbatore Long, Pusa Do Mausami, Arka Harit, VK-1-Priya, CO-1, MC-23, Pusa Vishesh, Punjab BG-14, NDB-1, Phule BG-6, Kalyanpur Sona, etc.
- **Bottle Gourd**: Pusa Summer Prolific Long, Pusa Summer Prolific Round, Punjab Long, Punjab Round, CO-1, Arka Bahar, Pusa Naveen, etc.
- **Cucumber**: Sheetal, Phule Shubhangi, Pusa Barkha, Pusa Uday, Kalyanpur Green, Swarna Poorna, Pant Khira-1, Swarna Sheetal, etc.
- **Tinda**: Arka Tinda, Tinda S-48, etc.
- **Long Melon**: IIHR-4, Karnal Selection, etc.
- **Musk Melon**: RM-43, Durgapura Madhu, Arka Rajhans, Arka Jeet, Pusa Madhuras, Lucknow Safeda, etc.
- **Pumpkin**: CO-1, CO-2, CM-14, etc.
- **Ridge Gourd**: Pusa Nasdar, CO-1, CO-2, etc.
- **Snake Gourd**: CO-1, CO-2, TA-19, etc.
- **Sponge Gourd**: Pusa Chikni, etc.
- **Summer Squash**: Punjab Chappan, Kaddu-1, Early Yellow Prolific, etc.
- **Watermelon**: Durgapura Meetha, Durgapura Kesar, etc.
- **Wax Gourd**: CO-1, KAU Local, etc.
- **Cauliflower**: Hisar-1, Pusa Him Jyoti, Snowball-16, Pusa Snowball K-1, Punjab Giant-26, Punjab Giant-35, Pant Gobhi-2, Pant Shubhra, Pusa Sharad, Pusa Snowball-2, Kashi Kunwari, Pusa Paushja, etc.
- **Cabbage**: Pride of India, etc.

- **Onion**: Pusa Ratnar, Pusa Riddhi, Hisar-2, etc.
- **Radish**: Pusa Chetki, CO-1, etc.
- **Turnip**: Pusa Swati, etc.
- **Carrot**: Ooty-1, etc.

19.3.2.5 Top Cross Progeny Selection

19.3.2.5.1 Recurrent Selection for General Combining Ability (GCA)

In recurrent selection for general combining ability (gca) a number of plants in the population are selected, selfed and crossed to an open-pollinated cultivar (tester). The next season the top crosses are tested; the lines with good gca are selected. The selfed seeds of the selected lines are used for intermating. It results in a new population for further selection. In essence this is a HS selection scheme. The expected genetic advance (gain), ΔG, is:

$$\Delta G = k \times \frac{\frac{1}{2}v_A}{\sqrt{v_{PHS}}}$$

19.3.2.5.2 Recurrent Selection for Specific Combining Ability (SCA)

In recurrent selection for specific combining ability (sca) a number of plants in the population are selected, selfed and crossed to homozygous tester (inbred tester). The next season the crosses are tested; the lines with higher sca are selected. The selfed seeds of the selected lines are used for intermating. It results in a new population for further selection. In this scheme, after a number of cycles of selection a number of inbreds will be available. Some of them with the tester may produce super hybrids.

An open-pollinated cultivar could be improved through 'selection and recombination' (i.e. recurrent selection) to the extent when its performance approaches that of a hybrid.

19.4 Inter-Population Improvement Methods

The inter-population selection scheme is also called reciprocal recurrent selection scheme because two populations are simultaneously improved. The goal of selection is to synthesize the superior varietal cross for immediate commercial use and also to produce superior hybrids from inbreds developed from these two improved populations. Even either of the two improved populations from the reciprocal recurrent selection programme can be used as a commercial open-pollinated cultivar. The steps involved in the inter-population selection schemes are the same as the family selection of intra-population selection scheme except that in this case we start with two populations instead of one.

19.4.1 Half-Sib Reciprocal Recurrent Selection (RRS-HS) (Comstock et al. 1949)

Suppose we have two populations, *viz.*, A and B. The half-sib (HS) progeny is developed in the manner given below:

Case-I: The selected individual plants in population A are used as ♂. Each is crossed to a number of selected plants used as ♀ in population B. Thus, HS progeny is developed in population B.

Case-II: Conversely, the selected individual plants in population B are used as ♂. Each is crossed to a number of selected plants used as ♀ in population A. Thus, HS progeny is developed in population A.

Thus, each population acts as a tester for the other. The ♂ plants in each population are selfed. Depending upon the performance of HS families, the S_1 seeds are used to intermate to produce a new population for starting another cycle of reciprocal recurrent selection. The different steps involved in the process of reciprocal recurrent selection are shown in Figure 19.5. This scheme is designed to utilize both gca and sca simultaneously. The test cross performance utilizes all types of gene actions and interactions (additive, dominance and epistasis). In the inter-population selection schemes, the expected genetic gain from selection refers to the improvement of the **population cross** (also called **'varietal cross'** or **'variety cross'**); it is the sum of expected genetic gain of each population with its tester. The expected genetic gain due to half-sib reciprocal recurrent selection is:

$$\Delta G = \frac{k_1\left(\frac{1}{4}\right)v_{A_{AB}}}{\sqrt{v_{p\bar{F}_{AB}}}} + \frac{k_2\left(\frac{1}{2}\right)v_{A_{BA}}}{\sqrt{v_{p\bar{F}_{BA}}}}$$

where V_{AAB} and V_{ABA} are additive genetic variances in populations A and B respectively, $\sqrt{V_{p\bar{F}_{AB}}}$ and $\sqrt{V_{p\bar{F}_{BA}}}$ are phenotypic standard deviations among families in populations A and B respectively, k_1 and k_2 are selection differentials in populations A and B respectively.

19.4.2 Full- Sib Reciprocal Recurrent Selection (RRS-FS) (Hallauer and Eberhart 1970)

Suppose we have two populations, *viz.*, A and B. FS families are developed in both populations keeping individuals in A as ♂ and individuals in B as ♀ and vice versa. The ♂ plants in both A and B are selfed. On the basis of performance of FS families, the S_1 lines are selected; they are intermated to produce new populations from which population cross can be obtained and tested for performance. The different steps involved in the process of this scheme are shown in Figure 19.5. The expected genetic gain due to selection would take the form:

$$\Delta G = \frac{k\left(\frac{1}{4}\right)\left(v_{A\,AB} + v_{A\,BA}\right)}{\sqrt{v_{p\bar{F}}}}$$

This selection scheme can be applied to a population producing two or more ears per plant. One ear is used for producing FS and the other one is selfed.

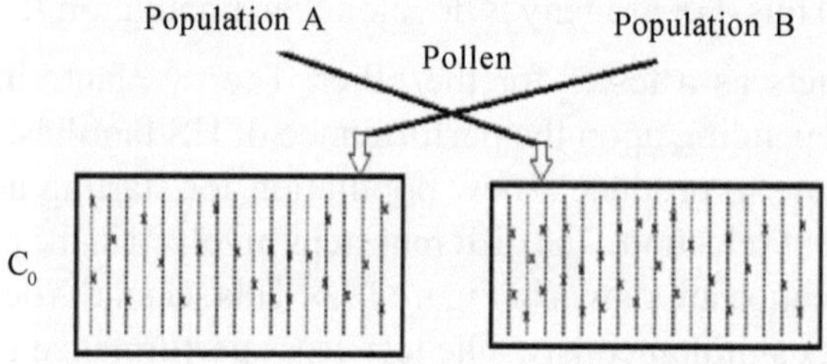

Steps:

1. Development of HS or FS in population A using ♂ from population B.
2. Selfing of ♂ and production of S_1 lines.
3. Selection of S_1 lines based on performance of HS or FS.
4. Intermating of selected S_1 lines for reconstitution of a population (C_1A).
5. Production of population cross hybrid $C_1A \times C_1B$ and comparison with $C_0A \times C_0B$.
6. Production of hybrids $S_1A \times S_1B$ and isolation of superior cross.

1. Development of HS or FS in population B using ♂ from population A.
2. Selfing of ♂ and production of S_1 lines.
3. Selection of S_1 lines based on performance of HS or FS.
4. Intermating of selected S_1 lines for reconstitution of a population (C_1B).
5. Production of population cross hybrid $C_1B \times C_1A$ and comparison with $C_0B \times C_0A$.
6. Production of hybrids $S_1B \times S_1A$ and isolation of superior cross.

Figure 19.5. HS and FS reciprocal recurrent selection in cross-pollinated crops

The phenotypic variance of FS is greater than HS at the same selection intensity. So, the genetic gain from RRS-FS will be lower than RRS-HS. However, as RRS-FS allows the testing of twice more plants for the same size of plots, the choice between RRS-HS and RRS-FS depends on the value of selection intensity and heritability. The choice of RRS-FS increases for lower selection intensity and lower heritability.

19.4.3 Modified Reciprocal Recurrent Selection-1(Modified RRS-1) (Paterniani 1967, 1973)

In this scheme, selected open-pollinated ears (HS) in both populations are harvested separately. They are planted ear-to-row with rows of ♂ from the other population as shown in Figure 19.6. The HS family rows are detasselled in each population. The open-pollinated ears from the HS families are harvested and composited. The different HSs are then evaluated. The remnant seeds of the selected HS families in a population are raised and intermated to produce an improved population. From the two improved populations a **population cross** (also called **'varietal cross'** or **'variety cross'**) can be made and evaluated. The expected genetic gain in this scheme is:

$$\Delta G = \frac{k_1\left(\frac{1}{16}\right)v_{A\,AB}}{\sqrt{v_{p\bar{F}_{AB}}}} + \frac{k_2\left(\frac{1}{16}\right)v_{A\,BA}}{\sqrt{v_{p\bar{F}_{BA}}}}$$

where $v_{A\,AB}$ and $v_{A\,BA}$ are additive genetic variances in populations A and B respectively, $\sqrt{v_{p\bar{F}_{AB}}}$ and $\sqrt{v_{p\bar{F}_{BA}}}$ are phenotypic standard deviations among families in populations A and B respectively, k_1 and k_2 are selection differentials in populations A and B respectively.

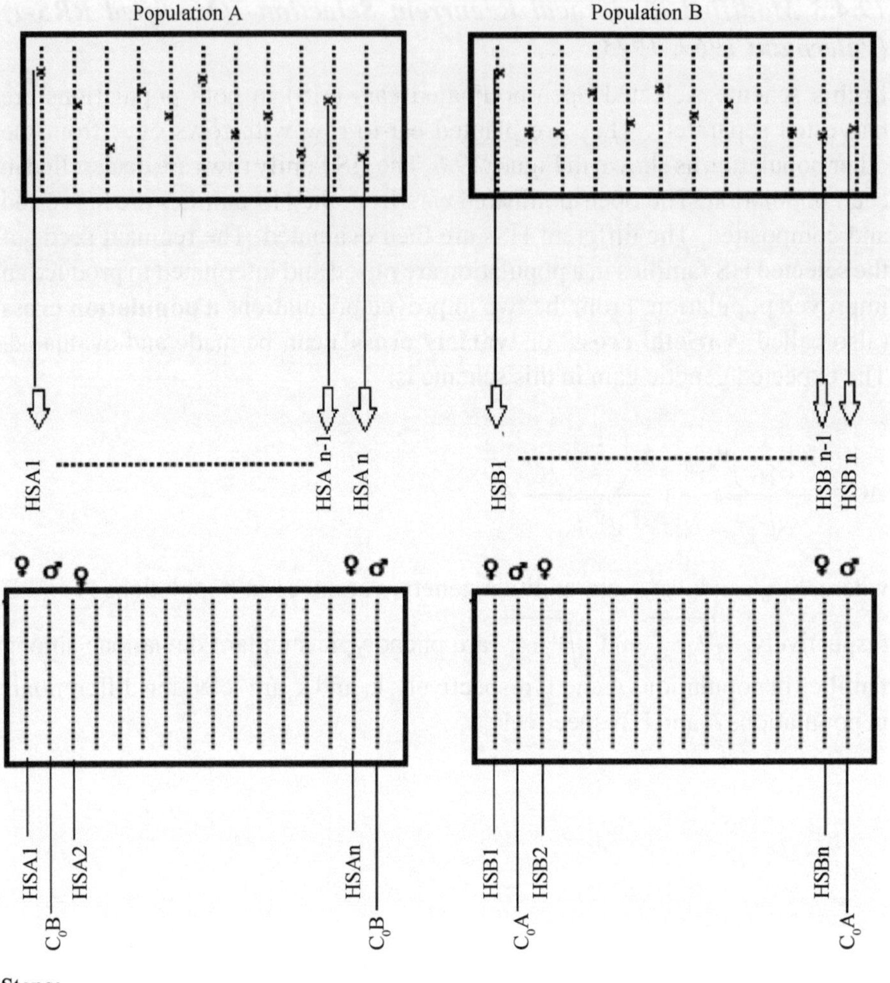

Steps:

1. Selection of open-pollinated ears (HS) in both populations.
2. Planting ear-to-row with rows of ♂ from the other population; detasseling of HS family rows in each population; development of HSA×C_0B progenies in population A, and HSB×C_0A progenies in population B.
3. Testing of HSA×C_0B and HSB×C_0A progenies; HSAs and HSBs are selected based on performance of HSA×C_0B and HSB×C_0A progenies, respectively.
4. Remnant seeds of the selected HSAs are raised and intermated to produce a new population C_1A. Similarly, remnant seeds of the selected HSBs are raised and intermated to produce a new population C_1B.
5. Production of improved population crosses C_1A×C_1B and C_1B×C_1A and its comparison with C_0A×C_0B and C_0B×C_0A, respectively.

Figure 19.6. Modified reciprocal recurrent selection-1 in cross-pollinated crops

19.4.4 Modified Reciprocal Recurrent Selection-2(Modified RRS-2) (Paterniani 1967, 1973)

This selection scheme is applicable in populations when two or more ears per plant are produced. Suppose we have two populations, namely, A and B. We need two blocks to plant them. In Block-I we plant both the populations row-wise and Population A is detasseled (i.e. made ♀). Likewise, in Block-II we plant both the populations row-wise and Population B is detasseled (i.e. made ♀). So, in Block-I Population B and in Block-II Population A are sources of pollens for open (also called random or uncontrolled) pollination. At flowering, bulked pollen from Population B (collected while detasseling) of Block-II is transferred to pollinate one of the two ears on plants of Population A used as ♀ in Block-I (selective or controlled pollination). Similarly, bulked pollen from Population A (collected while detasseling) of Block-I is transferred to pollinate one of the two ears on plants of Population B used as ♀ in Block-II (Figure 19.7). The other ear on the plants of Population A (used as ♀ in Block-I) as well as Population B (used as ♀ in Block-II) is open-pollinated (random or uncontrolled pollination). Thus, we have HS families developed through uncontrolled as well as controlled pollination in both Population A and Population B. On the basis of performance of HS families developed through controlled pollination, HS families developed through uncontrolled pollination are selected and intermated to produce an improved population. Now, population cross can be obtained using the two improved populations and evaluated. The expected genetic gain in this scheme is:

$$\Delta G = \frac{k_1\left(\frac{1}{8}\right)v_{A\,AB}}{\sqrt{v_{p\bar{F}_{AB}}}} + \frac{k_2\left(\frac{1}{8}\right)v_{A\,BA}}{\sqrt{v_{p\bar{F}_{BA}}}}$$

where $v_{A\,AB}$ and $v_{A\,BA}$ are additive genetic variances in populations A and B respectively, $\sqrt{v_{p\bar{F}_{AB}}}$ and $\sqrt{v_{p\bar{F}_{BA}}}$ are phenotypic standard deviations among families in populations A and B respectively, k_1 and k_2 are selection differentials in populations A and B respectively.

Appropriate variance components are not available to compare expected genetic gain from selection for inter-population selection methods. Open-pollinated ears are HSs produced by uncontrolled pollination.

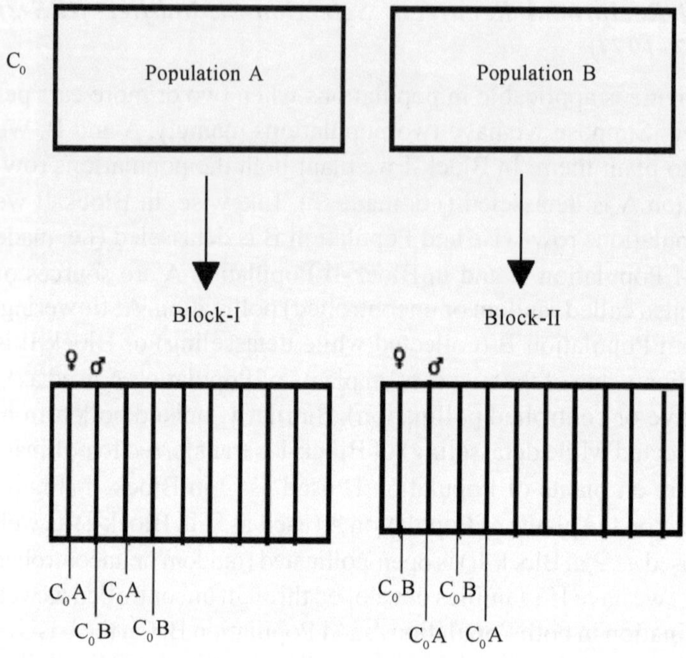

Steps:

1. Development of HS families through controlled and uncontrolled pollination in Population A and Population B:

   a) Bulked pollen from C_0B (collected while detasseling) in Block-II is transferred to pollinate one of the two ears on plants of C_0A in Block-I; thus, HS families through controlled pollination are produced. The other ear on the plants of C_0A in Block-I is open-pollinated and HS families through uncontrolled pollination are produced.

   b) Bulked pollen from C_0A (collected while detasseling) in Block-I is transferred to pollinate one of the two ears on plants of C_0B in Block-II; thus, HS families through controlled pollination are produced. The other ear on the plants of C_0B in Block-II is open-pollinated and HS families through uncontrolled pollination are produced.

2. Based on performance of HS families developed through controlled pollination, HS families developed through uncontrolled pollination are selected. This is done in both the populations.

3. Intermating of selected HS families (developed through uncontrolled pollination) to generate new base population C_1A. Similarly, C_1B is also produced.

4. Production of improved population crosses $C_1A \times C_1B$ and $C_1B \times C_1A$ and its comparison with $C_0A \times C_0B$ and $C_0B \times C_0A$, respectively.

Figure 19.7. Modified reciprocal recurrent selection-2 in cross-pollinated crops

19.5 Optimizing Gain from Selection in Population Improvement

The aim of a plant breeder is to make systematic progress in the mean expression of the character of interest from one cycle to the next. Achieving progressive gains in yield depends on several factors such as genetic variance, selection intensity, generations per cycle and field plot technique.

1. **Genetic Variance**: Additive genetic variance is critical to increase in gains per cycle. Additive genetic variance can be increased through increasing diversity in the entries used in population improvement.
2. **Selection Intensity**: When selection intensity is increased, the rate of gain due to selection is increased.
3. **Generations per Cycle**: Breeder's choice of the breeding system to use in a breeding programme is influenced by how rapidly each cycle of selection can be completed. If possible, using 2-3 generations per year can increase yield gains. Multiple generations per year can be possible by off-season nursery, or planting in the dry season providing irrigation.
4. **Field Plot Technique**: A breeder selects plants in the field. The field should be managed by adopting a suitable experimental design to reduce random variation. Whenever possible, a uniform field should be selected for field evaluation. Optimum cultural practices should be followed as much as possible to raise the crop. This will reduce variation between replications. Other factors to consider are plot size, number of plants per plot, number of replications per trial, and number of locations. If implemented properly, these factors reduce random variation.

19.6 Comparison of Methods of Intra-Population Improvement

The choice of which intra-population selection scheme to use depends on the relative theoretical efficiency, the ease with which it can be applied and the resource available. Considering the ratio of the amount of additive genetic variance and the number of generations per cycle $\left(\text{Ratio} = \frac{V_A}{\text{no. of generations/cycle}}\right)$, mass selection with control on ♂ has the highest ratio followed by mass selection with no control on ♂, modified ear-to-row and S_1 progeny testing. Mass selection is simple and inexpensive. But, if it is possible to grow two crops per year, FS or HS family selection might be the best. Again, HS family selection is more frequently used as it is rather simple. In crops where it is difficult to produce HS or FS, S_1 is the most efficient method of recurrent selection.

19.7 Comparison of Methods of Inter-Population Improvement

The expected gain is similar in RRS-HS and RRS-FS except for differences in the components of phenotypic variance. To give similar response selection differential (k) in RRS-FS would have to be 1.2 times greater than in RRS-HS.

At lower selection intensity, RRS-FS has an advantage over RRS-HS but this advantage decreases at higher selection intensity.

A comparison of RRS-FS with RRS 2-HS based on assumption of equal effective population size showed genetic gain in RRS 2-HS is higher than in RRS-FS but the reverse is true when the comparison is based on equal selection intensity. A comparison of RRS 1-HS and RRS 2-HS showed that RRS 2-HS utilizes more additive genetic variance than RRS 1-HS; so, it is a more efficient procedure but higher effective population size can be maintained in RRS 1-HS.

Thus, from the discussion on recurrent and reciprocal recurrent selection it can be concluded that in the initial stages of breeding programme at least simpler methods of intra-population selection (mass selection, FS, HS, S_1 or S_2 progeny) may be quite satisfactory. More complex methods of inter-population selection (RRS-HS, RRS-FS, RRS 1-HS, RRS 2-HS) may become advantageous in later stages after significant improvements have been achieved. In other words, one should utilize the additive genetic variance first, and then when the remaining genetic variation is largely of dominance and epistasis types, we should go for exploiting the non-additive genetic variation and thus isolation of lines for F_1 hybrid combination can be used to further enhance productivity.

19.8 Development of Synthetic Cultivars

19.8.1 Synthetic Cultivar Versus Germplasm Composite

There are two basic types of open-pollinated populations of crops–those produced by population improvement, and synthetics. Population improvement methods can be categorized into two–those that depend on purely phenotypic selection (mass selection) and those that involve selection with progeny testing. A **synthetic cultivar** (in brief **synthetics**) may be defined as an advanced generation of cross-fertilized (random mating in all combinations) seed mixture of parents (selected based on GCA) that may be strains, clones, or hybrids. Population improvement cultivars can be propagated indefinitely as such. But synthetic cultivars can be propagated for only a limited number of generations and then must be reconstituted from the parental stock.

Germplasm composites (in brief **composites**) is a broad term that refers to the mixing together of breeding materials on the basis of some agronomic character (e.g., yield potential, maturity, disease resistance) followed by random mating. There are many ways to put the breeding materials together to a composite. Germplasm composites are genetically broad-based and very complex. They can be cultivated commercially over a broad range of agroecological environments.

19.8.2 Desirable Features of Synthetic Cultivars

K. J. Frey summarized three major desirable features of synthetics as:

1. Yield reduction is less than a single or double cross. So, there is no need to use new seed each season.
2. A synthetic cultivar may become better adapted to the local production environment over time, as it is produced in successive generations in the region.
3. A synthetic cultivar is genetically heterogeneous. This makes it perform stably over changing environmental conditions. Again, due to this heterogeneity, both natural and artificial selection can modify the genotypic structure of synthetic cultivars. It means a plant breeder may achieve gain in performance by practising selection in Syn_2 and subsequent generations.

19.8.3 Applications

The possibility of commercial utilization of synthetic cultivars in maize was first suggested by Hayes and Garber in 1919. Synthetic cultivars have been of great value in the breeding for those cross-pollinated crops where pollination control is difficult.

19.8.4 Procedure of Development of Synthetic Cultivars

A synthetic cultivar is product of mixing lines (e.g., inbredlines, clones, open-pollinated cultivars, short-term inbredlines or other populations) that show good general combining ability (GCA). The following steps are involved (Figure 19.8) in its production:

1. Evaluation of parental lines for GCA
2. Production of the synthetic cultivar
3. Multiplication of the synthetic cultivar

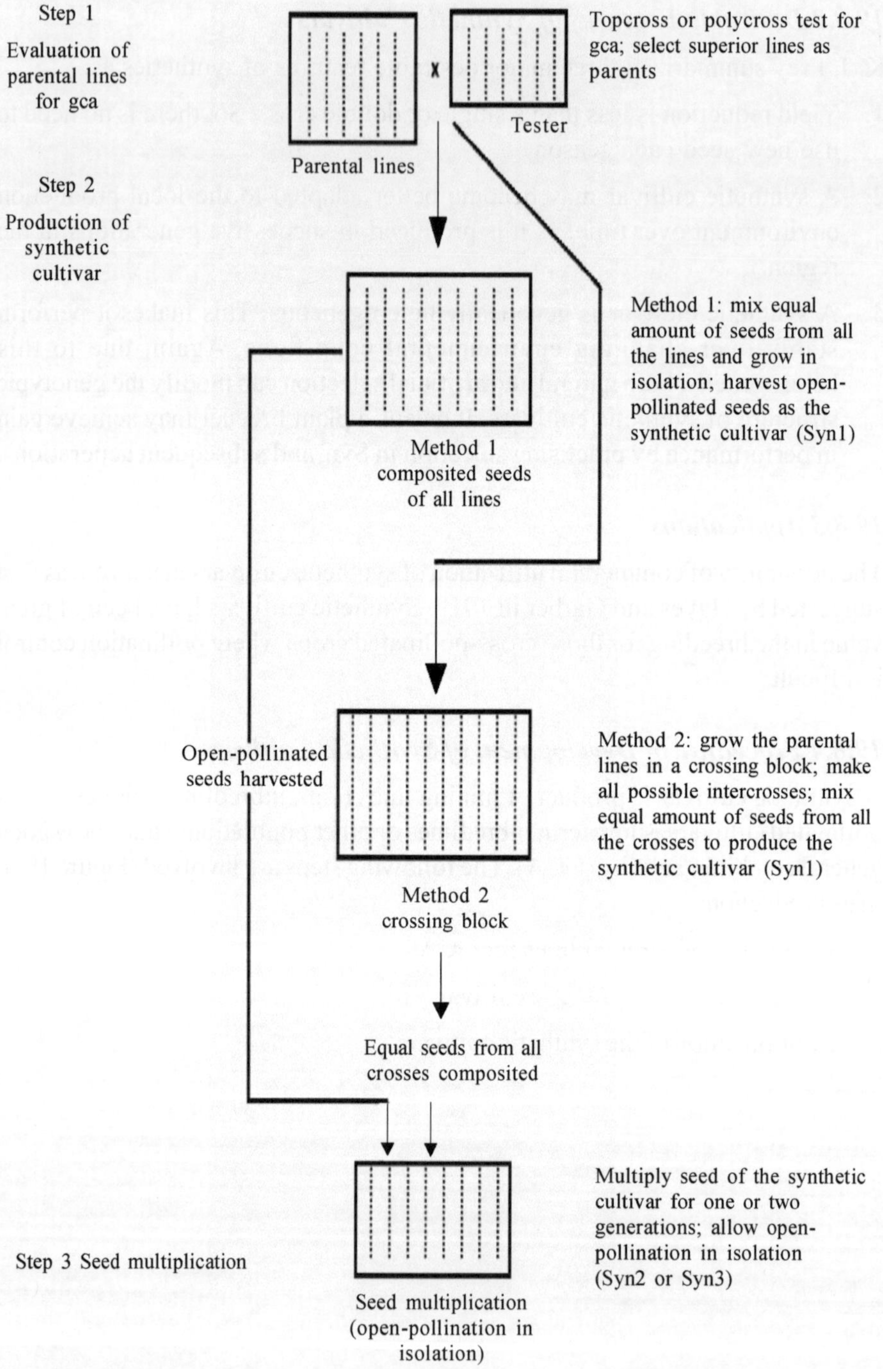

Figure 19.8. Steps in production of synthetic cultivar in cross-pollinated crops

19.8.4.1 Evaluation of Parental Lines for GCA

The GCA of parental lines are usually evaluated by topcross or polycross test. Topcross or polycross progenies of the selected lines are evaluated in a replicated yield trial. The performance of the progenies provides an estimate of GCA of the lines that produced them. The lines with high GCA are selected as parents (**Syn_0**) of the synthetic cultivar.

19.8.4.2 Production of the Synthetic Cultivar

A synthetic cultivar may be produced in the following two ways:

1. Equal amount of seeds from all the parental lines are mixed, planted in isolation and allowed to open-pollinate. The seed from this population is harvested in bulk; this is the **Syn_1** generation seed of the synthetic cultivar.
2. The parental lines are planted in isolation. All possible crosses among them are made by hand. Equal amount of seeds from all the crosses are composited to give the **Syn_1** generation seed of the synthetic cultivar.

The available experimental evidence suggests that both the methods produce comparable results.

19.8.4.3 Multiplication of the Synthetic Cultivar

Syn_1 seeds are planted in isolation and allowed to open-pollinate; this produces **Syn_2** generation seeds. Open-pollinated progeny of Syn_2 generation is called **Syn_3** and so on. In most crops, Syn_2 or Syn_3 seeds are distributed for commercial cultivation. But, in some crops, e.g., sugar beet, the Syn_1 seed itself is distributed for cultivation.

19.8.5 Performance of Synthetic Cultivars

The performance of a synthetic cultivar can be estimated as follows (Wright 1922):

$$Syn_2 = Syn_1 - \frac{Syn_1 - Syn_0}{n}$$

where Syn_2 is the expected performance of Syn_2 generation of the synthetic cultivar, Syn_1 is the performance of Syn_1 generation of the synthetic cultivar, Syn_0 is the mean performance of Syn_0 (i.e., parental lines) of the synthetic cultivar and n is number of parental lines of the synthetic cultivar.

The yield of Syn_2 is less than Syn_1; but there is no further decline in the performance of Syn_3, Syn_4 and the subsequent generations. The decline in performance of Syn_2 occurs due to the production of new gene combinations and a loss of heterozygosity due to the random mating in Syn_1 generation.

19.8.6 Merits of Synthetic Cultivars

1. The method is easy to implement.
2. It can be used to produce variability for hybrid breeding programmes.
3. Advanced genotypes of synthetic cultivars show little yield reduction from Syn_1. So, it is possible for farmers to save and use seed from the current season to plant the next season.
4. Synthetic cultivars offer a feasible means of utilizing heterosis in crop species where pollination control is difficult.
5. In variable environments, synthetic cultivars are likely to do better than hybrids.

19.8.7 Demerits of Synthetic Cultivars

1. The performance of synthetic cultivars is usually less than single or double cross hybrids.
2. The performance of synthetic cultivars is adversely affected by lines with poorer GCA.
3. Synthetics can be produced only in cross-pollinated crops whereas hybrids can be produced both in self- and cross-pollinated crops.

19.8.8 Achievements

Some examples of synthetic cultivars are:

- **Cauliflower**: Pusa Early Synthetic, Synthetic 78-1, Pant Gobhi-3, Pusa Synthetic, etc.
- **Cabbage**: Pusa Synthetic, etc.

19.9 Brief References

Acquaah 2012, Agrawal 1998, Brown, Caligari and Campos 2014, Chahal and Gosal 2002, Chaudhary 2014, Dabholkar 2006, Khan 2018, Kumar 2016, Priyadarshan 2019, Roy 2012a, Sharma JP 2009, Singh BD 2011, Singh and Shekhawat 2018, Singh Phundan 2011, Stoskopf, Tomes and Christie 2011.

Review Questions

Part A

Answer the following questions true or false.

1. In cross-pollinated crops, the aim of improvement is individual plants. True/ False

2. Additive genetic variance is critical to increase in gains per cycle. True/False
3. SCA is a critical factor in the development of synthetics. True/False
4. The yield of Syn_2 is greater than Syn_1.True/False
5. When both the parents are known, the progeny is called full-sib. True/False
6. Population cross is also called variety cross. True/False
7. Full-sib family selection comes under inter-population improvement methods. True/False
8. Mass selection is effective when the characters with low heritability are under selection. True/False
9. The inter-population selection scheme is also called reciprocal recurrent selection scheme. True/False
10. Synthetics can be produced both in self- and cross-pollinated crops. True/False

Part B

Answer the following questions.

1. Why are open-pollinated ears half-sibs?
2. Define synthetic cultivar.
3. Differentiate between synthetics and composites.
4. Full-sib selection will give more/less genetic gain than half-sib selection.
5. Ear-to-row selection is essentially a between-and-within-family selection; explain.
6. Mention two merits of mass selection.
7. Who proposed modified ear-to-row selection?
8. Mention one synthetic cultivar of cauliflower.
9. Why is selection called population improvement in cross-pollinated crops?
10. A synthetic cultivar is genetically heterogeneous/homogeneous.

Part C

Write a short note on each of the following topics.

1. Discuss the performance of synthetic cultivars.

2. What are the merits and demerits of mass selection?
3. Discuss the procedure of half-sib reciprocal recurrent selection.
4. Compare the methods of inter-population improvement.
5. How can we optimize gain from selection in population improvement?

20

Heterosis Breeding Development of Hybrid Cultivars

20.1 Introduction

A **hybrid cultivar** (also called **hybrid**) is the F_1 progeny of a planned cross between inbreds, cultivars, clones or populations. Depending upon the breeding approach, the hybrid cultivar may involve two or more parents. A critical requirement of hybrid production is that the parents are divergent. This divergence is the cause of superior performance of hybrids. The outstanding performance of crops is due to the exploitation of the phenomenon of **heterosis (hybrid vigour)** which is high when parents are divergent. Much of what we know about **heterosis breeding** (also called **hybrid breeding**) came from the work on maize hybrid cultivar development. However, commercial hybrids are now available in many crops including self-pollinated crops.

Starting of hybrid breeding programme is justified only when:

1. There exists a significant amount of heterosis over the better parent or the best available commercial cultivar.
2. Some means of easily obtaining hybrid seeds are available such as:
   a) Self-incompatibility.
   b) Male sterility.
   c) Single fruit producing enough hybrid seeds (e.g., tomato, brinjal, chilli, okra, tobacco, etc).
   d) Crop manifesting considerable heterosis and requiring relatively less hybrid seeds to plant a unit area (vegetable crops).
   e) High yield compensating for higher cost of small quantity of seed required to plant a unit area (vegetable crops).
   f) Different forms of sex expression (cucurbits, papaya).
3. Whether the seed or vegetative parts of hybrid is of economic value.

20.2 Types of Hybrids

We can have the following three types of hybrids:

1. Single Cross: A × B
2. Three-way Cross: (A × B) × C
3. Double Cross: (A × B) × (C × D)

If the number of parents is n, the number of different hybrids will be:

# of single crosses = n (n-1)/2

# of three-way crosses = n (n-1) (n-2)/2

# of double crosses = n (n-1) (n-2) (n-3)/8

The above formulas assume that no reciprocal crosses are made. Thus with 4 parents 6 single crosses, 12 three-way crosses and 3 double crosses can be made and with 10 parents 45 single crosses, 360 three-way crosses and 630 double crosses can be made.

20.3 Development of Hybrid Cultivars

The different of steps involved in the development of hybrid cultivars will be as follows:

i) Development of inbred lines

ii) Evaluation of inbred lines

iii) Production of hybrids

20.3.1 Development of Inbred Lines

Inbred lines are developed by continuous inbreeding of a plant population. The population could be an open-pollinated cultivar, a synthetic or composite cultivar or a single cross or double cross hybrid. Self-pollinated vegetables (e.g., tomato, brinjal, chillies, peas and beans) are maintained by selfing. So, every plant is an inbred line in these crops. However, in cross-pollinated vegetables (e.g., cole crops, root crops, bulb crops and leafy vegetables), plant populations are maintained by out-crossing. In this case, production of inbred lines involves continuous inbreeding for 3-5 generations.

Self-pollinated and often-cross-pollinated vegetables are marked by no to low inbreeding depression. So, selfing or inbreeding is not a problem in them. However, *cross-pollinated vegetables tolerate selfing only up to 2-3 generations* and inbreeding beyond these generations has to be supported by occasional random mating among full-sibs or half-sibs. The whole operation may take 6-7 generations. After this, all the plants will be homozygous. Good ones may be used to develop hybrids.

20.3.2 Evaluation of Inbred Lines

Random mating among all inbreds developed would not yield significantly higher than parental or original population. Therefore, it is necessary to evaluate them for making appropriate hybrid cultivar. There are three common methods that are used for evaluation of inbreds. The methods are briefly described below:

i) Phenotypic Performance

All inbreds are assessed for their yield and yield contributing traits. Various studies have indicated a positive correlation between performance of inbreds and hybrids developed from them. A replicated trial may be conducted, and poor performing lines are eliminated.

ii) Top-Cross Test

Top cross denotes the crossing of an inbred line with a broad genetic base tester (e.g., open-pollinated, synthetic or composite cultivar). This test gives an estimate of gca (general combining ability) of the inbred lines. Crossed seeds of inbred lines are evaluated in a replicated trial.

iii) Diallel Crossing

Diallel crossing refers to the mating of inbreds in all possible combinations. If there are n inbred lines to be evaluated, the number of cross combinations to be attempted would be $n(n-1)/2$. The cross combinations as such are evaluated over several locations for two years. The best ones are directly released as hybrid cultivars.

20.3.3 Production of Hybrids

Single cross hybrids have been commercially exploited in several vegetable crops because of their higher order uniformity for yield and yield contributing traits. More than 500 F_1 hybrid cultivars are in commercial use in major vegetables such as tomato, cucurbits, brinjal, okra, onion, carrot, cauliflower, capsicum etc.

Large scale seed production involves emasculation and pollination. Manual operations have been successful in generating adequate amount of seed in solanaceous vegetables such as tomato, brinjal and chilli. Male sterility has also been used in many vegetables for large scale seed production. In cucurbits, monoecious behaviour of plants has been utilized to facilitate seed production at commercial level. In cole crops, self-incompatibility has been used for F_1 hybrid seed production.

20.4 Hybrids in Horticultural Crops

A review by J. Janick (1996) indicates that hybrid seed is significantly used in horticultural production. A wide variety of mating systems are used in hybrid seed production of these species. These include hand emasculation (e.g., sweet pepper, tomato, brinjal), CMS (e.g., sugar beet, carrot, onion), self-incompatibility (e.g., cauliflower, broccoli), and monoecy (e.g., muskmelon, cucumber). The importance of hybrids is variable among species.

In ornamentals, F_1 hybrid seed is used in begonia (100% by emasculation), impatiens (100% by CMS), petunia (100% by CMS), seed geranium (100% by GMS), carnation (80% by GMS), and dianthus (70% by GMS)

20.5 Exploitation of Hybrid Vigour in Asexually Propagated Crops

20.5.1. Plants with Vegetative Propagation

Many horticultural plants are vegetatively propagated. The economic parts of many important food crops are non-seed, such as tubers (e.g., potato), stems (e.g., sugarcane) roots (e.g., cassava). Hybrid vigour can be effectively exploited in species that have the capacity to bear seed and yet propagated vegetatively. In such crops, the plant breeder only needs to create one superior genotype. The hybrid vigour and other traits assembled in the F_1 can be maintained indefinitely, as long as the genotype is propagated vegetatively. In horticulture, the superior genotype may be propagated by using techniques such as micropropagation, grafting, budding, sectioning and cutting.

20.5.2 Apomixis

Apomixis is the vegetative propagation through the seed. The seed in this instance is genetically identical to the female plant. Using the apomictic propagule is similar to reproducing the plant by other vegetative means. Hybrid vigour is fixed and expressed indefinitely through vegetative propagation.

20.6 Monoecy and Dioecy

Monoecious species bear male and female flowers on the same plants but on different parts of the plant. Environmental conditions (e.g., photoperiod and temperature) can influence sex expression making one plant more female or more male. In cucumber, short day and low night temperature promotes femaleness, while application of gibberellic acid promotes maleness. Breeders may manipulate the environment to produce hybrid seed.

20.7 Hybrid between Diploid and Tetraploid

We can develop hybrid between diploid and tetraploid and the resulting hybrid will be triploid as in case of seedless water melon. The tetraploid is synthesized through colchicine treatment of normal diploid.

Normal diploid (2n = 2x)

↓ Colchicine treatment

♀ Tetraploid (2n = 4x) × Diploid (2n =2x) ♂

↓

Triploid (2n = 3x)

The tetraploid line must be used as female parent to obtain triploid seed. The reciprocal cross results in empty seed. Both diploid and tetraploid lines are maintained through selfing. As triploid produces seedless watermelon because of parthenocarpy, the diploid plant cultivar should be interplanted with the seedless hybrid. The pollen from diploid plants pollinates the flower of triploid plants and this stimulates parthenocarpy which results in seedless watermelon.

20.8 Modified Breeding Method for Developing Hybrid in Fruit Crops (Roy 2012)

Fruit crops like mango, litchi, peaches, pears, apples which are multiplied through grafting/budding, are highly heterozygous and cross-pollinated in nature. Crossing between two parental lines (commercial cultivars) will result in segregating progenies with comparatively lower level of heterozygosity. Generally, the fruit breeder is crossing two cultivars and practicing selection in the crossed progeny and finally, selecting a genotype with an ideal plant type with improved or unique fruit quality. The selected genotype is then grafted/budded onto the root stock and thus multiplied. One thing is certain that in this method the probability of produced genotype being less heterozygous in comparison to parents is high and would thus show inbreeding depression and in turn less vigorous and will have quality very different from their parents. Now if only high heterozygosity is the requirement for successful fruit crop cultivars, then this method of breeding needs to be modified. In this suggested breeding method, two cultivars to be used in crossing programme be selfed first to generate populations of progenies with varying levels of homozygosity. These generated populations of genotypes will show inbreeding depression. Selection is practiced in each population of partially homozygous lines considering growth, flowering, quality and other traits and promising genotypes are selected from each population. Finally, selected genotypes from one population are crossed to selected genotypes from another population. In this way the heterozygosity is restored to the original level and there is every chance of increasing the level of heterozygosity in the selected individual. The genotypes developed can thus be called hybrids although **mistakenly cultivars developed from the standard two parent crosses are called hybrids by horticulturists.**

20.9 Merits of Hybrid Cultivars

1. The hybrid cultivars utilize hybrid vigour to the greatest extent.
2. Their produce, especially from single crosses, is much more uniform than those of open-pollinated, synthetic and composite cultivars.
3. They are produced in both self- and cross-pollinated crops.
4. Hybrid cultivars are easily maintained as parental inbreds/purelines and in case of asexually propagated crops, they are maintained indefinitely through vegetative propagation.

20.10 Demerits of Hybrid Cultivars

1. Farmers have to purchase new hybrid seed every year.
2. Hybrid seed production requires considerable technical skill. Hence, hybrid cultivar is costlier than synthetic/composite cultivars.
3. For exploitation of the full potential of hybrid cultivars, adequate inputs like fertilizers, irrigation etc. are essential. These may not be available to most Indian farmers.
4. Hybrid seed production may present several problems. Their solutions may increase cost of hybrid seed.
5. In case of most cross-pollinated species, isolation distance requirements are quite rigid. This makes hybrid seed production of such crops rather difficult in countries like India.

20.11 Achievements

Some examples of hybrids are given below:

Vegetables

- **Tomato:** Pusa Hybrid-8, KTH-1, KTH-2, Swarna Sampada, Arka Vardan, Swarna Baibhav, Solan Shagun, Pusa Divya, Pusa Hybrid-9 etc.
- **Brinjal:** Arka Navneet, Kashi Ganesh, Pusa Hybrid-5, Pant Brinjal Hybrid, Pusa Hybrid-6, Anand Brinjal Hybrid-1, BH-1, BH-2, Narendra Hybrid Brinjal-3, Swarna Ajay, Narendra Hybrid Brinjal-1, Kashi Sandesh etc.

- **Chilli:** CH-1, CH-2, Kashi Surkh, Arka Harita, Arka Sweta, Arka Meghna, Pusa Deepti, KTCPH-3 etc.
- **Okra:** Kashi Mahima, CO-2, Sheetala Uphar, Sheetala Jyoti, Hybrid Kashi Bhairav.
- **Cabbage:** KGMR-1, Sri Ganesh Gol, Nath-401, BSS-32, Nath-501 etc.
- **Cauliflower:** Pusa Hybrid-2, Pusa Kartik Sankar, DCH-541 etc.
- **Water Melon:** Durgapura Lal, Pusa Bedana, Arka Aishwarya, Arka Akash, Arka Madhura etc.
- **Musk Melon:** Pusa Rasraj, Punjab Hybrid-1, MH-10 etc.
- **Cucumber:** Pant Sankar Khira-1, Pusa Seedless-6, Pusa Sanyog etc.
- **Bitter Gourd:** Pusa Hybrid-1, Pusa Hybrid-2, NBGH-167 etc.
- **Bottle Gourd:** Pant Sankar Lauki-1, Pant Sankar Lauki-2, NDBH-4, Pusa Meghdoot, Pusa Manjari, Pusa Hybrid-3 etc,
- **Carrot:** Pusa Nayanjyoti, Hybrid-1 etc.

Fruits

- **Mango:** Mallika, Amrapalli, Pusa Arunima, Arka Anmol, Arka Puneet, Arka Aruna, Arka Neelkiran, Ratna, Sindhu, PKM-1, PKM-2 etc.
- **Guava:** Arka Amulya, Safed Jam, Kohir Safeda etc.
- **Litchi:** Sabohar Madhu, Sabohar Priya etc.
- **Papaya:** CO-3, CO-4, CO-7, IIHR-39 (Surya), IIHR-54, HPSC-3 etc.
- **Sapota**: CO-1, DHS-1, DHS-2, Hybrid -214, Hybrid-711 etc.

Ornamentals

Among ornamentals F_1 hybrids are available in *Antirrhinum*, *Ageratum*, *Begonia*, *Dianthus*, *Gerbera*, *Hollyhock*, *Impatiens*, marigold, *Petunia*, *Portulaca*, *Verbena* etc.

20.11 Brief References

Acquaah 2012, Agrawal 1998, Chahal and Gosal 2002, Dabholkar 2006, Kumar 2016, Roy 2012a, Sharma JP 2009, Singh BD 2011, Singh and Shekhawat 2018, Singh Phundan 2011, Stoskopf, Tomes and Christie 2011.

Review Questions

Part A

Answer the following questions true or false.

1. A hybrid is the F_2 progeny of a planned cross between inbreds, cultivars, clones or populations. True/False
2. Inbred lines are developed by continuous inbreeding of a plant population. True/False
3. Self-pollinated and often-cross-pollinated vegetables are marked by high inbreeding depression. True/False
4. We can develop hybrid between diploid and tetraploid. True/False
5. Farmers have to purchase new hybrid seed every year. True/False

Part B

Answer the following questions.

1. Monoecious species bear male and female flowers on the same/different plants.
2. What are the three steps of development of hybrids?
3. Mango is highly __________ and __________ in nature.
4. Farmers have to purchase new hybrid seed every year, why?
5. Mention two hybrids in cucumber.

Part C

Write a short note on each of the following topics.

1. Discuss the modified breeding method for developing hybrid in fruit crops.
2. Discuss merits and demerits of hybrids.
3. Types of hybrids.
4. Exploitation of hybrid vigour in asexually propagated crops
5. Hybrid between diploid and tetraploid.

21

Selection and Hybridization in Asexually Propagated Crops

21.1 Introduction

There are a number of field, vegetable, fruit and ornamental crops that are cultivated using asexual or vegetative parts such as stems, roots, modified flower etc. This is a heterogeneous group of plants and can be further categorized into four groups for ease in understanding the breeding methods applied to them. They are:

Group 1: Flowerless or Sterile

This group either does not bear flowers or if bears, they are sterile and do not set seed, e.g., ginger, mango ginger, turmeric, garlic, betel, many yams, etc.

Group 2: Apomicts

Apomixis has been reported in over 300 genera. But stable apomixis has been reported only in a few cultivated plants like meadow grass, citrus, mango, raspberry, etc.

Group 3: Flowers but Seed-Set Rare

In this group flowering takes place under specific situations but seed-set is limited, e.g., potato, sweet potato, sugarcane, etc.

Group 4: Normal Seed-Set but Propagated Vegetatively

This group consists of fruit trees, ornamentals, etc. which flower and fruit normally but due to long juvenile period, heterozygosity and desirable tree size, propagation is normally done through vegetative means. For example, mango, litchi, citrus, loquat, apple, pear, peaches, tea, and most ornamental plants.

21.2 Breeding Methods

The basic principles applicable to sexually propagated crops are equally well applicable to asexually propagated crops. However, there are some important differences which necessitate differential treatment, namely:

i) The extent of genetic variability, especially in crop species which are totally sterile, is rather limited to that already existing or that which may arise through natural (spontaneous) or induced mutations, and

ii) Once a desirable plant has been obtained, that is selected, regardless of heterozygosity, chromosome imbalances, sterility problems, etc., the same can be further propagated and multiplied through vegetative propagation for commercial use without genetic change.

The following breeding methods are employed for crop improvement of asexually propagated crops:

a) Clonal selection

b) Hybridization followed by clonal selection

c) Mutation Breeding

Mutation Breeding is discussed in the next chapter.

21.3 Clonal Selection

21.3.1 Clone

A clone may be defined as a group of plants derived from a single plant by vegetative propagation. In other words, all the vegetative progenies of a single plant make a clone. For example, all the trees of Dussehari mango available today can be traced back to a single plant of that cultivar in Dussehari village near Lucknow. Similarly, all the trees of Nagpur Seedless Orange can be traced back to Tree No.182 in Nagpur.

21.3.2 Characters of a Clone

1. Clonal population is homogeneous.
2. Individuals of a clone are heterozygous.
3. Clones are stable.
4. Mutation is the only means of creating variability.
5. Clone is propagated vegetatively.

21.3.3 Sources of Clonal Variation

The heritable variation in clones may occur through mutations, mechanical mixture and sexual reproduction.

21.3.3.1 Mutation

The heritable variation in clones may occur through mutations in any of the following ways:

a) Spontaneous Mutations: Clones are highly heterozygous and often complex hybrids. This seems to increase the chances of change within a somatic cell due to mutations that can occur due to chemical alterations in DNA at specific locations on the chromosome (point mutation), or gross structural changes in the chromosomes (deletions, duplications, inversions and translocations), or aneuploidy or polyploidy, or the plastids that independently determine plant characteristics. Variegated plants have a tendency to produce both normal and defective plastids.

b) Induced Mutations: Treating the growing points with a mutagenic agent, such as colchicine, bombarding seeds or scions meant for grafting with X-rays or radioactive emissions or other radiomimetic chemicals may induce mutations and chromosomal changes. Polyploidy may give rise to 'giant bud sports'.

21.3.3.2 Mechanical Mixture

Propagules of other clones may become mixed with those of a clone and give rise to genetic variation in clones.

21.3.3.3 Sexual Reproduction

Occasional sexual reproduction may occur in a clone. This would lead to segregation and recombination which will generate genetic variation. The sexual reproduction may also involve out-crossing with other clones; this would generate more and, possibly, novel genetic variation.

21.3.4 Clonal Degeneration

In general, **clones show a decline in vigour and yield with time; this is called clonal degeneration**. Clonal degeneration is neither due to the genotypic features of clones nor due to the vegetative/asexual reproduction. It may arise due to mutation, viral infection and bacterial diseases.

21.3.5 Procedure of Clonal Selection

This procedure is effective if variability exists in the natural clonal population. It is similar to pureline selection in self-pollinated crops. A simple description of clonal selection is given below (Figure 21.1):

1st Year: Space-plant variable population, i.e., mixture of several clones. Select 200-2000 superior plants. Traits with high heritability, e.g., plant height, flowering time, disease resistance etc. can be selected. Selection for traits with low heritability cannot be done at this stage.

2nd Year: Grow asexual progenies of each selected plant separately. Evaluate the clones visually for various desirable traits. Select 50-100 superior clones.

3rd Year: Conduct Preliminary Yield Trial (PYT) of selected clones along with check, select clones that are superior to check.

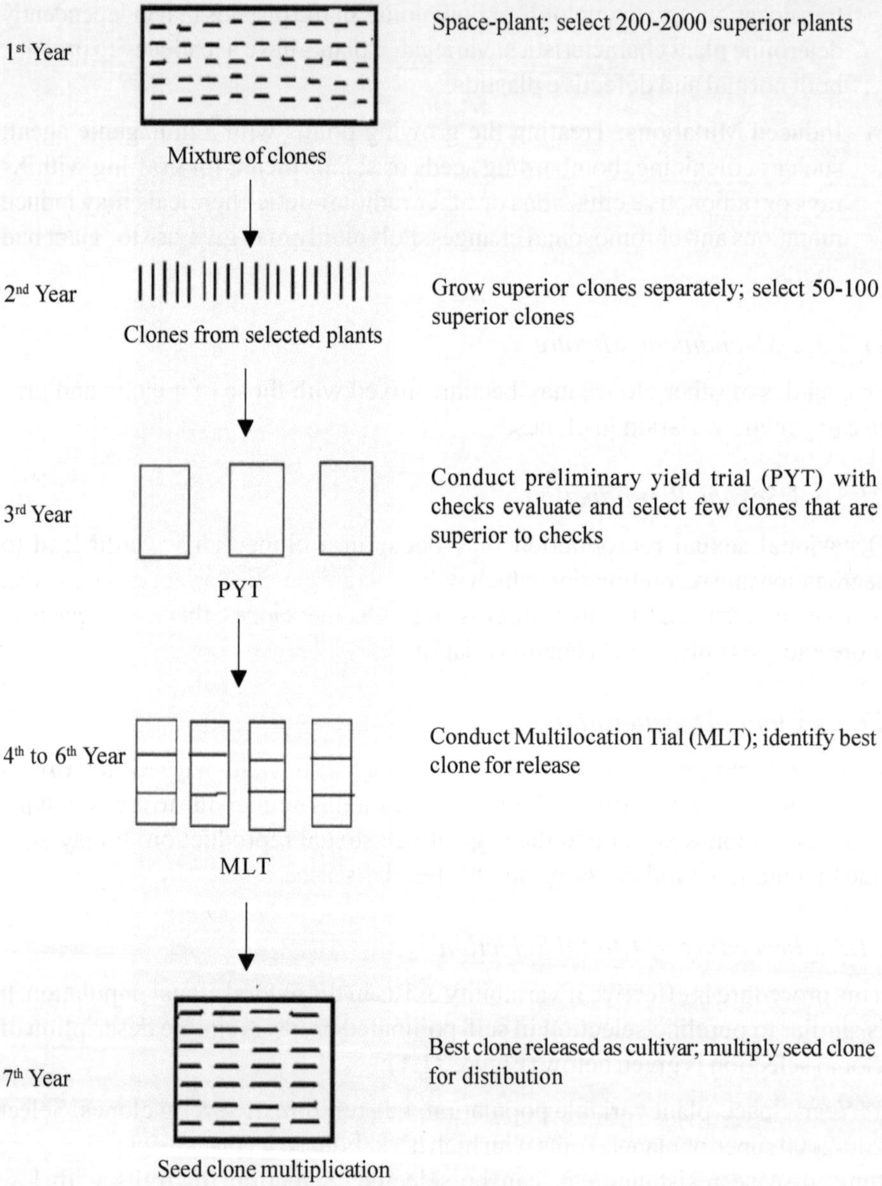

Figure 21.1. A generalized outline of clonal selection

4th to 6th Year: Conduct MLT of superior clones in all India coordinated trial along with best commercial check cultivars. Evaluate the clones for yield, quality, disease and insect resistance, etc. Identify superior clones to checks for release.

7 Year: Begin seed multiplication of identified clone. Then the clone is released and notified next year.

21.3.6 Merits of Clonal Selection

1. This is the only procedure to exploit genetic variation in asexually propagated crops (clonal crops).
2. It is useful in maintaining purity of clones.
3. It can be used to utilize the genetic variation created by hybridization or any other means.

21.3.7 Demerits of Clonal Selection

1. It can only utilize the variation present in the population.
2. The multiplication ratio is low.
3. It cannot create any new variability.

21.3.8 Achievements

In India, clonal selection has been successfully used for developing new cultivars in potato, sugarcane, banana, citrus and grapes. Some examples are:

- **Potato:** Kufri Red, Kufri Safed
- **Mango:** KO-11, KO-22, Neelam, Clone No-51, MA-1, Pusa Surya
- **Banana:** Bombay Green, Pride Monthan, High Gate
- **Grapes:** Pusa Seedless, HS 37-6
- **Ginger:** Suprabha, Suruchi
- **Turmeric:** Roma, Surama, Ranga, Rashmi
- **Sweet Potato:** Pusa Sunehari, Pusa Safed, Pusa Lal, CO-1
- **Tapioca**: CO-1, CO-2, CO-3, CO (TP)4
- **Yam (Greater Yam):** Sree Keerti, Sree Roopa
- **Yam (Lesser Yam):** Sree Latha
- **Taro:** Sree Rashmi, Sree Pallavi
- **Garlic:** G-41, Godavari Gulab, Agrifound White
- **Pointed Gourd:** FP-1, FP-3, FP-4
- **Elephant-foot Yam:** Kovvur

21.4 Hybridization followed by Clonal Selection

The natural genetic variability in **asexually propagated crops** (also called **clonal crops**) is exhausted by clonal selection. Hence, new genetic variability has to be created for further improvement of such crops. *Hybridization between clones is used to create variability where sexual reproduction is present.* In clonal crops where sexual reproduction is absent, mutation or biotechnological approaches have to be adopted to create genetic variation.

21.4.1 Procedure *(Figure 21.2)*

1st Year: Cross selected parents in pairs to produce single cross. Harvest F_1 seed.

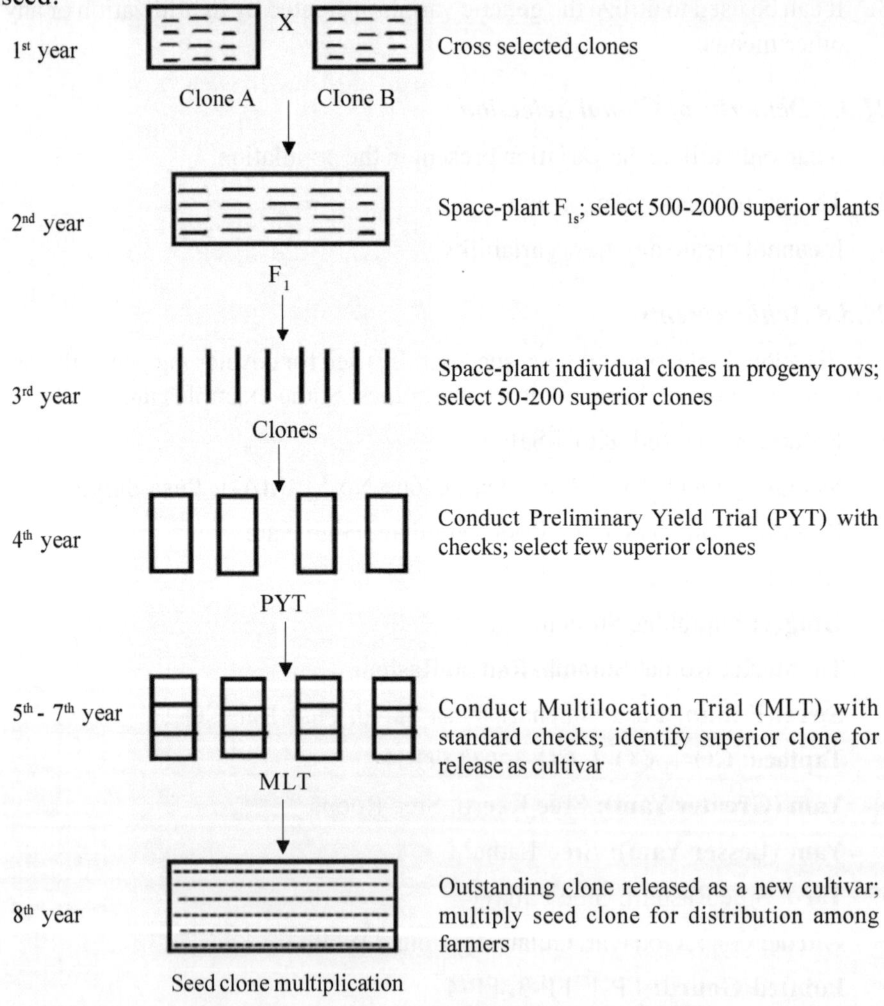

Figure 21.2. A generalized plan for hybridization followed by clonal selection

2nd Year: Space-plant and evaluate F_1s visually. Select 500-2000 vigorous and healthy plants.

3rd Year: Space-plant asexual progeny rows of selected plants. Each progeny is a clone. Evaluate the clones visually. Select about 50-200 superior clones.

4th Year: Conduct Preliminary Yield Trial (PYT) of selected clones with commercial cultivars as checks. Select clones superior to checks. Evaluate the superior clones for quality, disease and insect resistance, etc.

5th to 7th Year: Conduct Multilocation Trial (MLT) under coordinated project for the respective crop. Best commercial cultivars are included as checks. Quality, disease etc. tests are conducted. The clones superior to checks are identified for release as a cultivar.

8th Year: Seed multiplication begins for clones identified for release. Then, next year the identified clone will be released and notified and its certified seeds can be produced.

21.4.2 Achievements

Several cultivars of sugarcane (CO-541, COS-510, CO-1148, CO-1158, etc.) and potato (Kufri Sinduri, Kufri Kuber, Kufri Kundan, Kufri Chamatkar, etc.) have been developed by interspecific hybridization followed by clonal selection.

- **Tapioca:** H 97, H 165, Sree Visakham, Sree Prakash
- **Sweet Potato:** H 41, H 42, Rajendra Sakarkand-5

21.5 Brief References

Acquaah 2012, Agrawal 1998, Chaudhary 2014, Kumar 2016, Ram 2012, Shukla *et al.* 2020, Singh BD 2011, Singh Phundan 2011.

Review Questions

Part A

Answer the following questions true or false.

1. Asexually propagated crops are also called clonal crops. True/False
2. Individuals of a clone are heterozygous. True/False
3. Clones do not show any decline in vigour and yield with time. True/False
4. Clonal selection can create new variability. True/False
5. Suprabha is a cultivar of turmeric. True/False

Part B

Answer the following questions.

1. Define a clone.
2. What are the sources of clonal variation?
3. Mention two cultivars of ginger developed through clonal selection.
4. The natural genetic variability in asexually propagated crops is exhausted by ________.
5. Clonal population is homogeneous/heterogeneous.

Part C

Write a short note on each of the following topics.

1. Hybridization followed by clonal selection.
2. Sources of clonal variation.
3. Characters of a clone.
4. Procedure of clonal selection.
5. Categories of asexually propagated crops.

Part B
Non-Conventional Breeding Practices

22

Mutation Breeding Sexually and Asexually Propagated Crops

22.1 Introduction

Mutations are sudden heritable changes in the genotype of an organism. De Vries (1900) for the first time used the term mutation (mutare in Latin means 'to change') for the appearance of new types in evening primrose (*Oenothera*) plant. In 1901 he put forth the idea of producing mutations artificially for use in crop improvement. Alberto Pirovano (1922) was first to use X-rays and ultraviolet rays for inducing mutations in plants, but his work remained unknown. So, Muller (1927) became the first known to induce mutation in *Drosophila* fly using X-rays. Though Stadler (1928) had started working simultaneously on barley and maize and induced mutations, he was late in reporting mutation due to longer life cycle of these plants compared to *Drosophila*. Mutation may be **gene mutation** (the result of change in the nuclear gene or plasma gene) or **chromosomal mutation** (change in chromosome number, structure or chromosomal aberrations of various types).

22.2 Classification of Mutation

The process through which genetic changes get induced is called mutation and the changed (mutated) individual is called mutant. Mutants have variously been classified: **spontaneous** and **induced**, **natural** and **artificia**l based on their origin; **germinal** and **somatic** based on tissue involved; **chromosomal genic** and **cytoplasmic**, etc. The following classification put forth by Gaul (1961) is followed currently:

22.2.1 Macromutations

These are large mutations and can be recognized on a single plant basis. They may be (i) trans-specific resembling different species or even family and (ii) intra-specific resembling the same species.

22.2.2 Micromutations

These are mutations with small effects and can be recognized only when a group of 30 or more mutant plants are compared with a normal one. Micromutants differ from normals only quantitatively.

22.2.3 Spontaneous Mutations

These are naturally occurring mutations that arise automatically. They rise in nature continuously without any human control and create variability that forms the basis of conventional plant breeding methods. Their frequency is extremely low (10^{-6}). A number of factors cause spontaneous mutation, *viz*., genetic constitution, physiological conditions, nutrition, temperature, and naturally occurring mutagens.

22.2.4 Induced Mutations

Contrary to spontaneous mutations, these are induced by using various agents called mutagens. ***Mutagens** are the agents that induce mutation.*

22.3 Mutagen and Its Classification

Agents that induce mutations are called mutagens. Mutagens are of two types: physical and chemical mutagens.

22.3.1 Physical Mutagens

These are different kinds of radiations, e.g., X-rays, γ-rays, etc. Radiations are of two types: ionising and non-ionising radiations.

22.3.1.1 Ionising Radiations

These radiations produce ionisation as excitation in the atoms located in their path. When an atom loses or gains an electron, it becomes positively or negatively charged; this is called ionisation. An ionising radiation may produce either a dense or a sparse population of ions in its path. On this basis ionising radiations are termed as densely ionising or sparsely ionising radiations, respectively. In excitation, an atom does not lose an electron, but one of its electrons moves to an outer orbit representing a higher level of energy. Ionising radiations may be particulate, i.e., consist of atomic particles, or non-particulate, i.e., having no particles but only photons of high energy.

1. X-rays: These are electromagnetic radiation of 0.001-10Å wavelength produced by X-ray tube. X-rays are non-particulate, high energy, sparsely ionising and most penetrating of the various radiations.

2. γ-rays: They are similar to X-rays in physical and biological properties. Gamma-rays are produced from ^{60}Co, radium etc. through radioactive decay. Both x-rays and γ-rays are commonly used in mutation breeding programme.
3. Fast and Thermal Neutrons: Fast neutrons are produced during radioactive decay of heavier elements. When graphite or heavy water reduces their velocity, they become thermal or slow neutrons. They are uncharged, particulate, highly penetrating and densely ionising radiations.
4. α-Particles: Each -particle has 2 protons + 2 neutrons, and double positive charge. These are produced during radioactive decay. They are particulate, densely ionising and much less penetrating than neutrons.
5. β-rays: These are high-energy electrons produced during radioactive decay of ^{3}H, ^{12}P, ^{35}S, etc. They are particulate, negatively charged, sparsely ionising and have very little penetrating power as compared to X-rays.

22.3.1.2 Non-Ionising Radiation

Such a radiation does not cause ionisation but produces mutations. Ultraviolet (UV) rays (wavelength 100-3900Å) are the only example of this group. UV rays are produced by mercury vapour lamps, have very limited penetrating capacity (limited to 1-2 cell layers), and are commonly used to irradiate microorganisms and pollen grains. The most effective wavelength is 2540Å since DNA bases show the maximum absorption at this wavelength. UV has been rarely used for crop improvement.

22.3.2 Chemical Mutagens

There are certain chemicals that induce mutations. The chemical mutagens are grouped into the following categories: (i) alkylating agents, (ii) acridine dyes, (iii) base analogues and (iv) other chemicals.

1. **Alkylating Agents**: These chemicals transfer an alkyl (either methyl or ethyl) group to DNA, usually to the ring N of guanine. They cause DNA cross-linking, chromosome breaks, etc. similar to ionising radiation. Examples are ethylmethane sulphonate (EMS), methylmethane sulphonate (MMS), ethylene imine (EI), etc. These mutagens produce G: C $\rightarrow$ A: T transition. **Transition** is the replacement of an A by a G or of a C by a T, and vice-versa.
2. **Acridine Dyes**: Molecules of these chemicals become inserted between two base pairs of DNA (called **intercalation**) and usually produce deletions. Some examples of such mutagens are: acriflavine, proflavine, acridine orange, ethidium bromide, etc.

3. **Base Analogues**: These molecules are very similar to DNA bases and are readily incorporated into DNA during replication. During subsequent DNA replications they may pair with wrong DNA bases, thereby producing transitions. Examples of base analogues are: 5-bromouracil and 5-chlorouracil.
4. **Other Chemicals**: Many other chemicals are also mutagenic; some important ones are: nitrous acid, hydroxylamine, sodium azide (NaN_3), etc. The mechanisms of action of these chemicals are varied and different from each other.

22.4 Mutation Breeding

Mutation breeding refers to isolation and selection of desirable induced mutants in segregating generations that follow a mutagenic treatment of seeds or other plant parts. The selected mutants are used in breeding programmes and are also released for commercial cultivation depending on their usefulness.

22.4.1 Application and Uses of Mutation Breeding

Both spontaneous and induced mutations are usually undesirable, deleterious and sometime lethal. Hence the technique of mutation breeding should be used only under specific circumstances like:

i) If the crop to be improved is a vegetatively propagated one, mutation breeding can be preferably used to remove some specific defects.
ii) In fruit trees that have long life cycle, the use of conventional breeding methods is not easy. In these crops, mutation breeding can be used to improve some specific traits while retaining all the desirable ones of the old cultivar (like colour, flavour and taste of fruits).
iii) In crops where conventional breeding methods can be successfully used, the use of mutation breeding should be done only when the natural variability of the crop does not provide gene(s) for desired characteristics.
iv) Mutation breeding has a good scope when a desired gene is tightly linked with undesirable gene(s).
v) When recombination breeding reaches at stage where the biological system becomes helpless in further increasing the productivity, mutation breeding can work as a helping tool.
vi) When the breeder is willing to develop chimeras in an ornamental plant species, mutation breeding can be of great help in this regard. The development of chimeras in ornamentals is perhaps the most significant contribution made by mutation breeding.

vii) Mutation breeding can be helpful in rectifying a simply inherited defect in an otherwise promising cultivar, maintaining the superior genetic background of the cultivar.

viii) When the breeder desires to change some biochemical pathway in any crop, mutation breeding can make significant contribution.

ix) When there is negative association between disease resistance and some quality trait, one can resort to mutation breeding.

22.4.2 Main Features of Mutation Breeding

22.4.2.1 Selection of Mutagen

The frequency of spectrum of mutations (gene mutations, chromosomal aberrations) depends on the choice of mutagen and dose applied. Among physical mutagens, X-rays and γ-rays are widely used. The UV light may be effectively used with materials such as pollen. Fast and thermal neutrons cause more chromosomal aberrations. Chemical mutagens produce a higher rate of gene mutations. Table 22.1 presents a list of mutagens together with doses used by plant breeders in some select crops.

Table 22.1 Mutagens and recommended doses for inducing mutations in some selected crops.

Crop	Material	Recommended	Mutagen and Dose
A. Sexually Propagated Crops			
Pea	Dry seeds	X-rays	10 krad
		DES	2%
	Pollen	γ-rays	600 rad
Tomato	Dry seeds	EMS	0.8%
Pepper	Dry seeds	Fast neutrons	2.4 krad
		γ-rays	14-22 krad
B. Asexually Propagated Crops			
Potato	Tubers	EMS	100-500 ppm
Sweet Potato	Cuttings	γ -rays	20 krad
		EI	0.5%
Cassava (Tapioca)	Dry seeds	γ -rays	5-20 krad
	Stem	γ -rays	< 1.5 krad
	Pollen	γ -rays	2-6 krad
Coffee	Dry seeds	γ -rays	2.5-20 krad
Apple	One-year old dormant shoots	γ -rays	2.5 krad
Citrus	Dry seeds	γ -rays	1.0 -2.5 krad
	Buds	γ -rays	5 krad
Grape	Dormant buds	γ -rays	2.5-3.5 krad
		EMS	0.15-0.20%
Dahlia	Dormant tubers	X-rays	1-4 krad
Chrysanthemum	Rooted cuttings	X-rays	1200 rad
Carnation	Nodal stems	X-rays	1.5-2.0 krad

22.4.2.2 *Starting Material*

Choice of Genotype: The choice of genotype for mutagenic treatment should be based on familiarity with the available germplasm. The best cultivars are usually selected. In some instances, F_1 is used. The advantage of using F_1 is that two different genomes are exposed to a mutagen at the same time. This may increase gene recombination and thus reveal a wider spectrum of gene mutations.

Choice of Plant Material: In sexually (seed) propagated crops, the physical or chemical treatment of dry seeds is the most convenient and practical. In case of asexually (vegetatively) propagated crops, the material for treatment should come from disease free plant. The vegetative propagates such as bud woods, scions, cuttings, tubers, bulbs, or various explants of plant organs (shoot tips, meristems, epidermis, ovaries, nucellar tissues, etc.) are subjected to mutagenic treatment. These organs consist of many cells and after mutagenic treatment give rise to a chimeric structure consisting initially of genetically different cells in the form of a sector within a histogenic layer (mericlinal chimera). A mericlinal chimera occurring in the M_1V_1 generation may convert into periclinal chimera with a uniformly mutated cell layer. To favour this development and eventually obtain non-chimeric mutant plants, the following established techniques may be applied:

i) Pruning or cutting back the primary shoot and using the axillary buds in propagation.

ii) Repeated budding of those buds coming from preformed primordia and located generally in the basal and middle part of the M_1V_1 shoot.

iii) In-vitro multiplication of uniformly mutated plants through propagation of axillary buds.

iv) *Adventitious bud technique*: It is assumed that adventitious buds develop from a single cell and hence an adventitious bud developed on a mutagenized leaf, stem or root cutting is expected to give rise to a non-chimeric homohistont mutant. However, this technique is not applicable to all plant species.

22.4.2.3 *Nature of Induced Mutations*

The evidence for mutagenic specificity is very limited and control over the mutation spectrum is very limited. Some specificity of action is shown by different mutagens in the ratio of point mutations to chromosomal aberrations. A tendency to affect particular heterochromatic regions is typical of certain chemical mutagens. Thus, one particular mutagen tends to produce only part of the entire spectrum of possible genetic changes, and the type it produces is a reflection of its energy potential.

Densely ionizing radiations (neutrons) largely induce chromosomal alterations and a few point mutations. The more sparsely ionizing radiations (X-rays, γ-rays) induce a large number of point mutations. UV light has even lesser tendency to induce chromosomal alterations and a still higher proportion of point mutations. The chemical mutagens, *viz*., mustard gas and its derivatives produce some chromosomal alterations and some point mutations. The mutagenic nucleosides and other chemical mutagens with special effects on nucleic acids appear to produce mutations by special chemical affinities. Other factors, *viz*., metabolism of the treated plant, pretreatments (if any) or change in mineral nutrition may also influence or alter the genetic response to a particular mutagen.

22.4.2.4 Frequency of Desirable Mutations

The number of deleterious mutants is far higher than desirable ones. The recovery of desirable mutants at an economic cost is an important consideration. It is greatly influenced by the manner in which treated material is handled and the method of selection applied.

22.4.2.5 Chimera

A chimera or chimeric change occurs when an individual consists of two or more genetically different types of cells. Though heritable changes, these mosaics can only be maintained by vegetative propagation (not transferrable to progenies by sexual means). There are four basic types of chimeras:

1. **Sectoral**: This chimera is observed in growing shoot as two different tissues located side-by-side. The effect of this modification is that the stem develops with two distinct tissues on each half.
2. **Periclinal**: This type consists of two thin layers of different genetic makeup, one over the other.
3. **Mericlinal**: When an outer layer of different genetic tissue does not completely extend over the layer below, the chimera is called mericlinal.
4. **Graft Chimeras**: Unlike the first three chimeras that have genetic origin, a graft chimera is a non-heritable mixture of tissues that may occur after grafting is made.

   Whereas chimeras are undesirable in crop plants, they may be successfully exploited in horticulture.

22.4.3 Generalized Procedure of Mutation Breeding

22.4.3.1 Sexually Propagated Crops (Seed-Propagated Crops)

Step-1 Mutagenic Treatment: Seeds, shoot tips, pro-embryos, gametes, zygotes, single cells (*in-vitro* propagation) of selected genotypes are treated with mutagens, such as γ-rays, X-rays, fast neutrons, thermal neutrons or chemical mutagens. The choice of mutagen and the dose depends on the kind of material to be treated and the availability of mutagen.

Step-2 Grow M_1 Generation: Treated material is grown in isolation or bagged along with control as M_1 plants show some degree of sterility thereby increasing the potential out-crossing. Look for chimeric plants and non-chimeric heterozygous plants. Depending on the objective of breeding, decide whether to carry all M_1 harvested seeds, or a single seed per plant, or one or more reproductive organs (fruit, pod, capsule etc.) per M_1 plant and whether the M_2 generation is to be grown in bulk or in individual progenies. Harvest and keep the material accordingly.

Step-3 Grow M_2 Generation: Grow the material raised in M_1 generation. For diploid species, M_2 progenies of 20 plants should be sufficient to obtain a certain number of segregants homozygous for a mutant trait. For polyploids, and in general for quantitatively inherited characters, it is advantageous to allow more recombination to take place by delaying selection to the M_3 or M_4 generation.

Step-4 Grow M_3 Generation: Grow the material harvested in M_2 generation. Look for segregants. Verify mutants selected in M_2. For indirect use in a breeding programme crosses of desirable mutants with other breeding material may be made.

Step-5 Grow M_4 Generation: Selected mutants are included in a station trial for preliminary evaluation. Evaluate genetic stability of the selected mutants. For indirect use in a breeding programme crosses of desirable mutants with other breeding material may be made.

Step-6 to 9 Grow M_5 to M_8 Generation: Evaluate the selected stable mutants in multilocation trial (MLT). On the basis of performance, selected mutants may be directly released as a new cultivar or indirectly used for incorporating the desirable trait they carry into other promising breeding material.

Step-10 Release of mutant cultivar.

22.4.3.2 Asexually (Vegetatively) Propagated Crops

Step-1 Mutagenic Treatment: Plant organs such as shoot meristems, bulbs, tubers, cuttings, etc. of selected clones are treated with mutagens such as X-rays or γ-rays, or with chemical mutagens. The choice and dose of mutagen depend on the kind of material to be treated and the availability of the mutagen.

Step-2 Grow M_1V_1 Generation: Look for chimeric development from apical and axillary bud meristems (mericlinal, sectoral). Cut back the M_1V_1 shoot, bud grafting, etc.

Step-3 Grow M_1V_2 Generation: Grow M_1V_1 material. Mutant selection starts in M_1V_2 generation. Identify periclinal or uniformly mutated scion, branch, tree, etc.

Step-4 Grow M_1V_3 Generation: Grow M_1V_2 material. Verify genetic uniformity within a mutant clone. Further isolate somatic mutations. Vegetatively propagate mutant plants. Start PYT.

Step-5 to 10 Grow M_1V_4 to M_1V_9 Generation: In M_1V_4 and subsequent generations the stability and uniformity of the clone is assessed. Only uniform and stable clones should be planted for evaluation. The performance of mutant clones is evaluated. The improved mutants are released for commercial cultivation.

22.5 Achievements

A large number of cultivars have been released through mutation breeding. Some examples are:

Vegetables

- **Okra**: Anjitha, MDU-1, EMS-8, Parbhani Tillu, Panjab-8, etc.
- **Hyacinth Bean**: CO-10
- **Ridge Gourd**: PKM-1
- **Chilli**: MDU-1
- **Tomato**: CO-3, PKM-1, Pusa Lal Meeruti, S-12
- **Bitter Gourd**: MDU-1
- **Pea**: Hans
- **French Bean**: Pusa Parvati
- **Brinjal**: PKM-1
- **Cowpea**: CO-5, COCP-702, Cowpea-88, Amba, Shreshtha, Swarna
- **Turmeric**: BSR-1, CO-1

Fruits

- **Papaya**: Pusa Nanha
- **Mango**: Rosica

- **Grape**: Marvel Seedless
- **Banana**: Highgate, Motta Poovan
- **Orange**: Washington Navel
- **Grapefruit**: Marsh, Thompson

Ornamentals

- **Bougainvillea**: Arjuna, Jaya, Jayalakshmi, Pallavi, Suvarna, etc.
- **Chrysanthemum**: Agnisikha, Alankar, Anamika, Aruna, Asha, Basanti, Kanak, etc.
- **Dahlia**: Bichitra, Black Beauty, Happiness, Jayaprakash, Jyoti, Netaji, etc.
- **Gladiolus**: Shobha, Tambari
- **Hibiscus**: Anjali, Purnima
- **Tuberose (Polyanthes)**: Rajat Rekha, Swarna Rekha
- **Portulaca**: Five Petal, Jhumka, Karna Pali, Karna Phul, Lalita, Mukta, etc.
- **Rose**: Abhisarika H.T., Angara, Pusa Christina, Saroda, Sharada, etc.
- **Coleus**: Suphala

22.6 Brief References

Acquaah 2012, Agrawal 1998, Chaudhary 2014, FAO/IAEA 2018, Kumar 2016, Ram 2012, Roy D 2012a, Sharma and Sharma 2014, Shukla *et al*. 2020, Singh BD 2011, Singh Phundan 2011, Singh and Pawar 2006, Thirugnanakumar *et al*., 2014.

Review Questions

Part A

Answer the following questions true or false.

1. Mutations are sudden heritable changes. True/False
2. Two different tissues are located side-by-side in periclinal chimera. True/ False
3. The number of deleterious mutants is far lower than desirable ones. True/ False
4. The best cultivars are usually selected for mutagenic treatment. True/False
5. The frequency of spectrum of mutations depends on mutagen and its dose. True/False

Part B

Answer the following questions.

1. What is macromutation?
2. Differentiate between periclinal and mericlinal chimeras.
3. What are physical mutagens?
4. A graft chimera is a non-heritable/heritable mixture of tissues.
5. EMS stands for_______________.

Part C

Write a short note on each of the following topics.

1. Physical mutagens.
2. Selection of mutagen.
3. Procedure of mutation breeding in vegetatively propagated crops.
4. Nature of induced mutations.
5. Choice of plant material for mutation breeding.

23

Polyploidy Breeding Manipulating Chromosomes

23.1 Introduction

*An organism or individual having more than two basic sets of chromosomes is called **polyploid** and the condition is known as **polyploidy**.* Polyploidy is an important tool to increase genetic diversity by altering the chromosome number (2x) and consequently the number of genes within a single cell. Polyploidy has been one of the important mechanisms in the evolution of many cultivated crop plants.

23.2 Types of Polyploidy

23.2.1 Autopolyploidy

The condition that arises due to multiplication of the same chromosomes of a single species is known as **autopolyploidy** (also called '**simple polyploidy**' or '**single species polyploidy**'). Autopolyploids have generally larger leaves, stem and other vegetative parts and flower, fruit and seeds than their normal diploids.

Autopolyploids include

a) **Autotriploids (3x):** Having three copies of the same genome (basic set of chromosomes), e.g., triploid watermelon.

b) **Autotetraploids (4x):** Having four copies of the same genome, e.g., potato.

c) **Autopentaploids (5x):** Having five copies of the same genome.

d) **Autohexaploids (6x):** Having six copies of the same genome, e.g., sweet potato.

e) **Autoseptaploids (7x):** Having seven copies of the same genome.

f) **Autooctaploids (8 x):** Having eight copies of the same genome.

23.2.2 Allopolyploidy

The condition in which complete genomes of two or more species combine together is called **allopolyploidy**. An allopolyploid that arises by combining genomes of two diploid species is termed as **amphidiploid**. Allopolyploidy can be developed by interspecific crosses and fertility is restored by chromosome doubling with colchicine treatment.

Allopolyploids are of the following types

a) **Allotetraploids:** Having 2 copies each of two distinct genomes ($2x_1 + 2x_2$).

b) **Allohexaploids:** Having two copies each of three distinct genomes. ($2x_1 + 2x_2 + 2x_3$).

c) **Allooctaploids:** Having two copies each of four distinct genomes. ($2x_1 + 2x_2 + 2x_3 + 2x_4$).

23.3 Induction of Polyploidy

Polyploidy is induced by treatment with chemical known as colchicine. It is an alkaloid which is obtained from the seeds of *Colchicum autumnale* belonging to Liliaceae family. Colchicine is applied in a very low concentration (0.01 to 0.5%) because high concentration is highly toxic to the cells. For effective induction of polyploidy, usually concentration of 0.01% to 0.5% is used in different plant species. It is applied to growing tips, meristematic cells, seeds and axillary buds in aqueous solution or mixed with lanolin. The duration of treatment varies from 24-90 hours depending upon plant species.

Colchicine induces polyploidy by inhibiting formation of spindle fibres. The chromosomes do not line up on the equatorial plate and divide without moving to the poles due to lack of spindle fibres. The nuclear membrane is formed around them and the cells enter interphase. Thus, nucleus has double chromosome number. The colchicine-induced polyploidy is known as **colchiploidy**.

23.4 Autopolyploids

23.4.1 General Features of Autopolyploids

1. They have larger cell size than diploids.
2. Pollens are usually larger.
3. They are slower in growth and late in flowering.
4. They have larger and thicker leaves, larger flowers and, fruits are usually lesser in size.
5. They have reduced fertility.
6. Different species have different levels of optimum ploidy, e.g., triploidy in sugarbeet.
7. They have lower dry matter content.

23.4.2 Procedure of Breeding Induced Autotetraploids

23.4.2.1 Factors to Consider

Levan (cited by Dewey 1980) concluded that a crop most amenable to improvement through chromosome doubling should have the following features:

1. Low chromosome number
2. Be harvested for vegetative growth, and
3. Be cross-pollinated.

   Dewey added:

4. Have perennial habit, and
5. Have ability to reproduce vegetatively.

23.4.2.2 Selection of Material for Chromosome Doubling

A sufficient number of tetraploid lines should be synthesized from diverse diploid lines to insure allelic diversity in the breeding programme. Single and double crosses of the diverse 4x lines are required to produce cultivars with maximum hybrid vigour.

23.4.2.3 Colchicine Treatment

Colchicine is applied for inducing autopolyploidy, i.e., doubling the chromosome number. Proper aftercare of colchicine-treated material and isolation of the crop to avoid genetic contamination are the basic requirements. A selection strategy for each crop is imperative if plants with high fertility and ploidy level are desired.

23.4.2.4 Selection Strategies

Direct selection for ploidy level based on chromosome counts is rather laborious. The possibilities of indirect selection, such as stomatal length in C_0 and C_1 generations that could serve as a reliable criterion for selection at the ploidy level need to be explored. In vegetatively propagated crops heterozygosity of elite genotypes can be fixed instantly but in seed crops strategies based on maximizing heterozygosity within populations have to be implemented.

Schwanitz (1951) has emphasized that selection for smaller cell size and increased nuclear size and volume should be practiced for improved fertility of tetraploid strains. Selection for 5-6 generations is usually sufficient to achieve the desired type with tetraploidy and high fertility (seed-set).

23.4.2.5 Handling of Breeding Material

The other aspects of handling the material are similar to any other breeding procedures.

23.4.2.6 Application and Uses of Autopolyploids

1. **Triploid Sugarbeets:** Triploidy appears to be the optimum levels of polyploidy. Triploid roots are longer compared to diploid cultivars and also yield more sugar percent.
2. **Seedless Watermelon:** The yield of triploids and consumer acceptance has been good in Japan and the USA. However, it is necessary to interplant diploid plants among triploids as pollinators in a commercial field for fruit setting (1 diploid: 5 triploids). Triploids are produced by crossing tetraploid female with diploid male. However, the reciprocal cross is not successful.
3. **Triploid Cassava:** In cassava (tapioca) triploid Sree Harsha cultivar has been produced by crossing cultivated diploid with induced tetraploid. The triploid is characterized by vigorous growth, erect plant type, broad leaves and stout stem. Besides, it is significantly higher yielder than the diploids and has higher starch content (38-41%). Sree Harsha is the first triploid of cassava developed by polyploidy breeding and released for general cultivation.
4. **Triploid Banana:** Cultivated varieties of banana are triploids and seedless. Such bananas have larger fruits than diploids ones.
5. **Triploid Apples:** Some cultivars of apples are triploids that are propagated asexually by budding or grafting.
6. **Tetraploid Grapes:** Tetraploid grapes have been developed. They have larger fruits and fewer seeds per fruit than diploids.
7. **Triploid Tea:** Triploid tea clone TV29 has been released for commercial cultivation. It gives higher cured leaf yield and is more drought tolerant than the diploid cultivar. It is comparable to the diploid cultivar in leaf quality.
8. **Ornamentals:** Induced tetraploids have been most successful in ornamentals, e.g., marigold, petunia, snapdragon, portulaca, chrysanthemum, calendula and lily. In such crops novelty itself is a virtue. Autopolyploids may have bigger flower size, longer blooming period and relatively longer lasting flowers.

The outlook for utilization of autopolyploidy in crop species where vegetative parts are important is better, e.g., in medicinal plants.

23.5 Allopolyploids

The primary objective of allopolyploid (amphidiploid) breeding is to combine complementary characters of two different species into a new synthesized species. The progenitor species may be closely related or show a high level of divergence. The net result of interspecific hybridization and polyploidization is that the genomes of distinct species are combined in the allopolyploid.

23.5.1 Application and Uses of Allopolyploidy Breeding

Limited success has been achieved so far. Only triticale (wheat-rye amphidploid) and a tetraploid between Italian rye grass and perennial rye grass have been grown in farming systems to a rather limited extent. Poor success in allopolyploidy breeding is mainly due to plant's inability to stabilize meiotic behaviour to a level comparable to that found in natural allopolyploids.

23.6 Brief References

Agrawal 1998, Chaudhary 2014, Choudhary 2018, De and Bhattacharjee 2011, Khan 2018, Ramachandra 2014, Sharma 2009, Singh BD 2011, Singh Phundan 2011, Singh and Pawar 2006.

Review Questions

Part A

Answer the following questions true or false.

1. An organism having more than two genomes is called polyploid. True/False
2. Colchicine induces polyploidy by assisting formation of spindle fibres. True/False
3. Cultivated varieties of banana are tetraploids and seedless. True/False
4. Direct selection for ploidy level based on chromosome counts is easier. True/False
5. The colchicine-induced polyploidy is known as colchiploidy. True/False

Part B

Answer the following questions.

1. Define polyploidy.
2. What is the reason of poor success in allopolyploidy breeding?
3. Explain autopolyploidy.

4. Different species have _____________ levels of optimum ploidy.
5. What is the primary objective of allopolyploidy breeding?

Part C

Write a short note on each of the following topics.

1. General features of autopolyploids.
2. Triploid tapioca.
3. Procedure of breeding induced autotetraploids.
4. Induction of polyploidy.
5. Application and uses of autopolyploids.

24

Wide (Distant) Hybridization and Introgressive Breeding

24.1 Introduction

When crosses are made between two different species of a genus or between two different genera, they are known as **distant hybridization** (also called **wide hybridization**). It includes both interspecific and intergeneric hybridization. Thomas Fairchild (1717) was the first man to do distant hybridization. He produced a hybrid between two species of *Dianthus*, *viz.*, *D. caryophyllus* and *D. barbatus* (Carnation and Sweet William). Intergeneric hybrid produced by Karpechenko, a Russian scientist, in 1928 is Raphanobarssica. It is the amphidiploid of cross between radish (*Raphanus sativus*) and cabbage (*Brassica oleracea* var. *capitata*). Unfortunately, it had roots like cabbage and leaves like radish. Triticale was produced by Rimpau in 1890 itself. It is an amphidiploid obtained from a cross between wheat and rye (*Triticum* and *Secale*). Another example is *Saccharum* nobilisation involving *Saccharum officinarum*, *S. barberi*, *S. sinensis*, *S. spontaneum*. Wide hybridization has been used in the genetic improvement of some crops. It is an effective means of transferring desirable genes into cultivated plants from related species and genera.

24.2 Main Features of Interspecific Hybridization

Crossing or mating between two different species of the same genus is referred to as **interspecific hybridization** (also called **intrageneric hybridization**). The main features of interspecific hybridization are:

1. It is used when desirable character is not found within the species.
2. It is an effective method of transferring genes into cultivated plants from their related cultivated or wild species.
3. Interspecific hybridization is more successful in vegetatively propagated species than seed propagated species.
4. Interspecific hybridization gives rise to three types of crosses, *viz.*, (a) fully fertile, (b) partially fertile, and (c) fully sterile in different crop species.

5. Interspecific hybridization leads to introgression which refers to transfer of some genes from one species into the genome of another species.

24.3 Main Features of Intergeneric Hybridization

Intergeneric hybridization refers to crossing between two different genera of same family. Such crosses are rarely used in crop improvement because of various problems associated with them. The main features of intergeneric hybridization are given below:

1. Intergeneric hybridization is used when the desirable genes are not found in different species of the same genus.
2. This method is rarely used in crop improvement programmes and that too for transfer of some specific characters into cultivated species from allied genera.
3. Intergeneric hybridization has been generally used in asexually propagated species.
4. F_1 hybrids between two genera are always sterile. The fertility must be restored by chromosome doubling through colchicine treatment.

24.4 Barriers in Production of Distant Hybrids

In some cases, distant hybrids may be obtained without appreciable difficulty. But in majority of cases, F_1 hybrids may be obtained with variable degrees of difficulty and in many cases, hybrids may not be obtained with currently available technique. The difficulties encountered in the production of interspecific hybrids may be grouped into three broad categories:

1. Hybrid Sterility: Failure of zygote formation
2. Hybrid Inviability: Failure of zygote development
3. Hybrid Breakdown: Failure of F_1 seedling development

24.4.1 Failure of Zygote Formation

It may be due to failure of fertilization. The failure of fertilization may be due to (a) slow pollen tube growth, (b) style being longer than pollen tube, (c) pollen tube of polyploid species being thicker than diploid species and slower growth of pollen tube.

24.4.2 Failure of Zygote Development

In many cases, though fertilization takes place and zygote is produced, the development of zygote is blocked at various stages. It may be due to lethal genes, genotypic disharmony between two parental genomes, chromosome elimination, incompatibility of cytoplasm, and endosperm abortion.

24.4.3 Failure of F_1 Seedling Development

Due to chlorophyll deficiency seedling may die.

24.5 Techniques for The Production of Distant Hybrids

1. Generally the species with shorter style should be used as the female parent. When it is not possible, the style of the female is to be cut off.
2. In some cases, autopolyploidy may be helpful in interspecific hybridization. For example, diploid *Brassicas* would not cross; making them autotetraploids leads to easy crossing.
3. Whenever there is embryo abortion, embryo rescue technique can be adopted.
4. When species with different levels of ploidy are crossed, hybridization will be difficult. In such cases, higher ploidy level species is used as female for crossing. Secondly, the number of chromosomes of wild species of F_1 may be doubled to overcome sterility.
5. When two species cannot be crossed directly, a third species can be used as bridge species.
6. Use of growth regulators like IAA, 2,4-D at flowering stage increases seed set.

24.6 Role of Distant Hybridization in Crop Improvement

Wild species or wild genetic resources are potential sources of desirable genes for various traits of crop plants. Wild hybridization is an effective method of exploiting desirable traits from wild species for improvement of cultivated crop plants. Thus, the significance of wild species and distant hybridization are interlinked. Distant hybridization has played significant role in improving the crop plants for (1) resistance to biotic stress, (2) quality, (3) adaptation, (4) yield, (5) mode of reproduction, and (6) several other characters. The notable interspecific and intergeneric hybrids have been obtained in banana (A and B genomes), citrus (limequat, citromelos, tangerines and tangellos), papaya, guava, fig, annona, several temperate fruits, vegetables (solanaceous, leguminous, cucurbitaceous, and crucifers) and flowers (lillies, petunias, and orchids), etc.

24.7 Achievements

Some of the cultivars developed through interspecific hybridization are:

Cultivars

- **Tomato**: Pusa Red Plum, Hisar Anmol
- **Okra**: Punjab Padmini, Punjab 7, Parbhani Kranti, Arka Abhay, Arka Anamika
- **Potato**: Most of the modern cultivars
- **Bougainvillea**: Begum Sikander, Wajid Ali Shah, Chitra, Summer Time, etc.

New Crops

- **Citrus**: Tangelo (*C. paradisi* × *C. reticulata*), Tangor (*C. reticulata* × *C. sinensis*), Citrumelo (*C. paradisi* × *Poncirus trifoliata*), Citrange (*P. trifoliata* × *C. sinensis*), Citrangor (Citrange × *C. sinensis*), etc.
- **Annonas**: Arka Sahan (*Annona atemoya* × *A. squamosa*)
- **Prunus**: Pluots (*Prunus salicina* × *P. armeniaca*)
- **Rubus**: Red raspberry (*Rubus idaeus* × *R. strigosus*)

24.8 Brief References

De and Bhattacharjee 2011, Dinesh and Sankaran 2017, Ramachandra 2014, Sharma 2009, Singh Phundan 2011, Singh and Pawar 2006.

Review Questions

Part A

Answer the following questions true or false.

1. F_1 hybrids between two genera are sometimes sterile. True/False
2. Generally, the species with shorter style should be used as the female parent in distant hybridization. True/False
3. Interspecific hybridization is more successful in seed propagated species than vegetatively propagated species. True/False
4. Intergeneric hybridization is used when the desirable genes are not found in different species of the same genus. True/False
5. Bateson was the first man to do distant hybridization. True/False

Part B

Answer the following questions.

1. Define distant hybridization.

2. What is tangelo?
3. Growth regulators like IAA, 2,4-D at flowering stage increases/decreases seed set.
4. The fertility of sterile F_1 can be restored by chromosome __________.
5. What are the three barriers in production of distant hybrids?

Part C

Write a short note on each of the following topics.

1. Techniques for the production of distant hybrids.
2. Barriers in production of distant hybrids.
3. Main features of interspecific hybridization.
4. Role of wide hybridization in plant breeding.
5. Main features of intergeneric hybridization.

25

Marker Assisted Selection Assisting Conventional Methods

25.1 Introduction

The method of selection in which a gene/QTL (Quantitative Trait Locus) is selected on the basis of the genotype of the **molecular marker** (also called **DNA marker**) linked to this gene/QTL is known as **marker-assisted selection** (also called **marker-aided selection**). The idea of marker-assisted selection (MAS) was first advocated by Tanksley (1983). MAS can be used in combination with the conventional breeding methods or novel breeding schemes could be devised to take full advantage of the marker data. MAS has been widely used for introgression of the desired genes and QTLs using backcross programmes.

25.2 Marker-Assisted Backcrossing (MABC)

The backcross programme based on molecular marker is usually called **marker-assisted backcrossing** (MABC). MABC has three activities: (a) foreground selection, (b) background selection and (c) recombinant selection.

25.2.1 Foreground Selection

The use of molecular marker to select the gene/QTL being introgressed is known as **foreground selection**. The objective of foreground selection is to select the target gene, i.e., the gene being introgressed, without phenotypic evaluation for the trait governed by the target gene. Foreground selection for a gene/QTL will be highly desirable in a variety of situation, some of which are listed below:

1. The phenotypic evaluation of the character is time-consuming, costlier, burdensome or technically demanding.
2. The reliability of the phenotypic evaluation of the character is low.
3. The character is governed by the recessive allele of the gene.
4. The introgression of both vertical and horizontal resistance is required in a single cultivar.
5. Introgression of multiple QTLs of a character is necessary.

6. Introgression of genes/QTLs for multiple characters is necessary.
7. The character can be evaluated only after flowering.

When a marker is situated at more than 5 cM distance from the target gene, it is advisable to use an additional marker located on the other side of this gene for foreground selection. When one marker is located on one side of a locus and another marker is located on the other side of the locus, this pair of markers is called **flanking markers**. The use of flanking markers minimizes the risk of selection of the wrong allele of the gene/QTL on the basis of marker genotype, as a result increases the effectives of foreground selection.

25.2.2 Background Selection

It is important that the genome of recurrent parent (RP) is recovered in the new version of RP (RP^1) to the maximum extent possible. In a conventional backcross programme, a selection for the plant type of the RP is practiced during the backcross and the selfing generations to maximize the recovery of the RP genome. In this context, molecular marker may be used to increase the effectiveness of selection for the RP genome; this is referred to as **background selection**. *Background selection is based on markers distributed throughout the genome at sufficient density to ensure the selection for the entire RP genome, except for the genomic region containing the gene/QTL of interest.* The use of molecular markers for background selection increases the rate of RP genome recovery in the backcross progeny.

All the plants in a backcross generation must be screened for the markers used for the background selection before they flower so that this marker genotype information is used to select the plants for the next backcross. This is not feasible for many breeding programmes. Hence, in most MABC programmes the selection for RP genome is largely or exclusively based on the phenotypic evaluation for the RP plant type.

25.2.3 Recombinant Selection

Molecular markers can be used to minimize the risk of **linkage drag** by selecting against the donor parent (DP) genomic regions flanking the target gene/QTL. This application of molecular markers is termed as **recombinant selection**. Thus, recombinant selection is a special type of background selection that focuses on the genomic regions linked to and flanking the target gene/QTL. Further, recombinant selection is based on pair of markers that are located on either side of the target gene/QTL at < 5 cM. It is, however, preferable that these flanking markers are located within 1 cM of the gene being introgressed, particularly when the DP is an inferior germplasm accession, including a related wild species.

Recombinant selection is more convenient when it is done according to the two-step scheme. The marker M_1 is a gene-based marker located within the gene being introgressed and is used for foreground selection. The markers M_2 and M_3 are located within 1 cM of the target gene A, and are used for recombinant selection. The first step comprises identification of at least one plant in the BC_1 generation that shows recombination between the markers M_1 and M_2; such plant will have the marker genotype M_1m_2 (m_2 is the marker allele present in the RP). This plant is used to produce BC_2 generation. In this generation, all the plants will have the marker genotypes M_1m_2. Therefore, search is made for a plant that shows recombination between the markers M_1 and M_3. Such a plant will have the genotype M_1m_3 for these two markers, and this plant will be double recombinant, i.e., $m_2M_1m_3$. This plant is termed as double recombinant because it is produced as a result of two recombination events, *viz*., recombination between the markers M_1 and M_2, and between M_1 and M_3. Another scheme for recombinant selection is to select for the double recombinant in a single backcross generation.

25.2.4 A generalized Scheme for MABC

A generalized scheme for an effective MABC programme is outlined below (Figure 25.1):

1. The F_1 from the cross between selected DP and RP is backcrossed to RP to produce a sufficiently large BC_1 population.
2. In the BC_1 generation the plants are screened with M_1, M_2 and M_3. All BC_1 plants with either $m_2M_1M_3$, or $M_2M_1m_3$, or $m_2M_1m_3$ marker genotype are selected.
3. In case, a plant has the marker genotype $m_2M_1m_3$, it will be backcrossed to RP. Further, if more than one plant were obtained with other two genotypes, all the plants will be screened with the set of markers selected for background selection. The plant with the largest proportion of RP marker alleles will be selected and backcrossed to RP to produce a sufficiently large population.
4. In the BC_2 generation, all the plants are screened with M_1 and either M_2 or M_3. In case the BC_1 plant used for backcrossing was $m_2M_1M_3$, the marker M_3 will be used for screening. But the marker M_2 will be used if the selected plants in BC_1 were $M_2M_1m_3$. In case the BC_1 plant was $m_2M_1m_3$, only M_1 will be used to enable foreground selection for the desired allele of the target gene. All the plants that are recombinant for the concerned maker pairs used for screening are selected.
5. These selected BC_2 plants will be double recombinants, i.e., will be $m_2M_1m_3$. If there were more than one such plant, all of them will be screened with

the set of markers selected for background selection. The plant with the highest proportion of the RP marker alleles will be selected and used for either one or, preferably, more backcrosses.

6. All the backcross progeny will be subjected to foreground (using marker M_1) and background selections. The plant having M_1 allele and the highest proportion of RP marker alleles shall be identified and selfed.
7. Foreground and background selections will be carried out in the selfed generations and, in the end, a plant homozygous for introgressed allele (M_1M_1) and the highest proportion of RP genome is isolated.
8. The selected line is an NIL of RP (RP^1) modified with the introgression of the desired gene. This line will be evaluated for the introgressed trait and for agronomic performance and might be released for commercial cultivation if found to be desirable.

Parents RP x DP

$m_2m_1m_3$ ↓ $M_2M_1M_3$

F_1

- F_1 from selected RP and DP is backcrossed to the RP

↓

BC_1

- All the plants screened with markers M_1, M_2 and M_3
- Plants with marker genotype $m_2M_1M_3$, $M_2M_1m_3$ or $m_2M_1m_3$ identified
- If two or more plants with above genotypes are present, plant with highest proportion of RP marker alleles is selected and backcrossed with RP [If a BC_1 plant were $m_2M_1m_3$, it will be preferred to others]

↓

BC_2

- All the plants are screened with M_1 and M_2 or M_3 (If the selected BC_1 plant were $m_2M_1M_3$, M_3 will be used; if it were $M_2M_1m_3$, M_2 will be used; but if it were $m_2M_1m_3$, only M_1 will be used.)
- Plants with $m_2M_1m_3$ genotype are selected.
- If more than one plant has this genotype, the one having highest proportion of RP genome is selected and backcrossed with RP.

↓

BC_3

Foreground and background selections

Selfing

↓

BC_3F_2

- Foreground and background selections
- A line with the genotype $m_2m_2M_1M_1m_3m_3$ and highest proportion of RP genome is selected and used as RP'

Figure 25.1 A generalized scheme for marker-assisted backcrossing (MABC) RP: recurrent parent, DP: donor parent, RP': modified version of RP, i.e., RP with introgressed gene, F_1: first filial generation, BC_1, BC_2 and BC_3: first, second and third backcross generation, respectively.

25.3 Merits of Marker-Assisted Selection

1. Foreground selection is simpler and more convenient than phenotypic selection for many traits.
2. In case of introgression of recessive genes, the backcross can be made sequentially without any need for raising an F_2 generation.
3. The selected plants can be used for making crosses in the same season.
4. Recombinant selection minimizes the risk of linkage drag.
5. Background selection can be used in the backcross programme to ensure a high level of recovery of the RP genome.

25.4 Demerits of Marker-Assisted Selection

1. For many traits of interest tightly linked markers are not available.
2. Introgression of QTLs often does not produce the desired results.
3. Marker genotyping adds to the overall cost of the breeding programmes.

25.5 Brief References

Acquaah 2012, Janardhan 2012, Koh *et al.* 2015, Newbury 2003, Saxena and Balabhaskar 2014 b, Singh and Singh 2015, Singh Phundan 2014, 2015 and 2019, Singh and Shekhawat 2019, Wani and Malik 2014, Xu Y 2010.

Review Questions

Part A

Answer the following questions true or false.

1. Molecular marker is also known as DNA marker. True/False
2. Foreground selection is more complex than phenotypic selection. True/False
3. Recombinant selection increases the risk of linkage drag. True/False
4. For many traits of interest tightly linked markers are not available. True/False
5. The reliability of the phenotypic evaluation of the character is low. True/False

Part B

Answer the following questions.

1. QTL stands for ________.
2. What is background selection?

3. MABC has three/four activities.
4. What is recombinant selection?
5. Differentiate between foreground and background selection.

Part C

Write a short note on each of the following topics.

1. Foreground selection.
2. Background selection.
3. Recombinant selection.
4. Generalized scheme for MABC.
5. Merits of MAS

26

In-Vitro Breeding Crop Improvement Through Tissue Culture

26.1 Introduction

We have already discussed the basics of tissue culture techniques in Chapter 2. Plant tissue culture techniques in combination with recombinant-DNA technology are essential requirements for development of transgenic plants (will be discussed in next chapter). However, tissue culture techniques like anther, pollen, ovule and meristem culture can themselves be used for improvement of different traits or as an aid to conventional breeding methods. In recent years, isolated microspore culture has been preferred as a breeding tool and an experiment system for various genetic manipulations. Besides, using protoplast fusion and hybrid embryo rescue, distant hybridization has been successfully achieved in many crops. Protoplast fusion technique can be used for the transfer of cytoplasmic male sterility from one species to another in very short duration. Somaclonal variation generated during tissue culture can also be utilized for recovery of variants showing disease resistance or other desirable characters.

26.2 Utilization of Tissue Culture Techniques in Crop Breeding

26.2.1 Anther Culture and Doubled Haploids (DH)

Among several biotechnological tools available to crop breeders, anther and pollen culture holds special promise because of its potential application in crop improvement. Anther culture has the potential to produce haploid plants that can be useful in generating homozygous diploids (inbred lines) in very short time. These homozygous diploids have tremendous use in exploitation of hybrid vigour. Anther and pollen culture employs the same technique as that of tissue culture, only the explants used is either anther or microspores (pollens).

This method involves production of F_1 hybrids between two parents of desirable traits and then microspores (pollens) are cultured to obtain plantlets that are haploid and sterile. Application of colchicine is must to make them diploids, fertile and homozygous. Due to their inbred nature, the recessive deleterious as well as desirable quality traits are expressed in very first generation that makes the selection very easy.

26.2.1.1 Procedure to Develop Double Haploids (DH)

Double haploid is defined as a plant or line obtained by doubling chromosome number of a haploid plant. In many crops, plant breeders aim at developing superior homozygous lines. Hence, crosses must be advanced to F_5 or F_6 generations so that plants become reasonably homozygous to permit preliminary evaluation. But in many crops homozygous lines can be derived from F_1 generation itself. This is achieved by **double haploid technique** adopting the following procedure:

1. Extraction of haploid plants from F_1 using anther culture.
2. Chromosome doubling of such haploid plants to obtain completely homozygous double haploid plants using colchicine.
3. Seeds of individual double haploid plants are harvested separately which will be homozygous and evaluated separately.

This technique saves at least 3-4 years in the development of a new cultivar. Homozygosity is achieved in a single generation. But the availability of well-equipped tissue culture laboratory and dependable greenhouse facilities are a prerequisite for its success.

26.2.2 Embryo Culture and Wide Hybridization

Inter-specific hybridization is very effective method used to improve crops by transferring specific traits like biotic and abiotic stress resistance to crops from their wild relatives. Crossing barriers occur frequently when interspecific or intergeneric crosses are made. These barriers are the results of sexual incompatibility. Techniques like ovule culture and embryo rescue are employed to overcome these incompatibilities that cause endosperm failure or embryo abortion. Embryo culture can be applied successfully in crosses when pollinated flowers stay on plants for considerable time.

26.2.3 Protoplast Fusion and Somatic Hybrids

Protoplast (cell without cell wall) can be isolated and cultured at large scale. This protoplast is a very useful system for genetic manipulation. Protoplast culture employs a series of operations such as isolation of the protoplasts from the cells, culturing them on suitable medium inducing them to divide and then regenerating plantlets from them. Protoplasts can serve as a means of intensive *in-vitro* multiplication and can lead to extensive somaclonal variation. Their usefulness lies in their ability for **somatic hybridization** of close or distant relatives. The regeneration ability of fused protoplasts is low.

26.2.4 Somaclonal Variation

The genetic variation present in plant tissue/cell cultures or the plants regenerated from them is called **somaclonal variation**. Plant tissue and cell cultures are generally derived from somatic cells, and these cells divide mitotically during the culture period. Hence, all the cells present in a given culture are expected to have same chromosome constitution as well as genotype. However, chromosomal changes and genetic variation do arise in the tissue/cell cultures. These genetic changes may arise due to one or more of the following:

i) Gene mutations,

ii) Gene amplification,

iii) Gene deamplification,

iv) Increased transposition of transposable elements,

v) Mutations in the genomes of chloroplasts, and

vi) Changes in chromosome number or structure.

Somaclonal variants for different traits including yield are generally isolated by the screening procedure. In this approach, a large number of plantlets are generated from all cultures, and they are transferred into the field. The plants are selfed and the progeny are grown in, preferably, individual plant progeny rows. These progeny plants are evaluated for the various traits and compared with those of the parent cultivar. Any plant showing a desirable change is identified and selected. The selfed progeny from the plants are grown in individual plant progeny rows and desirable progenies/plants are selected. The selected lines are evaluated in replicated yield trial and superior variants may be released as new cultivars. This approach has been used to develop some useful cultivars in several crops, e.g., Vegisnac in carrot, Bio-13 in *Citronella*, Velvet Rose in *Geranium*, Scarlet in sweet potato. Alternatively, selection may be applied to cell cultures to select for variant cells, e.g., cells resistant to a toxin produced by a pathogen causing a disease. In case the toxin is involved in disease development, the toxin resistant cells would give rise to plants that would be resistant to the concerned pathogen. This approach can be used for selection for resistance to abiotic stresses like salt stress, cold stress, etc. as well. The technique shows some to considerable success in the selection of resistant cells, and the cell selection efforts have resulted in the development of some useful somaclonal variants, e.g., in case of *Mentha* sp.

26.3 Achievements

1. **Anther/pollen culture**: Haploids have been developed in *Asparagus officinalis, Beta vulgaris, Brassica oleracea, Capsicum annuum,*

Cucumis sativus, Solanum lycopersicum, Raphanus sativus, Solanum tuberosum, Nicotiana, Pelargonium, Anemone, Paeonia, etc.

2. **Embryo culture for wide hybridization**: Hybrid embryo rescue technique has been reported in (*Solanum lycopersicum*×*S. peruvianum*), (*Phaseolus vulgaris*×*P. angustissimus*), (*Vigna pubescens*×*V. unguiculata*), etc.
3. **Protoplast fusion and somatic hybrids**: Protoplast fusion has been reported in (*Allium ameloprasum* + *A. cepa*), (*Solanum lycopersicum* + *S. peruvianum*), (*S. lycopersicum* + *S. pennelli*), (*S. lycopersicum* + *S. richii*), (*Solanum melongena* cv Dourga + *S. khasianum*), (*S. melongena* + *S. nigrum*), (*S. melongena* + *S. torvum*), (*S. melongena* + *S. symbriifolium*), {*Citrus* (Tangelo) + *Murraya paniculata*}, {(*Citrus reticulata* × *Citrus paradisi*) + *Citrus jambhiri*}, (*Citrus sinensis* + *Citrus reticulata*), (*Citrus sinensis* + *Clausena lansium*), etc.
4. **Somaclonal variation:** Some cultivars have been developed using somaclonal variation in some crops, e.g., carrot, *Citronella*, *Geranium*, sweet potato, *Mentha*, etc.

26.4 Brief References

Agrawal 1998, Choudhary 2018, De and Bhattacharjee 2011, De Filippis 2014, Shukla *et al.* 2020, Singh and Shekhawat 2019, Thirugnanakumar *et al.* 2009.

Review Questions

Part A

Answer the following questions true or false.

1. Somaclonal variation generated during tissue culture cannot be utilized for recovery of variants. True/False
2. Protoplast is a cell without cell wall. True/False
3. Embryo culture cannot be applied successfully in crosses when pollinated flowers stay on plants for short time. True/False
4. Interspecific hybridization is very effective method used to improve crops by transferring specific traits from their wild relatives. True/False
5. Somaclonal variants for different traits are generally isolated by the screening procedure. True/False

Part B

Answer the following questions.

1. What is somaclonal variation?

2. Define double haploid.
3. Give two examples of protoplast fusion.
4. Crossing barriers occur frequently when ______ or ______ crosses are made.
5. Why do chromosomal changes and genetic variation arise in the tissue/cell cultures?

Part C

Write a short note on each of the following topics.

1. Procedure to develop double haploids.
2. Somaclonal variation.
3. Protoplast fusion and somatic hybrids.
4. Embryo culture and wide hybridization.

27

Transgenic Breeding Crop Improvement Through Genetic Engineering

27.1 Introduction

Crop improvement depends on genetic variation present in nature or created by hybridization followed by segregation and recombination. Again, gene transfers are possible only from the lines and species that are sexually compatible with the concerned species. Interspecific gene transfers pose many problems, including the loss of F_1 hybrids during embryo development due to abortion of the endosperm in the developing hybrid seeds. Embryo rescue technique was developed to rescue such interspecific hybrid embryos, and it has been widely used for this purpose. It is to be mentioned that embryo rescue can be used only in such cases where sexual hybridization is successful in producing zygotes that begin to develop into embryos.

Besides, the technique of somatic hybridization by protoplast fusion was developed to enable gene transfer from the species that do not produce zygotes on hybridization. This method has been extensively used and some useful lines have been derived. But this method is limited to utilization of genes present in plant species. Again, the results of somatic hybridization are unpredictable since many somatic hybrids fail to regenerate complete plants and the transfer of specific genes or groups of genes depends entirely on chance.

The **recombinant DNA technology** (also called **genetic engineering**) has permitted the transfer of genes from any organism into any organism. Besides, chemically synthesized genes can be transferred and integrated into genomes of various organisms to enable the expression of these genes. A gene transferred into an organism by genetic engineering is called a **transgene**. The individuals carrying transgenes are called as **transgenic organisms**. Hence, a **transgenic plant** can be defined as a plant that contains in its genome a transgene and expresses this transgene. Plant tissue culture and recombinant DNA technology are the basic requirements for the production of transgenic plants. The basics of these two techniques are discussed in Chapter 2.

27.2 Main Features of Transgenic Plants

1. Transgenic plants contain transgenes.
2. They are developed through tissue culture and genetic engineering.
3. They are developed bypassing sexual reproduction.
4. They are recovered at a very low frequency.
5. They are developed to solve specific problems in crop plants, e.g., susceptibility to biotic (diseases and insect-pests) and abiotic (drought, frost, salinity and metal toxicity) stresses, shelf-life of vegetables and fruits, development of male sterility, etc.

27.3 Transgenic Breeding

Transgenic breeding refers to genetic improvement of crops through the use of transgenes. Now-a-days, it is used for specific genetic improvement of different crops. It is expected to play an important role in crop improvement in the years ahead. The main advantages of transgenic breeding are:

1. **Rapid method of crop improvement**: By this technique, stable transgenic plants can be developed in 3-4 years whereas it takes 12-15 years to develop a new cultivar through conventional methods of breeding such as pedigree, bulk and backcross methods.
2. **Overcome crossing barriers**: It permits gene transfer between unrelated species and even unrelated organisms. It permits gene transfer even between plants and animals. For example, a freezing resistance gene has been transferred from fish to cultivated tomatoes.
3. **Evolution of new genotypes**: Sometimes it may lead to evolution of altogether new plant species, because it permits gene transfer between various plant species. Thus, it will affect the process of natural evolution.
4. **Application**: It can be used for the genetic improvement of both self- and cross-pollinated crops. Both seed and vegetatively propagated crops can be improved through this technique.
5. **Effectiveness**: It has been found effective for the genetic improvement of monogenic traits only. It has not been used, so far, for the genetic improvement of polygenic traits.

27.4 Procedure of Transgenic Breeding

Seven important steps are involved to develop transgenic plants. They are (i) gene identification and cloning, (ii) gene transfer, (iii) identification of transformed cells, (iv) regeneration, (v) gene expression, (vi) proper integration and (vii) field trials.

i) **Gene identification and cloning**: First the gene of interest is identified, isolated and cloned (made multiple copies).

ii) **Gene transfer**: There are four important methods of foreign gene (DNA) transfer in crop plants, namely, plasmid method, particle bombardment, direct DNA uptake and micro-injection. These methods are discussed in some details in Chapter 2. Currently, *Agrobacterium*-mediated gene transfer and bombardment of cells with plasmid coated particles are widely used for development of transgenic plants.

iii) **Identification of transformed cells**: After gene transfer, the transformed cells are identified. The presence of transgene is detected by several methods. The commonly used techniques of detecting transgene are (i) selectable marker gene, (ii) polymerase chain reaction (PCR), (iii) Enzyme-Linked Immunosorbent Assay (ELISA), (iv) Southern blot technique, (v) Northern blot technique and (vi) Western blot technique. Of these, the selectable marker gene is simple and widely used.

A selectable marker is attached to the target gene for easy identification. Generally, an antibiotic resistance gene is used as **selectable marker** to confirm whether new gene has been inserted in the host cells or not. The kanamycin resistance gene is used for identification of transformed cells. The lethal concentration of kanamycin is put in the plasmid mixed suspension of protoplast culture. The non-transformed cells will die and only transformed protoplast will survive. Thus, transformed cells are identified.

iv) **Regeneration**: The transgenic cells, calli or protoplasts are regenerated into whole plants by tissue culture.

v) **Gene expression**: The transgene inserted into host plant should have expression of the desired level.

vi) **Proper integration**: The gene should be integrated properly so that it is carried from generation to generation by usual means of reproduction.

vii) **Field trials**: First the transgenic plants are tested in laboratory for bio-safety such as allergenicity and toxicity with animals like rats, rabbits, poultry, goats, etc. Transgenic plants which have been passed by the regulatory authority are evaluated in multilocation trials for three years to test the performance of the gene of interest. Superior genotypes with stable performance are released for cultivation.

27.5 Achievements

Transgenic plants are reported in several crops. Some examples are given below:

- **Vegetables**: Tomato, brinjal, cauliflower, cabbage, etc.

- **Fruits:** Kiwi fruit, papaya, pecan nut, lime, rough lemon, mandarin, tangelo, strawberry, walnut, apple, banana, tamarillo, passion fruit, avocado, sweet orange, grape, peach, plum, apricot, pear, almond, trifoliate orange, citrange, cranberry, red raspberry, black berry.
- **Ornamentals**: Petunia, rose, chrysanthemum, carnation, *Dendranthema*, *Narcissus*, *Gladiolus*, *Lilium*, *Tulipa*, *Gerbera*, *Dendrobium*, *Antirrhinum*, *Anthurium*, *Eustoma*, *Pelargonium*, etc.

27.6 Brief References

Agrawal 1998, Choudhary 2018, De and Bhattacharjee 2011, Shukla *et al.* 2020, Singh and Shekhawat 2019, Singh Phundan 2013a, 2019.

Review Questions

Part A

Answer the following questions true or false.

1. Transgenic plants are recovered at a very high frequency. True/False
2. First the gene of interest is identified in transgenic breeding. True/False
3. Eight important steps are involved to develop transgenic plants. True/False
4. There are five important methods of foreign gene (DNA) transfer. True/False
5. A transgene cannot be detected by ELISA method. True/False

Part B

Answer the following questions.

1. Define transgene.
2. Stable transgenic plants can be developed in — years.
3. What is a transgenic plant?
4. Name two vegetables where transgenics have been developed.
5. The transgene inserted into host plant should have expression of the — level.

Part C

Write a short note on each of the following topics.

1. Advantages of transgenic breeding.
2. Main features of transgenic plants.
3. Procedure of transgenic breeding.

Part C
Mission-Oriented Breeding Practices

28

Breeding for Resistance to Biotic Stresses

28.1 Introduction

The stress that is caused by biological agents or factors such as diseases, insect-pests and parasitic weeds is called **biotic stress**. The term disease can be broken into two words: "dis" (means negative, reverse or opposite) and "ease" (means comfort, or freedom from pain or discomfort). Hence, *plant disease can be defined as any harmful alteration from the normal functioning of physiological processes*. Diseases may be bacterial, fungal or viral. A wide range of insects infests various crop plants. The extent of damage caused to crop plants by various organisms is usually in the order fungi > bacteria > viruses > nematodes = insects. But the relative importance of these organisms varies to a great deal in different crops. Breeding of cultivars resistant to diseases and/or insect-pests is the cheapest, safest and most dependable means of limiting the losses due to them. Resistant cultivars form an important component of integrated management of diseases and insect-pests.

28.2 Breeding for Resistance to Biotic Stresses-Diseases

Agrios (1970) has defined disease as: *"Disease is a series of invisible and visible responses of plant cells and tissues to a pathogenic microorganism or an environmental factor that result in adverse changes in form, function or integrity of plant and may lead to partial impairment or death of the plant or its parts."* The organism causing disease is called **pathogen**, while the plant affected by the disease is called **host**.

The principles underlying breeding for resistance to diseases are much the same as for other characters, except for one important difference, i.e., the diseases caused by pathogens are the product of interaction between two genetic systems, that of the host and that of the parasite. Both systems are capable of variation and evolution and affect the interaction between host and pathogen. Here the breeder is required to deal with the heritable traits of both the host plant and the pathogens invading it.

28.2.1 Productive and Resistance Breeding

In **productive breeding**, the plant genotype and its interaction with the environments are the two important considerations (G×E). But in **resistance breeding** there are three components, namely, genotype, environment and pathogen (G×E×P), known as "**disease triangle**" (Figure 28.1). Interaction of these three components produces resistant or susceptible reaction. Since the triangle is equilateral, all the three components have equal importance. Thus, resistance is observed when either weather is unfavourable or pathogen is avirulent or host has resistant genes. Conversely, virulent pathogen, favourable environment and susceptible host are essential for disease to occur. An avirulent race of a pathogen is the one which cannot cause the disease, and the one which can cause is virulent. A **pathogen race** or **pathotype** is said to be virulent if it is able to attack a host with specific resistance gene, and avirulent when it cannot attack. **Pathogenicity** is the ability of a pathogen to attack a host and is synonymous to **virulence.**

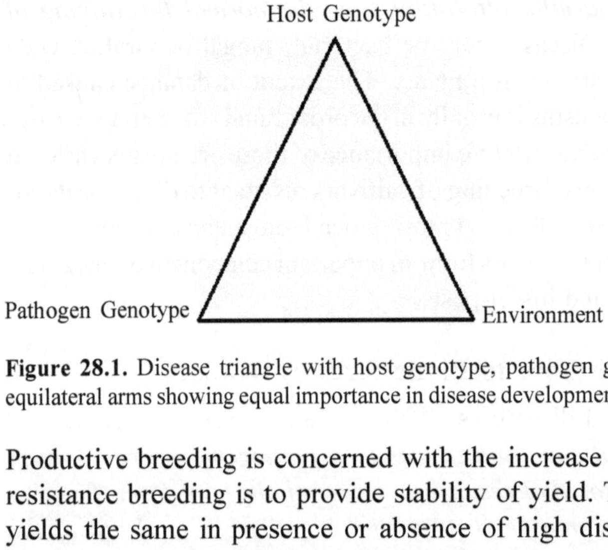

Figure 28.1. Disease triangle with host genotype, pathogen genotype and environment of equilateral arms showing equal importance in disease development.

Productive breeding is concerned with the increase of yield potential, while resistance breeding is to provide stability of yield. Thus, a resistant cultivar yields the same in presence or absence of high disease pressure. Artificial epiphytotics, i.e., artificial creation of high disease pressure is necessary to effect selection in resistance breeding while this is not necessary in productive breeding programmes. Yield potential once incorporated in a genotype becomes permanent feature while resistance reaction may be lost mainly due to development of virulent races.

28.2.2 Loss of Resistance

Resistant cultivars may become susceptible after sometime, which is called **break-down of resistance**. This may be due to any of the following reasons:

28.2.2.1 Change of Environment

Resistance genes function under given set and range of environment. Once this range is crossed, they do not function to provide resistance and the resistant cultivar becomes susceptible. For example, tomato cultivars carrying Cf1 gene show resistance to *Cladosporium fulvum* fungal disease in summer but not in winter.

28.2.2.2 Resistance Erosion

The host may lose the resistance gene either due to mutation, out-crossing with a susceptible cultivar or segregation. This type of loss is rare and known as **resistance erosion**.

28.2.2.3 Appearance of New Races or Biotypes

The disease-causing pathogen may create new races or insect-pests may create new biotypes. The resistant cultivar thus may become susceptible. This type of loss of resistance is very common.

28.2.3 Classification of Resistance

Each plant species is attacked by hundreds of types of pests and pathogens. Even a single plant is attacked by hundreds or thousands of individuals of a pathogen. Plants express their reaction and manage to survive using various mechanisms that can be classified variously:

A. Based on existence

1. *Preformed*: When the resistance is already present in the plant even in the absence of the pathogen, it is known as **preformed** (also called **axenic** or **passive**) **resistance**.
2. *Induced*: When the resistance is not present in the host plant in the absence of the pathogen, it is called **induced** (also called **apergic** or **active**) **resistance**.

B. Based on type of host response

1. *Immune*: No symptom develops in case of immune response.
2. *Tolerance*: Tolerance may be defined as the inherent or acquired capacity of the host to endure disease (Nelson,1973).
3. *Resistance*: Resistance is always partial and some symptoms do appear on the host, contrary to immunity, where no symptoms appear.

C. Based on genetic control

1. *Oligogenic*: **Oligogenic** (also called **major genic) resistance** is governed by a few genes whose individual effect is very clear.
2. *Polygenic*: **Polygenic** (also called **minor genic) resistance** is governed by many genes whose individual effects are not marked as contrasted in case of oligogenic resistance.
3. *Cytoplasmic*: Resistance of this type is governed by cytoplasmic factors.

D. Based on mechanism of resistance

1. *Mechanical*: **Mechanical** or **structural resistances** are rendered due to external or internal peculiarities of the host plant. The first line of defence against the pathogen is the surface. Cuticle, wax and thick hairs on plant parts and foliage check the germination and entry of the pathogen in the host tissue.
2. *Biochemical*: A series of defence reactions are evoked in the host following invasion by a pathogen. Some important reactions are as follows:
   - i) *Hypersensitivity*: Hypersensitivity is a potential defence reaction. It is due to the increased sensitivity of the host cells in the vicinity of the infection site and is expressed as necrotic spots. The necrotic tissue separates the normal living tissue of the host plant from the parasite and thereby checks its further growth.
   - ii) *Resistance to toxins*: Some pathogens produce toxins which kill the host or host tissue. Resistant hosts in such cases are non-sensitive to toxins.
   - iii) *Phytoalexins*: Phytoalexins may be defined as substances formed by the host plant in response to injury, physiological stimuli, infectious agents or their products that accumulate to levels which inhibit the growth of microorganism and repel insects and nematode (Day,1974). For example, pisatin in pea, phaseollin in beans, rishitin and phytotuberin in potato are the phytoalexins and provide resistance.

E. Based on degree and range of resistance

Van der Plank (1963,1968) introduced these concepts and classified resistance into two groups: vertical and horizontal.

1. *Vertical resistance*: **Vertical** (also called **differential** or **race-specific) resistance** is represented by the Y-axis once disease reaction is plotted in a histogram. Resistance gene functions against some races while it does not function against others.

2. *Horizontal resistance*: **Horizontal** (also called **uniform** or **race non-specific) resistance** corresponds to horizontal line of X-axis once the disease reaction is plotted in a histogram. This type of resistance provides resistance against all the prevalent races of a pathogen.

28.2.4 Genetics of Host-Pathogen Interaction

The first successful attempt to understand the genetics of host-pathogen interaction was made by Flor. He published his conclusion in 1956 in the form of a "gene-for-gene" hypothesis. This hypothesis states that "for each gene conditioning the resistant reaction in the host, there is a specific gene conditioning pathogenicity in the pathogen". Flor worked with linseed (flax) and its rust disease caused by *Melampsora lini* system. In this system resistance is dominant to susceptibility and avirulence is dominant to virulence. It means that resistance in linseed and avirulence in the rust are governed by dominant genes. In this situation resistant reaction is observed only when the host with a dominant gene is attacked by a pathogen with dominant gene. In other combination it will be a susceptible reaction.

28.2.5. Sources of Disease Resistance

Genes for disease resistance may be available from the following sources:

1. Improved cultivar
2. Germplasm collection
3. Related species
4. Induced mutation
5. Somaclonal variation
6. Unrelated organism

28.2.6 Methods of Breeding for Disease Resistance

The following methods have been used for breeding for disease resistance:

1. Selection
2. Introduction
3. Hybridization
   a) Pedigree method
   b) Backcross method without/with marker-assisted selection
4. Mutation breeding
5. Invitro breeding

a) Wide hybridization with embryo rescue
b) Somatic hybridization
c) Somaclonal variation

6. Transgenic breeding

These methods have been discussed in earlier chapters.

28.2.7 Merits of Breeding for Disease Resistance

1. Losses caused by diseases are minimized by using resistant cultivars.
2. Reduction in cost of production.
3. Reduces environmental pollution and health hazards by reducing use of chemicals.
4. Resistant cultivars are non-toxic to man, domestic animals and wild life.
5. Genetic resistance is only solution of some of the diseases such as wilt, rust, smut, nematode and bacterial blight.

28.2.8 Demerits of Breeding for Disease Resistance

1. It is a long-term process that takes 10-15 years to develop a cultivar.
2. In some cases, breeding for resistance to one disease leads to susceptibility to another pest.
3. It is a costly method.

28.3 Breeding for Resistance to Biotic Stresses–Insect Pests

All crops are attacked by insects. But the degree of damage caused and the number of insect species attacking different crops vary greatly. The average global loss due to insects is estimated as about 14% of the total yield of all the crops. Hence, breeding for insect resistance is an important objective, and insect resistant cultivars are usually an important component of pest management. There are many similarities between breeding for insect resistance and breeding for disease resistance. So only essential features of breeding for insect resistance are discussed here.

28.3.1 Biotype Differentiation

The different strains of an insect pest which differ from each other in their ability to attack different cultivars of their host species are called **biotypes**. It is similar to the physiological differentiation in plant pathogens, but less frequent.

28.3.2 Mechanism of Insect Resistance

There are four groups of mechanism for insect resistance: (1) non-preference, (2) antibiosis, (3) tolerance and (4) avoidance. In a given case of insect resistance, one or more of these mechanisms may be involved.

1. *Non-preference*: The host cultivars possessing **non-preference** are unsuitable for colonisation, oviposition or both by an insect-pest. In most cases of this type, the insect will not feed on the resistant cultivar even if alternative food is not available. But in some cases, the resistant cultivar is attacked by the insect although it is much less preferred than other cultivars. Non-preference may be due to morphological and/or biochemical reasons, and can be very effective. Again, the chances of such a resistance being durable are relatively greater.
2. *Antibiosis*: When insects feeding on a resistant cultivar show reduced growth and/or reproduction, it is known as **antibiosis**. In severe cases, the insects may even die. Antibiosis may involve morphological, physiological or biochemical features of the host plant.
3. *Tolerance*: When at the same level of insect attack a cultivar shows significantly lower reduction of yield than a susceptible one, it is called **tolerance**. It may be due to either a greater recovery or lower degree of damage to the tolerant cultivar.
4. *Avoidance*: Pest **avoidance** is the same as disease escape. Many environmental factors contribute to pest avoidance. It is not a case of true insect resistance. But often it is quite effective in minimizing losses due to insect pests.

28.3.3 Basis for Insect Resistance

Insect resistance may involve morphological, physiological or biochemical features of the host plant.

1. *Morphological features*: Several morphological traits produce insect resistance, e.g., hairiness of leaves, colour of plants, thickness and toughness of plant tissues, etc.
2. *Physiological features*: Some physiological features, e.g., osmotic concentration of cell sap, various exudates, etc. may be associated with insect resistance.
3. *Biochemical features*: Several biochemical factors are associated with insect resistance. These factors are more important than morphological and physiological features in producing non-preference and antibiosis.

28.3.4 Genetics of Insect Resistance

Insect resistance may be governed by oligogenes, polygenes or plasmagenes.

1. *Oligogenic resistance*: In this case one or few major genes govern resistance. Generally, the dominant allele controls resistance but in some cases, it is due to recessive allele. The differences between resistant and susceptible plants are clear-cut.
2. *Polygenic resistance*: Polygenic resistance is conferred by many genes, each contributing to the resistance. This type of resistance is usually complex. Horizontal resistance is usually polygenic.
3. *Cytoplasmic resistance*: This resistance is due to plasmagenes, e.g., root aphid in lettuce.

28.3.5 Sources of Insect Resistance

1. Cultivar
2. Germplasm collection
3. Related wild species
4. Unrelated organism

28.3.6 Methods of Breeding for Insect Resistance

The methods of breeding for disease resistance are also applicable to insect resistance breeding.

28.3.7 Problems in Breeding for Insect Resistance

1. Breeding for resistance to one insect may sometimes lead to susceptibility to another insect.
2. An insect-resistant cultivar may have inferior quality.
3. In many cases, insect resistance may be available only in related wild species posing many problems.
4. Tests for insect resistance are tedious, laborious and costly.
5. Breeding for insect resistance is a long- term programme.

28.4 Brief References

Agrawal 1998, Chaudhary 2014, Khan 2018, Ramachandra 2014, Singh BD 2011, Singh D *et al.* 2014, Singh Phundan 2013a.

Review Questions

Part A

Answer the following questions true or false.

1. The organism that causes disease is called pathogen. True/False
2. Related wild species may not be a source of disease resistance. True/False
3. Nelson put forth the "gene-for-gene" hypothesis. True/False
4. Avoidance is also a case of true insect resistance. True/False
5. The methods of breeding for disease resistance are not applicable to insect resistance breeding. True/False

Part B

Answer the following questions.

1. Define disease.
2. Differentiate between vertical and horizontal resistance.
3. Polygenic resistance is conferred by many/few genes.
4. No symptom develops in case of ______ response.
5. Oligogenic resistance is also called ______ resistance.

Part C

Write a short note on each of the following topics.

1. Genetics of host-pathogen interaction.
2. Basis for insect resistance.
3. Mechanism of insect resistance.
4. Methods of breeding for disease resistance.
5. Sources of disease resistance.

29

Breeding for Resistance to Abiotic Stresses

29.1 Introduction

An environmental factor that interferes with the complete expression of genotypic potential of a crop is called **stress**. The stresses may be biotic (pathogens, insect-pests, etc.) or abiotic. The main abiotic stresses are drought, mineral deficiency/toxicity and heat or cold. The losses due to abiotic stresses can be minimized by:

1. *Crop management*: The degree of stress can be reduced by suitable agronomic/soil management practices, e.g., irrigation to relieve drought stress, soil amendments to reduce salinity.
2. *Development of resistant cultivars*: This offers the cheapest, easiest to apply and most eco-friendly approach to reduce losses due to any stress.
   a) *Indirect breeding*: Evaluation of materials under stress often helps in identification of a resistant line that may become an important commercial cultivar.
   b) *Direct breeding*: It is deliberate breeding of materials for stress resistance. When a genetically variable material is grown under the stress environment, selection can be based on survival, yield and other traits contributing to stress resistance.

29.2 Breeding for Drought Resistance

Drought is the most prevalent abiotic stress. It is the inadequacy of water availability to a crop; it hinders the expression of full genetic yield potential.

29.2.1 Mechanisms of Drought Resistance

Drought resistance is the mechanism causing minimum loss of yield in drought situation in relation to maximum yield in drought-free situation. It is grouped into three categories:

1. *Drought escape*: A susceptible cultivar avoids drought.
2. *Dehydration avoidance*: Ability to retain a relatively higher level of hydration under drought.

3. *Dehydration tolerance*: Ability to suffer less damage due to water stress in the cell.

29.2.2 Sources of Drought Resistance

1. Cultivar
2. Landraces
3. Wild relatives
4. Transgenes

29.2.3 Selection Criteria

The selection criteria for drought should have the following properties:

1. They should be easy to score/estimate.
2. They should have high heritability.
3. They should have large genetic variability.
4. They should have significant correlation with drought.
5. They should exhibit a positive correlation with drought under moisture stress.

29.2.4 Genetics of Drought Tolerance

Information on genetics of drought tolerance is lacking in crop plants. The available information reveals that morphological and physiological traits contributing towards drought tolerance show different inheritance pattern (monogenic and polygenic) and gene action (additive and non-additive) in various crops. Root characters are controlled by polygenic inheritance. The long root and high root number are governed by dominant alleles and thick root tip by recessive alleles. Root-shoot ratio is also governed by dominant and recessive alleles.

29.2.5 Breeding Approaches

In breeding programmes, creation of the correct drought environment is critical for success. Drought tolerance/resistance can be managed by adopting approaches such as conventional, molecular and transgenic.

29.2.5.1 Conventional Breeding Approach

a) *Choice of Selection Environment*: There are three strategies that have been proposed to address the issue of choice of selection environments (Mitra 2001). They are:

   1. *Selection of plants under optimum growing conditions*: The selection of plants has been recommended for high yield under optimum (water-

stress-free) conditions. The maximum genetic potential of yield is expected to be realized under optimum growing condition. A high positive correlation exists between performance under optimum and stressed conditions.

2. *Selection of plants in stressful environment*: The selection and breeding under actual drought condition is also recommended. The intensity of drought is highly variable from year to year. As a result, environmental selection pressure on breeding materials changes drastically from generation to generation limiting the progress of breeding programme.
3. *Selection of plants under optimum and stressful environments*: An alternative approach to selection of plants under optimum and stressful environments would be to improve drought tolerance/resistance in high yielding genotypes through incorporation of morphological and physiological mechanisms of drought tolerance/resistance. In contrast, improving the yield potential of an already resistant material may be more promising approach, provided there is genetic variation within such material. Simultaneous selection (in non-stressed environment for yield and in drought conditions for stability) of genotypes can be carried out to achieve the desired goal of evolving high yielding and drought tolerant/ resistant genotypes.

b) *Breeding Methods*: In general, pedigree and bulk methods are used in self-pollinated crops and recurrent selection in cross-pollinated crops. However, if transfer of a few traits relating to drought resistance to a high yielding genotype is the aim, backcross is the appropriate breeding method. On the other hand, biparental mating (half-sib and full-sib) maintains the broad genetic base as well as provides the scope to evolve the desired genotypes for drought tolerance/resistance.

29.2.5.2 Molecular Breeding Approach

Genetic improvement can be assisted by using molecular markers. The identification of molecular markers linked to drought resistance gene(s) provides several opportunities for drought tolerance/resistance strategies. The use of molecular markers will facilitate to develop drought tolerant/resistant genotypes more effectively as their expressions are independent of environmental effects. After identification of molecular markers linked with yield or other morphological traits related to drought tolerance, these markers could be used in marker-assisted selection (MAS) for drought tolerance.

29.2.5.3 Transgenic Breeding Approach

The production of drought tolerant crops through genetic engineering is considered as an alternative approach to increase crop production under drought stress. This requires the identification and isolation of key genes underlying drought stress and introducing them into crops. Key genes (transgenes) observed to express under drought stress in other plants have been employed to produce transgenic plants.

29.3 Breeding for Resistance to Mineral Stresses

The stresses caused by mineral salts are second in importance after drought. Mineral stresses may be grouped into the following three categories:

1. Mineral deficiency
2. Mineral toxicity
3. Accumulation of an excess amount of soluble salts in the root zone

29.3.1 Salinity Resistance

Salt-affected soils may be subjected to reclamation or they may be used for growing salt-tolerant crops and cultivars. Salinity stress generates water stress, mineral toxicity stress and disturbances in the mineral nutrition of the plants. Hence, salinity resistance may arise from resistance to water stress, resistance to mineral toxicity or a combination of both.

29.3.1.1 Selection Criteria for Salinity Resistance

Salinity resistance may be estimated in terms of cell survival, germination, dry matter accumulation, leaf death or senescence, leaf ion content, root growth, osmoregulation and yield. Of these cell survival, leaf death or senescence and osmoregulation seem to be more reliable as selection criteria.

29.3.1.2 Sources of Salinity Resistance

1. Cultivar
2. Germplasm collection
3. Related species
4. Somaclonal variants
5. Transgenes

29.3.1.3 Breeding Approaches for Salinity Resistance

The breeding approaches are conventional, molecular and transgenic breeding. Three basic genetical approaches are commonly used to improve salt tolerance. They are exploitation of natural genetic variation through direct selection in salt stress environments, mapping QTL and subsequent marker-assisted selection, and development of transgenic plants to introduce novel genes or to change the tolerance levels of existing genes to improve the degree of salt tolerance.

29.3.2 Resistance to Mineral Deficiency

Resistance to mineral deficiency may arise due to:

1. Mineral redistribution
2. Efficient mineral uptake
3. Increased mineral transport
4. Increased root/shoot ratio
5. Increased root hair density or length

Resistance to mineral deficiency is monogenic in several cases but polygenic in some cases. The sources of resistance are cultivars, landraces, mutants and selected species. Selection criteria are visible deficiency symptoms, mineral contents of plant tissues, biochemical assays, and yield. The mineral deficiency environment can be created in the field or in greenhouse using soil-filled pots or hydroponics.

29.3.3 Resistance to Mineral Toxicity

Mineral toxicity is related chiefly to Al and Mn toxicity in acid soils. Al-toxicity may be observed at any pH below 5.5. Resistance to Al-toxicity arises in different ways: increased in pH of rhizosphere, exclusion at root, compartmentation in roots, accumulation in older leaves, intracellular compartmentation, etc. Mn-toxicity often occurs with Al-toxicity. Resistance to Mn-toxicity may be due to reduced transport from root to shoot, oxidation of Mn^{2+} to Mn^{4+} and possibly tolerance to Mn.

Mineral toxicity resistance is usually polygenic, but in some cases monogenic. The sources of resistance are cultivars adapted to acid soils and cultivars bred on acid soils. The mineral toxicity environment may be provided in the field, in pots filled with problem soil, or in nutrient solution-based hydroponics/sand cultures. The selection criteria may be shoot dry matter, root length, root weight, root deformation or discolouration, or yield. In case of Al-toxicity, root staining with haematoxylin is a good assay for Al-exclusion at root.

29.4 Breeding for Resistance to Temperature Stress

A temperature higher or lower than the optimum range of temperatures for normal crop growth constitutes temperature stress. There are three types of temperature stress: heat stress, chilling stress and freezing stress.

29.4.1 Resistance to Heat Stress

Heat stress is generated by temperatures higher than optimal range. Heat resistance is the ability of some genotypes to perform better than others under heat stress. Heat resistance mechanisms may be grouped as heat avoidance and heat tolerance. Heat avoidance is the ability of a genotype to dissipate the radiation energy primarily by transpiration, and due to reflective properties of leaves. Heat tolerance is the ability of some genotypes to perform better than others at elevated temperatures. Heat tolerance is usually under polygenic control; both additive and dominance gene effects are important. In most crops, heat tolerance is available within breeding germplasm. Selection can be carried out in the field in the region concerned or in greenhouses and growth chambers. Some assays, e.g., membrane stability test, can be done even in test tubes.

29.4.1.1 Selection Criteria for Heat Tolerance

Selection for heat tolerance may be based on the following:

1. Growth under heat stress
2. Yield under heat stress
3. Flower, fruit, seed, etc. formation, including pollen fertility
4. Seed germination under heat stress
5. Recovery after heat stress
6. Sensitivity of the photosynthetic process
7. Membrane stability after heat shock

29.4.2 Resistance to Chilling Stress

Chilling stress is imposed by temperatures above zero but below the optimal range. Tropical species are typically chilling sensitive. Chilling tolerance involves membrane-lipid unsaturation, reduced sensitivity of photosynthesis, increased chlorophyll accumulation, and improved germination, fruit and/or seed set and pollen fertility. Chilling tolerance is usually under polygenic control, and may be present in well-adapted breeding populations, germplasm collections, cold tolerant mutants, somaclonal variants and related wild species. The selection may be carried out in the field environment or in controlled environment, e.g., in a growth chamber. Selection for chilling tolerance may be based on germination, growth under chilling stress, chlorophyll loss due to chilling, membrane stability, photosynthesis, seedling mortality, seed and/or fruit set, and pollen fertility.

29.4.3 Freezing Resistance

Freezing stress is produced when plants are subjected to sub-zero temperatures. Dormant state is conducive to freezing resistance, while resistance in actively growing tissues is rare. Freezing resistance is a complex trait. It involves variety of characters. It involves two distinct strategies, namely, freezing avoidance and freezing tolerance. Freezing avoidance is the ability of tissues and organs to avoid ice formation at sub-zero temperatures. In contrast freezing tolerance results due to osmotic adjustment, bound water, plasma membrane stability, cell-wall properties and cold-responsive proteins. Freezing tolerance is preferred to freezing avoidance, particularly in regions where temperature may drop below – 41°C. The sources of freezing tolerance may be cultivars, germplasm collections, mutants, related wild species and transgenes. Selection for this trait may be done in the field environment, or in a controlled environment. The selection criteria may be field survival, freezing test, freezing of isolated crowns (in cereals) and osmoregulation.

29.5 Brief References

Khan 2018, Singh BD 2011, Singh D *et al.* 2014.

Review Questions

Part A

Answer the following questions true or false.

1. Heat stress is generated by temperatures higher than optimal range. True/False
2. Chilling stress is imposed by temperatures close to 0°C. True/False
3. Salinity stress generates water stress. True/False
4. Dormant state is not conducive to freezing resistance. True/False
5. Mineral toxicity resistance is usually monogenic. True/False

Part B

Answer the following questions.

1. What is drought?
2. Define heat tolerance.

3. Freezing resistance is a simple/complex trait.
4. Sources of salinity resistance are __________, __________, __________ and __________.
5. What is the difference between heat avoidance and heat tolerance?

Part C

Write a short note on each of the following topics.

1. Breeding approaches for drought resistance.
2. Selection criteria for heat tolerance.
3. Freezing resistance.
4. Resistance to chilling stress.
5. Resistance to mineral deficiency.

Colour Plates

Chapter 5: Molecular Genetic Basis II: Association Mapping

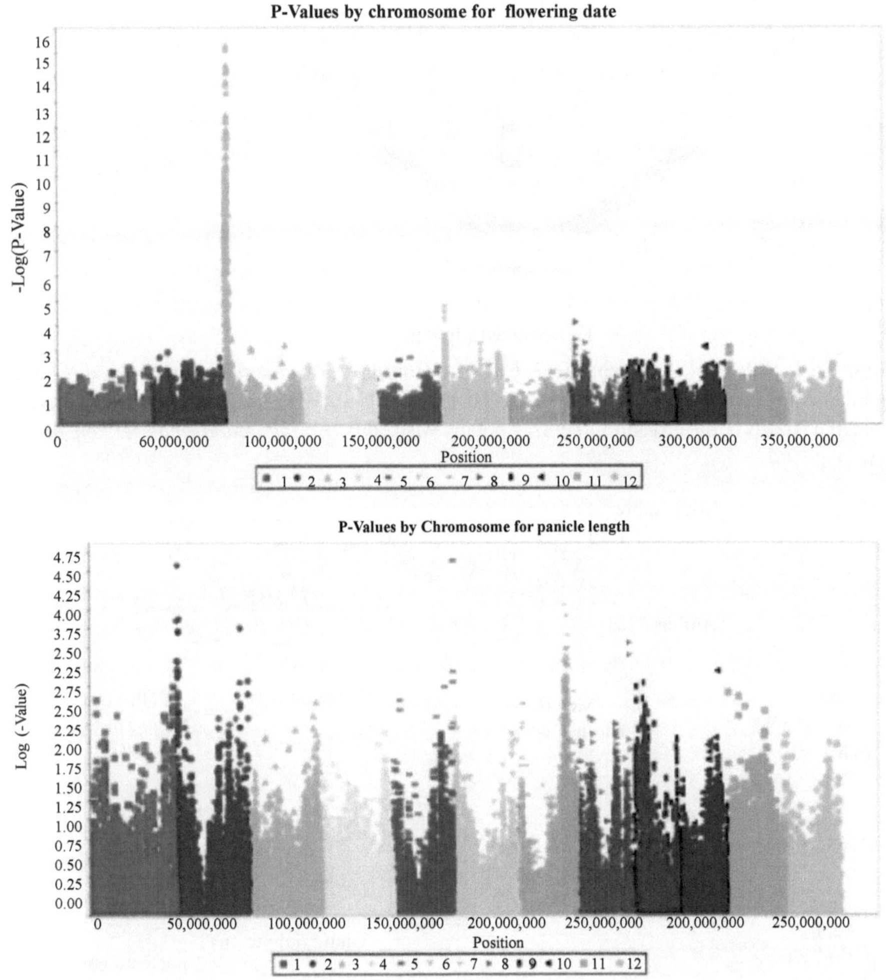

Figure 5.2. AM results for agronomic traits on all 12 chromosomes (2012 dry season data), analysed from Begum et al. (2015). Manhattan plots for flowering date (a) and panicle length (b). Analysis was performed using TASSEL software Version 4.1.34 (Bradbury et al., 2007). Results indicate a major locus for flowering time on chromosome 3 and multiple QTLs for panicle length, typical of quantitative traits. *(For black and white version, please visit page 84)*

Chapter 6: Mode of Reproduction in Relation to Horticultural Crop Breeding

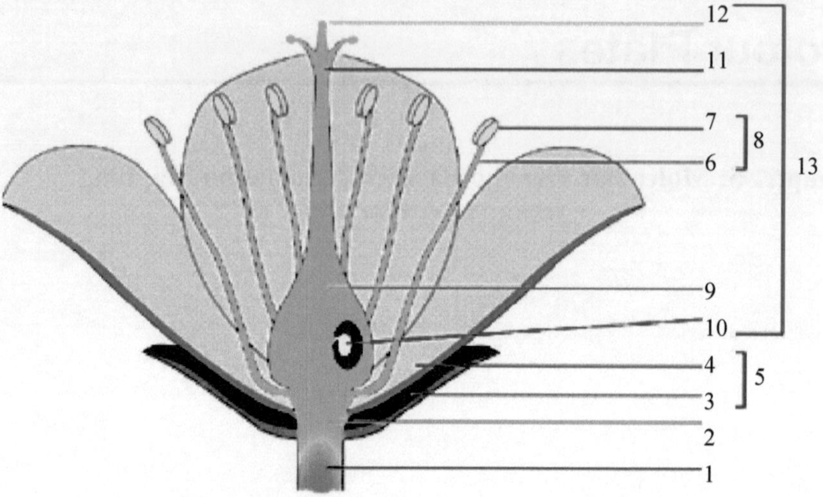

Generalized Flower

Figure 6.1. 1. Pedicel, 2. Receptacle, 3. Calyx, 4. Corolla, 5. Accessory or non-essential whorls, 6. Filament, 7. Anther, 8. Stamen, 9. Ovary, 10. Ovule, 11. Style, 12. Stigma, 13. Reproductive or essential whorls (*Courtesy*: http://courses.botany.wisc.edu) *(For black and white version, please visit page 96)*

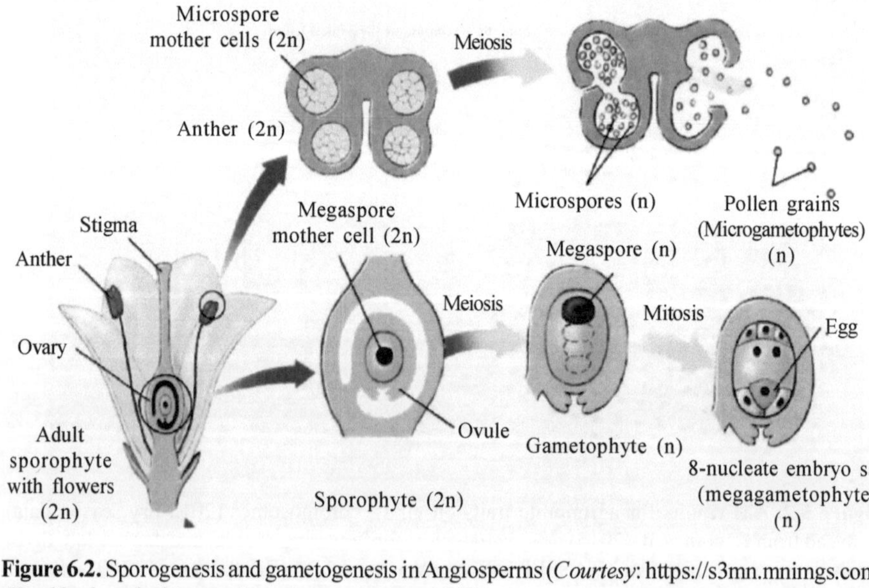

Figure 6.2. Sporogenesis and gametogenesis in Angiosperms (*Courtesy*: https://s3mn.mnimgs.com) *(For black and white version, please visit page 100)*

References

Suggested Further Readings

1. **Acquaah G. (2012).** *Principles of Plant Genetics and Breeding*, 2nd ed. Wiley-Blackwell.

2. **Brown J., Caligari P. and Campos H. (2014).** *Plant Breeding*, 2nd ed. Wiley-Blackwell.

3. **Orton T.J. (2019).** *Horticultural Plant Breeding*. Academic Press-Elsevier Inc., USA.

4. **Priyadarshan P.M. (2019).** *Plant Breeding: Classical to Modern.* Springer-Nature, Singapore.

5. **Sharma A.K. and Sharma R. (2014).** *Crop Improvement and Mutation Breeding*, Print 2019. Scientific Publishers (India), Jodhpur.

6. **Singh B.D. (2018).** *Plant Breeding: Principles and Methods*, 11th ed, Reprint 2019. Kalyani Publishers, New Delhi-Ludhiana.

7. **Singh B.D. and Singh A.K. (2015).** *Marker-Assisted Plant Breeding: Principles and Practices*, 1st South Asian Ed 2017. Springer (India) Pvt Ltd, New Delhi.

8. **Singh D., Singh N.P. and Singh P. (2014).** *Breeding Crop Plants for Stress Resistance.* Kalyani Publishers, New Delhi-Ludhiana.

9. **Singh P. (2013a).** *Essentials of Plant Breeding*, 5th ed, Reprint 2014. Kalyani Publishers, New Delhi, Ludhiana.

10. **Thirugnanakumar S., Manivanan K., Prakash M., Narasimman R. and Vasline Y.A. (2009).** *In Vitro Plant Breeding*, Agrobios (India), Jodhpur.

Detailed References

A

1. **Abdurakhmonov IY 2011.** *Plant Breeding.* InTechJaneza Trdine 9, 51000 Rijeka, Croatia. www.intechopen.com

2. **Abraham Z 2017.** *Fruit Breeding.* Agri Horti Press, New Delhi.

3. **Acquaah G 2012.** *Principles of Plant Genetics and Breeding*, 2nd ed. Wiley-Blackwell.

4. **Agrawal RL 1998.** *Fundamentals of Plant Breeding and Hybrid Seed Production*, Reprint 2017. Oxford and IBH Publishing Co Pvt Ltd, New Delhi (A Unit of CBS Publishers & Distributors Pvt Ltd).

5. **Allard RW 1960.** *Principles of Plant Breeding.* John Wiley & Sons Inc, New York.

6. **Allard RW 1999.** *Principles of Plant Breeding,* 2nd ed, Reprint 2013. Wiley India Pvt Ltd, New Delhi.

7. **Andrade FH, Sala RG, Pontaroli AC, León A and Castro S 2015.** Integration of biotechnology, plant breeding and crop physiology: Dealing with complex interactions from a physiological perspective (Chapter 19). In: Sadras, VO and Calderini, D (ed) *Crop Physiology: Applications for Genetic Improvement and Agronomy*, 2nd ed. Elsevier Inc, Amsterdam, Netherlands, pp 487-503. DOI: 10.1016/B978-0-12-417104-6.00019-4

8. **Ashraf M and Harris PJC 2005.** *Abiotic Stresses: Plant Resistance through Breeding and Molecular Approaches*, Indian Print 2006. International Book Distributing Co, Lucknow.

B

9. **Babcock EB 2013.** *Genetics and Plant Breeding*, 2nd ed, Indian Print 2013. Agrobios (India), Jodhpur.

10. **Babu KN, Divakaran M, Raj RP, Anupama K, Peter KV and Sarma YR 2015.** Biotechnological approaches in improvement of spices: A review (Chapter 25). In Bir Bahadur *et al* (ed) *Plant Biology and Biotechnology Vol II: Plant Genomics and Biotechnology*. Springer India, pp 487-516.

11. **Babu R, Nair SK, Prasanna BM and Gupta HS 2004.** Integrating marker-assisted selection in crop breeding – prospects and challenges. *Current Science* 87(5): 607-619.

12. **Barr AR 2009**. Marker-assisted selection in theory and practice (Chapter 19). In Ceccarelli S, Guimaraes EP and Weltzien E (ed) *Plant Breeding and Farmer Participation.* FAO, Rome, pp 479-517.

13. **Bernard A, Lheureux F and Dirlewanger E 2018**. Walnut: past and future of genetic improvement. *Tree Genetics and Genomes* 14:1-28.

14. **Breseghello F and Coelho ASG 2013.** Traditional and modern plant breeding methods with examples in rice (*Oryza sativa* L.). *Journal of Agricultural and Food Chemistry* 61: 8277-8286. dx.doi.org/10.1021/jf305531j

15. **Brown J, Caligari P and Campos H 2014.** *Plant Breeding*, 2nd ed. Wiley-Blackwell.

16. **Brumlop S and Finckh MR 2011**. *Applications and Potentials of Marker Assisted Selection (MAS) in Plant Breeding,* BfN-Skripten 298, Final report of the F+E project "Applications and Potentials of Smart Breeding"(FKZ 350 889 0020) On behalf of the Federal Agency for Nature Conservation,December 2010, Bundesamt für Naturschutz (BfN),Federal Agency for Nature Conservation, Konstantinstrasse 110, 53179 Bonn,Germany. http://www.bfn.de/0502_skripten.html

C

17. **Ceccarelli S, Guimaraes EP and Weltzien E 2009.** *Plant Breeding and Farmer Participation*. FAO, Rome.

18. **Chahal GS and Gosal SS 2002.** *Principles and Procedures of Plant Breeding: Biotechnological and Conventional Approaches,* Reprint 2014. Narosa Publishing House, New Delhi.

19. **Chandrasekharan SN and Parthasarathy SV 1975.** *Cytogenetics and Plant Breeding*, 3rd Ed, Revised and Enlarged by Krishnaswamy N and Hrishi N. P Varadachary & Co, Madras (Chennai).

20. **Chaudhari HK 1971.** *Elementary Principles of Plant Breeding,* 2nd ed, Reprint 1978. Oxford and IBH Publishing Co Pvt Ltd, New Delhi.

21. **Chaudhary RC 2014.** *Introductory Principles of Plant Breeding,* 2nd ed. Oxford and IBH Publishing Co Pvt Ltd, New Delhi (A Unit of CBS Publishers & Distributors Pvt Ltd).

22. **Chopra VL 1989.** *Plant Breeding: Theory and Practice.* Oxford and IBH Publishing Co Pvt Ltd, New Delhi.

23. **Choudhary BR 2018**. *Breeding of Vegetable Crops*. Kalyani Publishers, New Delhi-Ludhiana.

24. **CIMMYT 1997.** *Introduction to RFLP Mapping and Plant Breeding Applications.* CIMMYT, Mexico.

25. **Collard BCY, Jahufer MZZ, Brouwer JB and Pang ECK 2005.** An introduction to markers, quantitative trait loci (QTL) mapping and marker-assisted selection for crop improvement: The basic concepts. *Euphytica* 142: 169-196.

D

26. **Dabholkar AR 1999.** *Elements of Biometrical Genetics*, 2nd ed. Concept Publishing Co, New Delhi.

27. **Dabholkar AR 2006**. *General Plant Breeding*. Concept Publishing Co, New Delhi.

28. **Dar ZA, Meena PD and Kumar A 2013.** *Principles of Plant Breeding*. Pointer Publishers, Jaipur.

29. **Das LDV 2000.** *Problems Facing Plant Breeding*, Reprint 2012. CBS Publshers and Distributers, Pvt Ltd, New Delhi.

30. **Datta SK 2005.** *Role of Classical Mutation Breeding in Crop Improvement.* Daya Publishing House, Delhi.

31. **De LC and Bhattacharjee SK 2011**. *Ornamental Crop Breeding*. Aaviskar Publishers, Distributers, Jaipur.

32. **De Filippis LF 2014.** Crop improvement through tissue culture (Chapter 12) In Ahmad P, Wani MR, Azooz MM and Tran LSP (ed) *Improvement of Crops in the Era of Climatic Changes,* Vol 1. Springer, New York, pp 289-346.

33. **Dias JS and Ortiz R 2014.** Advances in transgenic vegetable and fruit breeding. *Agricultural Sciences* 5:1448-1467.

34. **Dinesh MR and Sankaran M 2017.** *Distant Hybridization in Horticultural Crops.* Daya Publishing House, New Delhi.

35. **Dudley JW 1997.** Quantitative genetics and plant breeding. *Advances in Agronomy* 59: 1-23.

36. **Dutta AC 1974.** *Botany for Degree Students*, 4th ed, Reprint 1976. Oxford University Press, Calcutta (Kolkotta).

E

37. **Evans DM, Gillespie NA and Martin NG 2002**. Biometrical genetics. *Biological Psychology* 61: 33-51. www.elsevier.com/locate/biopsycho

F

38. **Falconer DS 1960.** *Introduction to Quantitative Genetics*, Reprint 1967. Oliver and Boyd Ltd, London.

39. **Falconer DS 1989.** *Introduction to Quantitative Genetics*, 3rd ed, ELBS Ed 1989, Longman, UK.

40. **FAO/IAEA 2018.** *Manual on Mutation Breeding*, 3rd ed. Spencer-Lopes, MM, Forster, BP and Jankuloski, L (eds.), Food and Agriculture Organization of the United Nations. Rome, Italy.

41. **Ford-Lloyd B and Jackson M 1986.** *Plant Genetic Resources: An Introduction to their Conservation and Use*. Edward-Arnold (Publishers), London.

42. **Frey KJ 1981.** *Plant Breeding II,* 1st Indian Ed 1982. Kalyani Publishers, New Delhi-Ludhiana.

G

43. **Guimarães EP, Ruane J, Scherf BD, Sonnino A and Dargie JD 2007.** *Marker-Assisted Selection: Current Status and Future Perspectives in Crops, Livestock, Forestry and Fish,* Reprint 2009. FAO, Rome.

44. **Gupta SK 2000.** *Plant Breeding: Theory and Techniques,* Reprint 2008. Agrobios (India), Jodhpur.

45. **Guzhov Y 1989.** *Genetics and Plant Breeding for Agriculture.* MIR Publishers, Moscow.

H

46. **Halfhill MD and Warwick SI 2008.** Mendelian genetics and plant reproduction (Chapter 2). In Stewart CN, Jr (ed) *Plant Biotechnology and Genetics: Principles, Techniques, and Applications.* John Wiley & Sons Inc, New Jersey, pp 21-46.

47. **Hayes HK, Immer FR and Smith DC 1955.** *Methods of Plant Breeding,* 2nd ed, Indian Reprint 2012. Biotech Books, New Delhi.

48. **Hayes HK and Garber RJ 2008.** *Breeding Crop Plants*, 2nd ed, Indian Reprint 2008. Asiatic Publishing House, Delhi.

49. **Hospital F 2003.** Marker-assisted breeding (Chapter 2). In Newbury HJ (ed) *Plant Molecular Breeding.* CRC Press, USA, pp 30-59.

50. **http://www.icar.org.in/en/horticulture.htm** (accessed on October 9, 2017)

J

51. **Jain JP 1992.** *Statistical Techniques in Quantitative Genetics,* 2nd ed, Hindustan Publishing Corporation, Delhi.

52. **Janardhan KV 2012.** *Crop Improvement: A Biotechnological Perspective.* Kalyani Publishers, New Delhi-Ludhiana.

53. **Jiang GL 2013a**. Plant marker-assisted breeding and conventional breeding: challenges and perspectives. *Advances in Crop Science and Technology* 1: e106. doi:10.4172/2329-8863.1000e106

54. **Jiang GL 2013b.** Molecular markers and marker-assisted breeding in plants (Chapter 3). In Andersen SB (ed) *Plant Breeding from Laboratories to Fields.* InTechOpen. Pp 45-83. http://www.intechopen.com/books/plant-breeding-fromlaboratories-to-fields.

K

55. **Kadirvel P, Senthilvel S, Geethanjali *S*, Sujatha M and Varaprasad KS 2015.** Genetic markers, trait mapping and marker-assisted selection in plant breeding (Chapter 4). In Bir Bahadur et al (ed) *Plant Biology and Biotechnology: Volume II: Plant Genomics and Biotechnology.* Springer India, pp 65-88. DOI 10.1007/978-81-322-2283-5_4

56. **Kalloo G 1988.** *Vegetable Breeding*, Combined ed Vol I, II & III, 1st Indian Reprint1994. Panima Educational Book Agency, New Delhi.

57. **Kamthan A, Chaudhuri A, Kamthan M and Datta A 2016.** Genetically modified (GM) crops: milestones and new advances in crop improvement. *Theoretical and Applied Genetics* 129 (7): DOI 10.1007/s00122-016-2747-6

58. **Kang MS 2002.** *Quantitative Genetics, Genomics and Plant Breeding*. CAB International, UK.

59. **Kataria SK 2011.** *Plant Breeding (Theory and Techniques)*. Educational Publishers and Distributors, Delhi.

60. **Khan H 2018.** *Basics of Plant Breeding*. Daya Publishing House, New Delhi.

61. **Khan MA 2013.** *Plant Breeding.* Biotech Books, Delhi.

62. **Knapp SJ 1998.** Marker-assisted selection as a strategy for increasing the probability of selecting superior genotypes. *Crop Science* 38:1164-1174.

63. **Koh H-J, Kwon S-Y and Thomson M 2015.** *Current Technologies in Plant Molecular Breeding: A Guide Book of Plant Molecular Breeding for Researchers.* Springer Science+Business Media, Dordrecht.

64. **Koornneef M and Stam P 2001.** Changing paradigms in plant breeding. *Plant Physiology* 125: 156-159.

65. **Kozjak P and Meglic V 2012.** Mutagenesis in plant breeding for disease and pest resistance (Chapter10). In Mishra R (ed) *Mutagenesis*. InTech, pp 195-219 http://dx.doi.org/10.5772/50332

66. **Kuligowska K, Lutken H, Hegelund JN and Muller R 2015.** Future perspectives of in vitro culture and plant breeding. *Acta Horticulurae* 1083: 27-34.

67. **Kumar N 2018.** *Breeding of Horticultural Crops: Principles and Practices*, 3rd ed. New India Publishing Agency, New Delhi.

68. **Kumar PA, Mohapatra T, Sharma TR, Bhattacharya R, Dash PK, Gupta NC and Solanke AKU 2011.** Biotechnology and crop improvement. *Indian Journal of Agricultural Sciences* 81 (9): 787–800.

69. **Kumar R 2016.** *Vegetable Breeding*. New Vishal Publications, New Delhi.

70. **Kumaresan V 2009.** *Plant Breeding*, Reprint 2014.Saras Publication, Periavilai, Kanyakumari.

71. **Kumari RU and Gowthami R 2015.** *Textbook on Plant Genetic Resources Conservation.* Kalyani Publishers, New Delhi-Ludhiana.

72. **Kumpatla SP, Buyyarapu R, Abdurakhmonov IY and Mammadov JA 2011.** Genomics-Assisted Plant Breeding in the 21st Century:Technological Advances and Progress(Chapter 7). In Abdurakhmonov IY (ed) *Plant Breeding.* InTechOpen Janeza Trdine 9, 51000 Rijeka, Croatia, pp 131-184.www.intechopen.com

73. **Kute NS and Aher AR 2013.** *Principles of Plant Breeding*. Agri-Biovet Press, New Delhi.

74. **Kute N and Shinde G 2016**. *Principles of Biometrical Genetics*. Daya Publishing House, New Delhi.

L

75. **Lamkey KR and Lee M 1993**. Quantitative genetics, molecular markers, and plant improvement. In Imrie BC and Hacker JB (ed) *Focused Plant Improvement: Towards Responsible and Sustainable Agriculture*. Proc. 10th Australian Plant Breeding Conf, Gold Coast,18-23 April 1993. Organising committee, Australian Convention and Travel Service: Canberra pp 104-115.

76. **Lefebvre V and Chevre AM 1995**.Tools for marking plant disease and pest resistance genes: a review. *Agronomie* 15: 3-19.

77. **Liu WY, Kang JH, Yang HB *et al* 2014.** Combined use of bulked segregant analysis and microarrays reveals SNP markers pinpointing a major QTL for resistance to *Phytophthora capsici* in pepper. *Theoretical and Applied Genetics* 127:2503-2513.

M

78. **Mackay I 2011**. *Statistical Methods and Design in Plant Breeding and Genomics: Course Manual*. Armidale Animal Breeding Summer Course 2011, organized by the University of New England, Armidale, Australia 7-11 February 2011.

79. **Mather K 1949**. *Biometrical Genetics: The study of continuous variation*. Methuen, London.

80. **Mather K and Jinks JL 1982.** *Biometrical Genetics: The Study of Continuous Variation*, 3rd ed. Chapman and Hall, London.

81. **Misra KK, Mishra NK and Chand S 2017.** *Fruit Breeding: Approaches and Applications*. Biotech Books, New Delhi.

82. **Mohanan KV 2010.** Essentials of *Plant Breeding*. PHI Learning Pvt Ltd, New Delhi.

83. **Mukherjee BK 1995.** *The Heterosis Phenomenon*. Kalyani Publishers, New Delhi-Ludhiana.

84. **Muallem T and Bekeko Z 2016.** Advances in quantitative trait loci, mapping and importance of markers assisted selection in plant breeding. *International Journal of Plant Breeding and Genetics*, 10 (2): 58-68. DOI: 10.3923/ijpbg.2016.58.68

85. **Murovec J and Bohanec B 2011**. Haploids and doubled haploids in plant breeding (Chapter 5). In Abdurakhmonov IY (ed) *Plant Breeding*. InTech Janeza Trdine 9, 51000 Rijeka, Croatia, pp 87-106. www.intechopen.com

N

86. **Nadarajan N and Gunasekaran M 2005.** *Quantitative Genetics and Biometrical Techniques in Plant Breeding*, Reprint 2014. Kalyani Publishers, New Delhi-Ludhiana.

87. **Narain P 1990.** *Statistical Genetics*. Wiley Eastern Ltd, New Delhi.

88. **Newbury HJ 2003**. *Plant Molecular Breeding*. Blackwell Publishing Ltd, UK.

O

89. **Onaga G and Wydra K 2016a**. Advances in plant tolerance to abiotic stresses (Chapter 9). In Abdurakhmonov IY (ed) *Plant Genomics*. InTechOpen, DOI:10.5772/64350

90. **Onaga G and Wydra K 2016b**. Advances in plant tolerance to biotic stresses (Chapter 10). In Abdurakhmonov IY (ed) *Plant Genomics*. InTechOpen, DOI:10.5772/64351

91. **Orton TJ 2019.** *Horticultural Plant Breeding*. Academic Press-Elsevier Inc., USA.

92. **Osei MK, Asante MD, Agyeman A, Adebayo MA and Adu-Dapaah H 2014.** Plant breeding: a tool for achieving food sufficiency (Chapter 11). In Nandwani D (ed) *Sustainable Horticultural Systems*, Sustainable Development and Biodiversity 2. Springer International Publishing Switzerland, pp 253-274.

P

93. **Panigrahi J, Mishra RR, Sahu AR, Rath SC, Seth S and Mishra SP 2013.** Marker-assisted breeding for simple inherited traits conferring stress resistance in crop plants. *The Bioscan* Special Issue, 3: 217 – 233.

94. **Peter KV, Bonny B and Pradeepkumar T 2015.** Horticulture science and profession (Chapter 1). In Peter KV (ed) *Basics of Horticulture* (2nd rev ed). New India Publishing Agency, New Delhi, pp 1-10.

95. **Poehlman JM 1987.** *Breeding Field Crops,* 3rd ed. AVI Publishing Co Inc, Westport, Connecticut, pp 1-289.

96. **Poehlman JM and Borthakur DN 1969.** *Breeding Asian Field Crops*. Oxford and IBH Publishing Co, New Delhi, pp 1-100.

97. **Priyadarshan PM 2019.** *Plant Breeding: Classical to Modern.* Springer-Nature, Singapore.

R

98. **Ragimekula N, Varadarajula NN, Mallapuram SP, Gangimeni G, Reddy RK and Kondreddy HR 2013**. Marker assisted selection in disease resistance breeding. *Journal of Plant Breeding and Genetics* 1(2): 90-109.

99. **Rajani and Joshi S 2017**. The role of horticultural crops in enhancing nutrient security. *International Journal of Current Microbiology and Applied Sciences* 6 (9): 311-316.

100. **Ram HH 2012.** *Vegetable Breeding: Principles and Practices*, 3rd Ed, Reprint 2015. Kalyani Publishers, New Delhi-Ludhiana.

101. **Ram M 2014.** *Plant Breeding Methods*. PHI Learning Pvt Ltd, New Delhi.

102. **Ramachandra RK 2014**. *Principles of Plant Breeding*, Reprint 2020. Narendra Publishing House, Delhi.

103. **Reddy M and Rao A 2016.** *Plant Breeding in Horticulture.* Pacific Books International, New Delhi.

104. **Roy B and Basu AK 2009.** *Abiotic Stress Tolerance in Crop Plants: Breeding and Biotechnology.* New India Publishing Agency, New Delhi.

105. **Roy D 2012a.** *Plant Breeding: A Biometrical Approach.* Narosa Publishing House Pvt Ltd, New Delhi.

106. **Roy D 2012b.** *Biometrical Genetics: Analysis of Quantitative Variation.* Narosa Publishing House Pvt Ltd, New Delhi.

S

107. **Saeed T and Shahzad A 2016.** Basic principles behind genetic transformation in plants (Chapter 13). In Shahzad *et al* (ed) *Biotechnological Strategies for the Conservation of Medicinal and Ornamental Climbers*. Springer International Publishing, Switzerland, pp 327-350.

108. **Sakiyama NS, Ramos HCC, Caixeta ET and Pereira MG 2014.** Plant breeding with marker-assisted selection in Brazil. *Crop Breeding and Applied Biotechnology* 14: 54-60.

109. **Sattler MC, Carvalho CR and Clarindo WR 2016**. The polyploidy and its key role in plant breeding *Planta* 243(2): 281-296. DOI 10.1007/s00425-015-2450-x

110. **Saxena A 2014**. *Principles of Plant Breeding*. Sonali Publications, New Delhi.

111. **Saxena PK and Balabhaskar KR 2014a**. *Introduction to Plant Breeding*. Campus Books International, New Delhi.

112. **Saxena PK and Balabhaskar KR 2014b.** *Molecular Plant Breeding*. Campus Books International, New Delhi.

113. **Saxena PK and Balabhaskar KR 2014c.** *Modern Plant Breeding: Principles and Procedures*. Campus Books International, New Delhi.

114. **Saxena PK and Balabhaskar KR 2014d.** *Plant Breeding in Organic Agriculture.* Campus Books International, New Delhi.

115. **Sharma AK and Sharma R 2014**. *Crop Improvement and Mutation Breeding*, Print 2019. Scientific Publishers (India), Jodhpur.

116. **Sharma JP 2009.** *Principles of Vegetable Breeding*, Reprint 2014. Kalyani Publishers, New Delhi-Ludhiana.

117. **Sharma JR 1994.** *Principles and Practice of Plant Breeding*. Tata McGraw-Hill Publishing Co Ltd, New Delhi.

118. **Sharma JR 1998.** *Statistical and Biometrical Techniques in Plant Breeding.* New Age International (P) Ltd, New Delhi.

119. **Shashidhar HE, Kanbar A, Toorchi M, Raveendra GM, Kundur P, Vimarsha HS, Soman R, Kumar NG, Bekele BD and Bhavani P 2013**. Breeding for drought

resistance using whole plant architecture – conventional and molecular approach (Chapter 6). In Andersen SB (ed) *Plant Breeding from Laboratories to Fields.* InTechOpen, pp151-166. www.intechopen.com

120. **Shimelis H and Laing M 2012.** Timelines in conventional crop improvement: pre-breeding and breeding procedures. *Australian Journal of Crop Science* 6 (11):1542-1549.

121. **Shu QY, Forster BP and Nakagawa H 2011**. *Plant Mutation Breeding and Biotechnology*. Jont FAO/IAEA Programme.

122. **Shukla Anil K, Shukla Arun K, Noor Mohamed MB, Singh A and Tiwari D 2020.** *Fruit Breeding: Approaches and Achievements*, 2nd ed. New India Publishing Agency, New Delhi.

123. **Siddiqui BA and Khan S 1999.** *Breeding in Crop Plants: Mutations and In vitro Mutation Breeding*, Reprint 2008. Kalyani Publishers, New Delhi-Ludhiana.

124. **Simmonds NW 1979.** *Principles of Crop Improvement*, ELBS ed 1981, Longman Group Ltd, UK.

125. **Singh BD 1998.** *Biotechnology*, Kalyani Publishers, New Delhi-Ludhiana.

126. **Singh BD 2011.** *A Text Book of Plant Breeding*, 3rd ed, Reprint 2016. Kalyani Publishers, New Delhi-Ludhiana.

127. **Singh BD 2018.** *Plant Breeding: Principles and Methods*, 11th ed, Reprint 2019. Kalyani Publishers, New Delhi-Ludhiana.

128. **Singh BD and Shekhawat NS 2018.** *Molecular Plant Breeding*. Scientific Publishers (India), Jodhpur.

129. **Singh BD and Shekhawat NS 2019.** *Plant Breeding in 21st Century*. Scientific Publishers (India), Jodhpur.

130. **Singh BD and Singh AK 2015.** *Marker-Assisted Plant Breeding: Principles and Practices*, 1st South Asian ed 2017. Springer (India) Pvt Ltd, New Delhi.

131. **Singh D, Singh NP and Singh P 2014.** *Breeding Crop Plants for Stress Resistance.* Kalyani Publishers, New Delhi-Ludhiana.

132. **Singh NP 2015.** *Basic Concepts of Fruit Science*, 2nd ed, CBS Publishers & Distributors Pvt Ltd, New Delhi.

133. **Singh Parameshwar 2013.** *Concepts in Plant Breeding.* Anmol Publications Pvt Ltd, New Delhi.

134. **Singh Phundan 2011.** *Plant Breeding (For Undergraduate Students)*, Reprint 2015. Kalyani Publishers, New Delhi-Ludhiana.

135. **Singh Phundan 2013a.** *Essentials of Plant Breeding*, 5th ed, Reprint 2014. Kalyani Publishers, New Delhi-Ludhiana.

136. **Singh Phundan 2013b.** *Introduction to Biotechnology*, 2nd ed, Reprint 2016. Kalyani Publishers, New Delhi-Ludhiana.

137. **Singh Phundan 2014.** *Plant Breeding: Molecular and New Approaches*, 3rd ed. Kalyani Publishers, New Delhi-Ludhiana.

138. **Singh Phundan 2015.** *Molecular Plant Breeding*. Kalyani Publishers, New Delhi-Ludhiana.

139. **Singh Phundan 2016.** *Plant Breeding Related Legislations*. Daya Publishing House, New Delhi.

140. **Singh Phundan 2019.** *Principles of Crop Improvement*. Daya Publishing House, New Delhi.

141. **Singh Phundan and Narayanan SS 2013.** *Biometrical Techniques in Plant Breeding*, 5th ed. Kalyani Publishers, New Delhi-Ludhiana.

142. **Singh PK 2014.** *Manual on Vegetable Breeding.* Daya Publishing House, New Delhi.

143. **Singh RK and Chaudhary BD 1985.** *Biometrical Methods in Quantitative Genetic Analysis*, 3rd ed., Reprint 2014, Kalyani Publishers, New Delhi-Ludhiana.

144. **Singh S and Pawar IS 2005.** *Theory and Application of Biometrical Genetics.* Reprint 2010. CBS Publishers and Distributors, Pvt Ltd, New Delhi.

145. **Singh S and Pawar IS 2006.** *Genetic Basis and Methods of Plant Breeding,* Reprint 2014. CBS Publishers and Distributors, Pvt Ltd, New Delhi.

146. **Stoskopf NC, Tomes DT and Christie BR 2011.** *Plant Breeding: Theory and Practice*, 3rd ed, Print 2019. Scientific Publishers (India), Jodhpur.

147. **Suslow TV, Thomas BR and Bradford KJ 2002.** Biotechnology provides new tools for plant breeding. *Agricultural Biotechnology in California Series Publication No.8043*. University of California, Division of Agriculture and Natural Resources. http://anrcatalog.ucdavis.edu.

T

148. **Thirugnanakumar S, Anandan A and Vasline YA 2014.** *Basics of Mutation Breeding*. New India Publishing Agency, New Delhi.

149. **Thirugnanakumar S, Manivanan K, Prakash M, Narasimman R and Vasline YA 2009.** *In Vitro Plant Breeding*, Agrobios (India), Jodhpur.

150. **Thirugnanakumar S, Saravanan K, Senthilkumar N, Anandan A and Eswaran R 2012.** *Quantitative Genetics and Crop Breeding*. New India Publishing Agency, New Delhi.

151. **Thomson MJ, Ismail AM, McCouch SR and Mackill DJ 2010.** Marker assisted breeding (Chapter 20). In Pareek A, Sopory SK, Bohnert HJ and Govindjee (ed) *Abiotic Stress Adaptation in Plants: Physiological, Molecular and Genomic Foundation.*Springer Science + Business Media BV, pp 451–469. DOI 10.1007/978-90-481-3112-9_20

152. **Tinker NA 2008.** Plant breeding (Chapter 3). In Stewart CN, Jr (ed) *Plant Biotechnology and Genetics: Principles, Techniques and Applications*. John Wiley & Sons Inc, New Jersey, pp 47-82.

153. **Tohidfar M and Khosravi S 2015.** Transgenic crops with an improved resistance to biotic stresses: A review. *Biotechnology, Agronomy, Society and Environment*. 19(1): 62-70.

154. **Trehan K 1990.** *Biotechnology*, Reprint 1991. Wiley Eastern Ltd, New Delhi.

155. **Trevan MD, Boffey S, Goulding KH and Stanbury P 1987.** *Biotechnology: The Biological Principles,* Reprint 1991. Tata McGraw-Hill Publishing Co Ltd, New Delhi.

156. **Tyagi ID 2015.** *Principles of Plant Breeding,* 2nd ed. Jain Brothers, New Delhi.

U

157. **Uddin MS and Cheng Q 2015.** Recent application of biotechniques for the improvement of mango research (Chapter 12). In Poltronieri P and Hong Y (ed) *Applied Plant Genomics and Biotechnology*. Woodhead Publishing, UK, pp 195-212.

V

158. **Vencovsky R, Ramalho MAP and Toledo FHRB 2012.** Contribution and perspectives of quantitative genetics to plant breeding in Brazil. *Crop Breeding and Applied Biotechnology S2*: 7-14.

159. **Verdeprado H, Kretzschmar T, Begum H, Raghavan C, Joyce P, Lakshmanan P, Cobb JN and Collard BCY 2018.** Association mapping in rice: basic concepts and perspectives for molecular breeding, *Plant Production Science* DOI: 10.1080/1343943X.2018.1483205

160. **VIB 2016**. *The Past, Present and Future of Plant Breeding.* VIB, Belgium. www.vib.be

161. **Visarada KBSR, Meena K, Aruna C, Srujana S, Saikishore N and Seetharama N 2009**. Transgenic breeding: perspectives and prospects. *Crop Science* 49: 1555-1563.

162. **Vindhiyavarman R and Gunasekaran M 2012.** *Participatory Plant Breeding: Theory and Practices.* Kalyani Publishers, New Delhi-Ludhiana.

W

163. **Wani SH and Malik CP 2014.** *New Plant Breeding Techniques*. Aaviskar Publishers, Distributers, Jaipur.

164. **Wijerathna-Yapa A 2017.** Transgenic plants: resistance to abiotic and biotic stresses. *Journal of Agriculture and Environment for International Development.* 111(1): 245-275.

X

165. **Xu Y 2010.** *Molecular Plant Breeding*. CAB International, UK.

166. **Xu Y and Crouch JH 2008.** Marker-assisted selection in plant breeding: from publications to practice. *Crop Science* 48: 391-407.

Y

167. **Yadav RK and Krishna R 2012.** *Experimental Science of Genetics* and *Plant Breeding.* Kalyani Publishers, New Delhi-Ludhiana.

168. **Yadav S and Sharma KD 2016**. Molecular and morphophysiological analysis of drought stress in plants (Chapter 10). In Rigobelo EC (ed) *Plant Growth.* InTechOpen, pp149 -173. DOI:10.5772/65246.

Z

169. **Zeven AC and de Wet JMJ 1982**. *Dictionary of Cultivated Plants and Their Regions of Diversity*. PUDOC, Wageningen.